Lecture Notes in Mathematics

A collection of informal reports and seminars
Edited by A. Dold, Heidelberg and B. Eckmann, Zürich

322

Nonlinear Problems in the Physical Sciences and Biology

Proceedings of a Battelle Summer Institute,
Seattle, July 3-28, 1972

Edited by Ivar Stakgold, Northwestern University,
Evanston, IL/USA
Daniel D. Joseph, University of Minnesota,
Minneapolis, MN/USA
and David H. Sattinger, University of Minnesota,
Minneapolis, MN/USA

Springer-Verlag
Berlin · Heidelberg · New York 1973

AMS Subject Classifications (1970): 34D30, 34K99, 35B20, 35B25, 35B35, 35J99, 35K55, 35P99, 35Q10, 35R10, 45G99, 58F10, 82A05, 82A30, 82A75, 73Pxx, 76D05, 86A05, 92A15, 35J60, 47H15, 58E05, 76Exx, 80A30, 93D05

ISBN 3-540-06251-3 Springer-Verlag Berlin · Heidelberg · New York
ISBN 0-387-06251-3 Springer-Verlag New York · Heidelberg · Berlin

Offsetdruck: Julius Beltz, Hemsbach/Bergstr.

PREFACE

For four weeks in July 1972, the Battelle Research Center in Seattle was host to a Summer Institute devoted to the mathematical analysis of nonlinear problems in the physical and biological sciences. As has become customary in these Battelle conferences, nearly equal representation was sought from mathematics and from the areas of application. Participants were selected for their research contributions and their ability to interact in more than a superficial manner with colleagues in other disciplines. The Summer Institute was marked by a lively interchange of ideas stimulated by four principal series of lectures dealing both with mathematical topics at the center of nonlinear analysis and with various areas of application such as biology, statistical mechanics, hydrodynamics, and chemical reaction engineering.

To capture in print the friendly and invigorating spirit of the Summer Institute is admittedly an impossible task. We have therefore confined ourselves to converting the lecture series and the individual talks into the scientific papers collected in these Proceedings. If a single theme emerges it is that disparate branches of science generate common mathematical problems of nonlinear analysis. For example, the spontaneous generation of new forms of motion in hydrodynamics, of new states of equilibrium in chemically reacting mixtures, and of new structures in biological systems can all be viewed mathematically as manifestations of instability and bifurcation.

The Proceedings contain a mix of new or previously inaccessible research results and of more expository articles that are, however, always close to the frontiers of current research. We hope that this volume will be a useful source book for scientists interested in the applications of mathematical analysis to nonlinear problems.

There remains the pleasant task of acknowledging the friendly and efficient organizational talent of the Battelle staff. The facilities and arrangements provided by the Center were uniformly excellent.

Ivar Stakgold
Daniel Joseph
David Sattinger

CONTENTS

[illegible]

PARTICIPANTS

Organizers

Norman W. Bazley, Battelle-Geneva Research Center, 1227 Carouge - Geneva, Switzerland.

Frederick J. Milford, Battelle Memorial Institute, Columbus, OH 43201.

Ivar Stakgold, Northwestern University, Evanston, IL 60201.

Principal Lecturers

Donald S. Cohen, California Institute of Technology, Pasadena, CA 91109.

Hirsh Cohen, IBM, Yorktown Heights, NY 10598.

David H. Sattinger, University of Minnesota, Minneapolis, MN 55455.

Colin J. Thompson, University of Melbourne, Parkville, Victoria 3052, Australia.

Discussion Leaders

Daniel D. Joseph, University of Minnesota, Minneapolis, MN 55455.

Joel L. Lebowitz, Yeshiva University, New York, NY 10033.

J. F. G. Auchmuty, University of Indiana, Bloomington, IN 47401.

Friedrich Busse, UCLA, Los Angeles, CA 90024.

Bernard D. Coleman, Carnegie-Mellon University, Pittsburgh, PA 15213.

Ellis H. Dill, University of Washington, Seattle, WA 98195.

Paul C. Fife, University of Arizona, Tucson, AZ 85721.

Bruce A. Finlayson, University of Washington, Seattle, WA 98195.

Wilfred M. Greenlee, University of Arizona, Tucson, AZ 85721.

Richard Handelsman, University of Illinois, Chicago, IL 50680.

Tosio Kato, University of California at Berkeley, Berkeley, CA 94720.

Oscar E. Lanford, III, University of California at Berkeley, Berkeley, CA 94720.

J. B. McGuire, Florida Atlantic University, Boca Raton, FL 33432.

J. B. McLeod, Oxford University, Oxford, England.

William R. McSpadden, Battelle Northwest, Richland, WA 99352.

Grégoire Nicolis, Université Libre de Bruxelles, 1050 Brussels, Belgium.

W. E. Olmstead, Northwestern University, Evanston, IL 60201.

George C. Papanicolaou, Courant Institute, New York University, New York, NY 10012.

George H. Pimbley, Los Alamos Scientific Laboratory, Los Alamos, NM 87544.

M. Reeken, Battelle-Geneva Research Center, 1227 Carouge - Geneva, Switzerland.

G. F. Roach, University of Strathclyde, Glasgow, Scotland.

Peter D. Robinson, University of Bradford, Bradford, England.

R. B. Simpson, University of Waterloo, Waterloo, Ontario, Canada.

Juergen Strott, Battelle-Frankfurt, 6000 Frankfurt/Main 90, Germany.

Charles Walter, University of Texas, M. D. Anderson Hospital, Houston, TX 77025.

LYAPUNOV METHODS AND EQUATIONS OF PARABOLIC TYPE

J. F. G. Auchmuty
Department of Mathematics
State University of New York, Stony Brook
and
Indiana University, Bloomington

In this paper we describe the use of Lyapunov's second method to study the stability and instability of equilibrium solutions of nonlinear parabolic equations. First, we shall define Lyapunov functionals in a manner which is appropriate for evolution equations on a Banach space (and which is different from the usual definitions for ordinary differential equations). In the second section these Lyapunov functionals are used to obtain results on the stability of solutions of evolution equations. The third section is a study of certain classes of nonlinear parabolic equations. For these equations Lyapunov methods provide a rigorous justification of certain "energy principles." In particular, we shall show that the equilibrium solutions which minimize a certain functional are stable and that many other equilibrium solutions are unstable.

Some results on the use of Lyapunov methods for nonlinear partial differential equations have already been obtained by Zubov (9) and Chafee and Infante (3). Lyapunov methods have been used for nonlinear parabolic equations in chemical engineering: see Aris (2) and the references therein.

1. LYAPUNOV FUNCTIONS AND EVOLUTION EQUATIONS

Let X be a Banach space and X^* be its dual space with respect to a pairing given by the inner product on a Hilbert space H. Assume that

$$X \subset H \subset X^*$$

and that the imbeddings are continuous.

Consider the Cauchy problem for the autonomous operator equation

$$\frac{du}{dt} = L\,u + N(u) \tag{1.1}$$

$$u(0) = \xi \ , \tag{1.2}$$

where L and N are continuous maps of X into X^* and L is linear.

We shall assume that (1.1)-(1.2) has a unique solution $u(t,\xi)$ defined for $0 \leq t < \infty$ which is a strongly continuous curve in X.

A functional $V : X \to R$ is a Lyapunov function associated with (1.1)-(1.2) provided it satisfies the following two conditions;

(V_1) V is strongly continuous on X and bounded on bounded subsets of X.

(V_2) $\frac{dV}{dt}(u(t,\xi)) \leq 0$ along any solution curve $u(t,\xi)$.

A Lyapunov function V is said to be a strict Lyapunov function provided it satisfies

(V_3) $\frac{dV}{dt}(u(t,\xi)) < 0$ along any solution curve $u(t,\xi)$ unless $u(t,\xi) \equiv \xi$ for $t > 0$.

In (V_2) and (V_3), the derivatives are just ordinary derivatives of real-valued functions of the real variable t.

A point ξ in S is a critical point of V provided

$$\frac{dV}{dt}(u(t,\xi)) = 0 \qquad \text{for } t \geq 0 \ .$$

A point ξ in X is an equilibrium solution of (1.1)-(1.2) provided

$$u(t,\xi) = \xi \qquad \text{for } t \geq 0 \ .$$

Any equilibrium solution of (1.1)-(1.2) is a critical point of V. However, V must be a strict Lyapunov function for the converse to hold.

These definitions are different from those usually introduced for ordinary differential equations. There it is usually assumed that the function V has a given sign, while here we are requiring that $\frac{dV}{dt}$ always have a given sign.

Example. Consider equations (1.1)-(1.2). Assume that L is symmetric and N is potential. (An operator $N : X \to X^*$ is potential if there exists a functional $F : X \to R$ which is Gâteaux differentiable and obeys

$$F'(u) = N(u) \quad \text{for all } u \text{ in } X \text{ ; see Vainberg (8).)}$$

Take $V(u) = -1/2(Lu,u) - F(u)$, where $(\; , \;)$ represents the pairing between X and X^* . Then

$$\frac{dV}{dt}(u(t,\xi)) = -(Lu, \frac{du}{dt}) - (N(u), \frac{du}{dt})$$

$$= -(\frac{du}{dt}, \frac{du}{dt}) \; .$$

Thus V obeys (V_3). If one now imposes conditions on L , N so that (V_1) is obeyed, one finds that V is a strict Lyapunov functional.

2. LYAPUNOV FUNCTIONS AND STABILITY

In this section we shall use Lyapunov functions to obtain criteria for the stability, asymptotic stability or instability of solutions of (1.1)-(1.2). We shall restrict our attention to questions involving stability to perturbations of the initial conditions. Assume that p is some norm on X . Henceforth, a p-neighborhood or a p-ball is a neighborhood or a ball in X , defined by the norm p .

A solution $u(t,\xi)$ of (1.1)-(1.2) is p-stable if, given any $\epsilon > 0$, there exists $\delta > 0$ and $T \geq 0$ such that

$$p(\xi-\eta) < \delta \text{ implies that } p(u(t,\xi)-u(t,\eta)) < \epsilon \text{ for } t \geq T \; .$$

The solution $u(t,\xi)$ is p-asymptotically stable if it is p-stable and

$$\lim_{t \to \infty} p[u(t,\xi) - u(t,\eta)] = 0$$

for all η such that $p(\xi-\eta)$ is sufficiently small.

If $u(t,\xi)$ defines a strongly continuous curve in X , then let

$$\Gamma_\xi = \{u(t,\xi); \; 0 \leq t < \infty\} \; .$$

Let Ω_ξ be the set of strong ω-limit points of Γ_ξ . That is, w is in Ω_ξ , provided there exists a sequence $\{t_n\}$ such that

$t_n \to \infty$ and $u(t_n, \xi)$ converges strongly to w in X.

For general curves there is no guarantee that this set Ω_ξ is non-empty. So we shall use the following hypothesis which usually holds for solution curves of parabolic equations (ω). The set Ω_ξ is non-empty whenever Γ_ξ is bounded in X.

Theorem 1. *Let η be a critical point of a Lyapunov function V of (1.1)-(1.2). Suppose there exists an $r > 0$ such that*

$$V(u) - V(\eta) \geq \varphi(p(u-\eta)) \quad \text{for} \quad p(u-\eta) < r \ , \tag{2.1}$$

where $\varphi(s)$ is a continuous, non-negative and monotone function obeying $\varphi(0) = 0$ and $\varphi(s) \neq 0$ for $s \neq 0$. Then η is a p-stable equilibrium solution of (1.1)-(1.2).

Proof. Choose ξ so that $p(\xi-\eta) < r$. From (V_2) and (2.1) one has

$$\varphi(p(u(t,\xi) - \eta) \leq V(u(t,\xi)) - V(\eta) \leq V(\xi) - V(\eta) \ .$$

Thus $p(u(t,\xi) - \eta) \leq \varphi[V(\xi) - V(\eta)]$ and η is p-stable. The assumptions on φ, together with (2.1), imply that η is an equilibrium solution.

A functional V on X is said to be strongly convex at w with respect to p, provided there exists $r > 0$, $\nu > 0$ and $c > 0$ such that

$$p(u-w) < r \ \text{ implies that } \ V(u) - V(w) \geq c \ p(u-w)^\nu \ .$$

Taking $\varphi(s) = cs^\nu$ in (2.1), one sees that the following corollary holds.

Corollary 1. *Suppose V is a Lyapunov functional for (1.1)-(1.2) and V attains a local minimum at a point η in X. If V is strongly convex at η with respect to p, then η is a p-stable equilibrium solution of (1.1)-(1.2).*

Theorem 2. *Let V be a strict Lyapunov functional for (1.1)-(1.2) and assume that η is an isolated critical point of V, (isolated with respect to the topology induced by p). If (ω) and (2.1) hold, then η is a p-asymptotically stable equilibrium solution of (1.1)-(1.2).*

Proof. Choose r_1 sufficiently small so that $r_1 < r$ in (2.1) and η is the only critical point in a p-ball of radius r_1 about η. Choose ξ so that $p(\xi-\eta) < r_1$. Then $V(u(t,\xi))$ is a monotone decreasing function of t and

$$V(u(t,\xi)) \geq V(\eta) \quad \text{for} \quad t \geq 0 .$$

Assume that $V(u(t,\xi))$ converges to α as t tends to ∞. If $\alpha = V(\eta)$, then the result follows from (2.1).

If $\alpha > V(\eta)$, choose w in Ω_ξ. Then from (V_1), $V(w) = \alpha$. There exists a sequence $\{t_n\}$ such that $u(t_n,\xi)$ converges strongly to w. Thus $u(t_n+T,\xi)$ converges strongly to $u(T,w)$ since (1.1) is autonomous. Therefore, $V(u(t_n+T,\xi)$ converges to $V(u(T,w))$. But $V(u(T,w) < V(w) = \alpha$ since w is not an equilibrium point. This is impossible; therefore $\alpha = V(\eta)$.

Corollary 1. Let V be a strict Lyapunov functional of (1.1)-(1.2). Assume that V attains a local minimum at a point η in X and V is strongly convex with respect to p at η. If (ω) holds for all ξ in some p-neighborhood of η, then, η is a p-asymptotically stable equilibrium solution of (1.1).

This is proved in a manner similar to the proof of the corollary to Theorem 1.

Now we shall prove some results on instability. A solution $u(t,\xi)$ of (1.1)-(1.2) is p-unstable if it is not p-stable.

Theorem 3. Suppose V is a strict Lyapunov functional of (1.1)-(1.2) for all ξ in a p-neighborhood S of a point η in X. Let η be a p-isolated critical point of V and assume that (ω) holds for all ξ in S. Then η is a p-unstable equilibrium solution provided $V(u) - V(\eta)$ can be negative for u in any p-neighborhood of η.

Proof. Choose ξ so that $V(\xi) - V(\eta) < 0$ and $p(\xi-\eta)$ is sufficiently small. Then $V(u(t\ \xi))$ is monotone decreasing for all $t \geq 0$ from (V_3). Suppose that $\alpha = \lim_{t\to\infty} V(u(t,\xi))$. If $\alpha = -\infty$, then $u(t,\xi)$ is unbounded from (V_1). If $\alpha = -\infty$, choose w in Ω_ξ. If w is in S, then it is an equilibrium solution of (1.1) and $w \neq \eta$ since

$$V(w) = \alpha < V(\xi) < V(\eta) .$$

Thus η is p-unstable since η is isolated by assumption and $w \neq \eta$.

When w is not in S , then again η is p-unstable.

Corollary 1. Suppose V and η are defined as in Theorem 3 and that η is not a local minimum for V . Then η is an unstable equilibrium solution of (1.1).

This corollary is just a restatement of the theorem. The corollaries to each of these theorems provide a mathematical form for the physical "principle" that equilibrium solutions which locally minimize certain "energy" functionals are stable, while those that are just critical points and not minima are unstable.

3. APPLICATIONS OF LYAPUNOV METHODS

The results of the last two sections can be used to study the stability of solutions of a wide variety of nonlinear partial differential equations. The main problem usually is to find an appropriate Lyapunov functional. In this section we shall study the stability of solutions of some nonlinear second order elliptic boundary value problems. For these problems there is a natural Lyapunov function, namely, the variational functional for the solutions of the equilibrium problem.

Let Ω be a smooth bounded domain in R^n . Consider the initial-boundary value problem

$$\frac{\partial u}{\partial t} = \sum_{i,j=1}^{n} \frac{\partial}{\partial x_i} \left(a_{ij}(x) \frac{\partial u}{\partial x_j}\right) + f(x,t) \qquad (x,t) \text{ in } \Omega \times (0,\infty) \tag{3.1}$$

subject to

$$u(x,t) = 0 \qquad (x,t) \text{ in } \partial\Omega \times (0,\infty) \tag{3.2}$$

and

$$u(x,0) = \xi(x) \qquad \text{for } x \text{ in } \Omega . \tag{3.3}$$

Here $\partial\Omega$ is the boundary of Ω , and we shall assume that the equation is uniformly parabolic so that there exists a $c > 0$ such that

$$\sum_{i,j=1}^{n} a_{ij}(x)\alpha_i\alpha_j \geq c \sum_{i=1}^{n} \alpha_i^2 \qquad \text{for all } x \text{ in } \Omega$$

and α in R^n .

Under appropriate regularity conditions on the coefficients $a_{ij}(x)$, on the boundary conditions and on the domain Ω, the elliptic part of equation (3.1) may be represented by a linear operator L. The domain of L is some subspace of the real Sobolev space $W_0^{1,2}(\Omega)$, and one has

$$(Lu,v) = -\int_\Omega \sum_{i,j=1}^n a_{ij}(x) \frac{\partial u}{\partial x_i} \frac{\partial v}{\partial x_j} dx = (Lv,u) \qquad \text{for all } u,v \text{ in } W_0^{1,2}(\Omega) .$$

$(,)$ represents the duality between X and X^*; here and henceforth $X = W_0^{1,2}(\Omega)$. X is a Banach space with respect to the norm

$$\|v\|_X^2 = \int_\Omega \sum_{i=1}^n \left(\frac{\partial u}{\partial x_i}\right)^2 dx ,$$

and there should exist constants $c_0 > 0$ and $c_1 > 0$ such that

$$c_0\|u\|_X^2 \leq -(Lu,u) \leq c_1\|u\|_X^2 \qquad \text{for any } u \text{ in } X . \tag{3.4}$$

When $n = 1$, $f(x,s)$ need only obey the following condition:

(F1) $f(x,s)$ is separately continuous in both x and s on $\Omega \times R^1$.

When $n \geq 2$, it will also be assumed that (F2) holds.

(F2) If $n > 2$, there exists a constant k and a function $m(x)$ in $L^q(\Omega)$ such that

$$|f(x,s)| \leq m(x) + k|s|^r$$

for all x in Ω and s in R^1. Here $q = 2n/(n+2)$ and $0 \leq r < (n+2)/(n-2)$.

When $n = 2$, there exist constants c,k such that

$$|f(x,s)| \leq c + k\, e^{s^2} \qquad \text{for all } x \text{ in } \Omega \text{ and } s \text{ in } R^1 .$$

Under these conditions the operator $N : X \to X^*$, defined by

$$N(u)(x) = f(x,u(x)) ,$$

is completely continuous: see Hempel (4).

With these definitions of L, N and X, equations (3.1)-(3.3) may be written in the form of (1.1)-(1.2). It will be assumed that these equations have a solution $u(t,)$ which is a strongly continuous function from $[0,\infty)$ into X.

The Lyapunov functional appropriate to this problem is

$$V(u) = -1/2(Lu,u) - \mathcal{F}(u) . \tag{3.5}$$

Here

$$\mathcal{F}(u) = \int_{\Omega} F(x,u(x))\, dx \tag{3.6}$$

and

$$F(x,v) = \int_0^v f(x,s)\, ds . \tag{3.7}$$

Lemma. Under the above assumptions on L, N and X, the functional V(u), defined by (3.5)-(3.7), is a strict Lyapunov functional for (3.1)-(3.3).

Proof. The term $-1/2(Lu,u)$ obeys (V_1) from the estimates on (Lu,u).

From standard measure-theoretic arguments, one may show that conditions (F1)-(F2) imply that $\mathcal{F}(u)$ obeys (V_1).

Also

$$\frac{dV}{dt}(u(t,\xi)) = -(Lu, \frac{du}{dt}) - (N(u), \frac{du}{dt})$$

$$= -(\frac{du}{dt}, \frac{du}{dt}) .$$

Thus V is actually a strict Lyapunov functional.

A function v in X is an equilibrium solution of (3.1)-(3.3) provided

$$Lv + N(v) = 0 . \tag{3.8}$$

The following theorems show how the Lyapunov functional defined by (3.5)-(3.7) can be used to describe the stability of equilibrium solutions of (3.1)-(3.3) in a number of special cases.

Theorem 4. Assume that L and N are as defined above and that $f(x,s)$ is a monotone decreasing function of s for any x in Ω. Then there is a unique equilibrium solution of (3.1)-(3.3), and it is asymptotically stable in the X-norm.

Proof. V(u) attains a unique minimum u_0 on X since V is strongly convex in u on X under these assumptions. u_0 is an equilibrium solution of (3.1)-(3.3) since (3.8) is the Euler equation for the functional V.

Also condition (ω) holds for all ξ in a neighborhood of u_0 in X. Thus

u_0 is asymptotically stable in X from the corollary to Theorem 2.

Now consider the case where

$$f(x,s) = \lambda[c(x)s + g(x,s)] \quad . \tag{3.9}$$

Assume that (F1) and (F2) hold, that $c(x) \geq k > 0$ on Ω and that c is in $L^{\infty}(\Omega)$. g also obeys the following conditions:

(i) $g(x,-s) = -g(x,s)$ and $g(x,s) = 0$ if and only if $s = 0$ for each x in Ω .

(ii) $s^{-1}g(x,s)$ is a decreasing function of $|s|$, and

$$\lim_{s \to 0} s^{-1}g(x,s) = 0 \text{ and } \lim_{s \to \infty} s^{-1}g(x,s) < -c(x) \text{ uniformly for all } x$$

in Ω .

(iii) $g(x,s)$ is twice continuously differentiable with respect to s almost everywhere in Ω .

The usual examples of functions which obey these conditions are smooth functions which are odd in s and are strictly concave for $s > 0$. For such functions one need only check that the last part of (ii) holds.

A number of authors have studied the problem of obtaining equilibrium solutions of (3.1)-(3.3) when λ is positive and f satisfies the above conditions. In particular, Simpson and Cohen (7) have studied the positive solutions of this problem, while Hempel (4) has obtained quite detailed results on the existence of non-trivial equilibrium solutions under some additional assumptions on g .

In the case where L is the one-dimensional Laplacian, Chafee and Infante (3) have studied the existence and stability of equilibrium solutions of (3.1)-(3.3) with f obeying conditions somewhat different from the above.

To prove the following theorem, one needs some terminology and results from the theory of the calculus of variations on Banach spaces. For this, the reader is referred to Vainberg (8). It will also be assumed that the Schauder theory as in (1), (6) or (7) may be applied to (3.8).

<u>Theorem 5</u>. <u>Assume that</u> L <u>and</u> N <u>are defined as before</u>, <u>that</u> $f(x,s)$ <u>is</u>

given by (3.9) and that c, g obey the given conditions. Then for any $\lambda \geq 0$, $V(u)$ attains a global minimum at $u = u_+(\lambda)$ and $u = u_-(\lambda)$. One has $u_+(\lambda,x) \geq 0$ on Ω and

$$u_-(\lambda,x) = - u_+(\lambda,x) \quad . \tag{3.10}$$

$u_+(\lambda)$ and $u_-(\lambda)$ are, respectively, maximal and minimal equilibrium solutions of (3.1)-(3.3), and they are X-stable solutions of (3.1)-(3.3). There exists a constant $\lambda_1 > 0$ such that the zero solution is stable for $\lambda < \lambda_1$ and unstable for $\lambda > \lambda_1$.

Proof. We shall first show that V attains a global minimum. V is weakly lower semicontinuous on X since $- 1/2(Lu,u)$ is strongly continuous and convex on X and $\mathcal{F}(u)$ is weakly continuous.

From (ii) one sees that for any $\epsilon > 0$ and any $\lambda \geq 0$, there exists an $s_0 > 0$ such that

$$f(x,s) \leq \epsilon s \quad \text{for} \quad s \geq s_0 \quad \text{and} \quad f(x,s) \geq \epsilon|s| \quad \text{for} \quad s < -s_0 \quad .$$

Thus $F(x,s) = \int_0^s f(x,r)dr \leq Cs_0 + \frac{\epsilon}{2} s^2$ if $s > s_0$, and similarly

$$F(x,s) \leq Cs_0 + \frac{\epsilon}{2} s^2 \qquad \text{if} \quad s < -s_0 \quad .$$

Therefore one has

$$V(u) = - 1/2(Lu,u) - \int_\Omega F(x,u(x))dx$$

$$\geq 1/2\, c_0 \|u\|_X^2 - c_2 - 1/2\, \epsilon \int |u(x)|^2 dx \ ,$$

using the above estimates and those for $-(Lu,u)$, where c_2 is a constant.

Choosing ϵ sufficiently small, one sees that $V(u)$ is coercive on X. Thus from the basic theorem in the calculus of variations, V attains a global minimum on X.

Suppose that this global minimum is attained at $u = u_0(\lambda)$.

Let $u_+(\lambda,s) = |u_0(\lambda)(x)|$. Then one sees that

$$V(u_+(\lambda)) = V(u_0(\lambda))$$

since every term in $V(u)$ is quadratic or symmetric about $u = 0$. Similarly

$$V(u) = V(-u) \quad \text{for any } u \text{ in } X .$$

Thus if $u_-(\lambda,x)$ is defined by (3.10), it also minimizes V on X .

For sufficiently small λ $V(u)$ is minimized by $u(x) \equiv 0$ for all x in Ω . Let $\lambda_1 > 0$ be the infimum of the set of λ's such that $u_+(\lambda\ x)$ is not identically zero on Ω . Then λ_1 is the least eigenvalue of the equation

$$Lu + \lambda c(x)u = 0 \text{ in } \Omega \qquad u(x) = 0 \text{ on } \partial\Omega .$$

The fact that there are maximal and minimal equilibrium solutions of (3.1)-(3.3) follows from condition (ii). Choose M so large that

$$f(x,M_0) < 0 \text{ for all } x \text{ in } \Omega .$$

Then the function $v(x) = M$ for $M \geq M_0$ is an upper solution of the equilibrium equations in the sense of Amann (1) or Sattinger (6).

Similarly, $w(x) = -M$ is a lower solution of the equilibrium equations whenever $M \geq M_0$. Thus if u is any solution of (3.8), then

$$-M_0 \leq u(x) \leq M_0 \text{ for all } x \text{ in } \Omega . \tag{3.11}$$

Using the iteration methods defined in (1) or (6) and the fact that the function which is identically zero on Ω is a solution of (3.8), one obtains maximal positive and minimal negative solutions of (3.8). These two solutions differ only in sign.

Simpson and Cohen have shown that if there is a positive solution, then it is unique when g obeys condition (ii). Thus if $u_+(\lambda,x)$ is not identically zero on Ω , it is the maximal positive solution of (3.8). Similarly for $u_-(\lambda,x)$. From the corollary to Theorem 1, one finds that $u_+(\lambda)$ and $u_-(\lambda)$ are X-stable solutions of (3.1)-(3.3).

Let u be an equilibrium solution of (3.1)-(3.3) and write $\emptyset(\alpha) =$

$V(u+\alpha v) - V(u)$. Then

$$\phi(\alpha) = -1/2(L(u+\alpha v),u+\alpha v) + 1/2(Lu,u) - \int_\Omega [F(x,(u+\alpha v)(x))-F(x,u(x))]dx$$

$$= -1/2\alpha^2(Lv,v) - \int_\Omega [F(x,u+\alpha v(x)) - F(x,u(x)) - \alpha f(x,u(x))v(x)]dx$$

$$= -1/2\alpha^2\{(Lv,v) - \int_\Omega \frac{\partial f}{\partial u}(x,u(x))v(x)^2dx\} - \int_\Omega r(x,\alpha)dx$$

where $r(x,\alpha) = F(x,(u+\alpha v)(x)) - F(x,u(x)) - \alpha f(x,u(x))v(x) - 1/2\alpha \frac{\partial f}{\partial u}(x,u(x))v(x)^2$.

If u,v are continuous and bounded on Ω, then from condition (iii) and the Lebesgue dominated convergence theorem, one has

$$\lim_{\alpha \to 0} \int_\Omega \alpha^{-2} r(x,\alpha)dx = 0 .$$

Thus

$$\lim_{\alpha \to 0} 2\alpha^{-2}\phi(\alpha) = (Lv,v) - \int_\Omega \frac{\partial f}{\partial u}(x,u(x))v(x)^2dx . \qquad (3.12)$$

However $\frac{\partial f}{\partial u}(x,u(x)) = \lambda[c(x) + \frac{\partial g}{\partial u}(x,u(x))]$.

From (ii) and (iii) one finds that $\frac{\partial g}{\partial s}(x,s)$ is a non-positive continuous function of s with $\frac{\partial g}{\partial s}(x,s) = 0$.

If $u(x) = 0$ for all x in Ω, then one sees that it is stable provided $\lambda < \lambda_1$. It is unstable when $\lambda > \lambda_1$, since from (3.12) one finds that ϕ may be negative in any neighborhood of $\alpha = 0$, for an appropriate choice of v. Condition (ω) holds for any solution of this equation.

One actually can get some more information about the stability of equilibrium solutions of (3.1)-(3.3) when f is given by (3.9). Suppose $u(\lambda,x)$ is a non-trivial, isolated, equilibrium solution with $\lambda > \lambda_1$ and

$$\sup_{x \in \Omega} |u(\lambda,x)| - m .$$

Then if

$$\frac{\partial g}{\partial s}(x,m) \geq (\frac{\lambda_1}{\lambda} - 1)c(x) \qquad \text{everywhere on } \Omega ,$$

One finds that $\phi(\alpha)$ may be negative for α near 0, with some v, and $u = u(\lambda)$. Such $u(\lambda)$ must be unstable. This says that equilibrium solutions of (3.8) which bifurcate from the zero solution for $\lambda > \lambda_1$ must be unstable for some finite distance along the new branch.

The preparation of this paper was supported in part by NSF Grant No. GP-29273.

REFERENCES

1. Amann, H., On the existence of positive solutions of nonlinear elliptic boundary value problems, Indiana Univ. Math. J. 21, 125-146 (1971).

2. Aris, R., On stability criteria of chemical reactor engineering, Chem. Eng. Sci. 24, 149-169 (1969).

3. Chafee, N. and Infante, E.F., A bifurcation problem for a nonlinear partial differential equation of parabolic type, to appear.

4. Hempel, J. A., Multiple solutions for a class of nonlinear boundary value value problems, Indiana Univ. Math. J. 20, 983-996 (1971).

5. Keller, H. B. and Cohen, D.S., Some positone problems suggested by nonlinear heat generation, Jour. Math. Mech. 16, 1361-1376 (1967).

6. Sattinger, D. H., Monotone methods in nonlinear elliptic and parabolic boundary value problems, Indiana Univ. Math. J. 21, 979-1000 (1972).

7. Simpson, R. B. and Cohen, D. S., Positive solutions of nonlinear elliptic eigenvalue problems, J. Math. Mech. 19, 895-910 (1970).

8. Vainberg, M. M., Variational methods for the study of nonlinear operators, Holden-Day, San Francisco 1964.

9. Zubov, V., Methods of A. M. Lyapunov and their application, Noordhoff, 1964.

MULTIPLE SOLUTIONS OF NONLINEAR PARTIAL DIFFERENTIAL EQUATIONS*

Donald S. Cohen

Department of Applied Mathematics
California Institute of Technology

Table of Contents

---

*This work was partially supported by the U.S. Army Research Office (Durham) under Contract DAHC-04-68-C-0006 and the National Science Foundation under Grant No. GP 18471.

Chapter I

THE EQUATIONS OF CHEMICAL REACTOR THEORY

In Chapters I-IV we shall be concerned exclusively with the equations describing tubular and continuous stirred tank chemical reactors. In their simplest non-dimensional form the equations for the tubular reactor and one reacting species are

(1.1) $u_t = \frac{1}{P} u_{xx} - u_x - \beta(u-u_c) + f(u,w), \quad 0 < x < 1, \quad t > 0,$

(1.2) $w_t = \frac{1}{P} w_{xx} - w_x + g(u,w), \quad 0 < x < 1, \quad t > 0,$

where u represents dimensionless temperature, w represents dimensionless concentration, and f(u, w) and g(u, w) are nonlinear rate functions of chemical kinetics which we shall specify later. We shall confine our analysis to the commonly occuring Arrhenius rate function, but our analysis applies equally well to the more general rate functions of chemical kinetics. All other symbols appearing in equations (1.1)-(1.3) are prescribed physical constants whose ranges of values we shall give later. (For a list of references in the chemical engineering literature where good derivations and explanations of the governing equations and the associated boundary conditions are given, see D. S. Cohen [1].

We shall investigate two distinct problems involving (1.1) and (1.2), namely, reactors with recycle and reactors without recycle. For the tubular flow reactor without recycle the appropriate boundary conditions are

(1.3) $u_x(0,t) - Pu(0,t) = 0, \quad t \geq 0,$

(1.4) $w_x(0,t) - Pw(0,t) = 0, \quad t \geq 0,$

(1.5) $u_x(1,t) = 0, \quad t \geq 0,$

(1.6) $w_x(1,t) = 0, \quad t \geq 0.$

In the case where there is a recycle line, the boundary conditions become

(1.7) $\frac{1}{P} u_x(0,t) - u(0,t) + Ru(1,t) = -(1-R)u_0, \quad t \geq 0,$

(1.8) $\frac{1}{P} w_x(0,t) - w(0,t) + Rw(1,t) = -(1-R), \quad t \geq 0,$

(1.9) $u_x(1,t) = 0, \quad t \geq 0,$

(1.10) $w_x(1,t) = 0, \quad t \geq 0.$

Here R, with $0 \leq R \leq 1$, represents the fraction of the effluent which is recycled, and u_0 represents the temperature of the feed stream. (See C. R. McGowin and D. D. Perlmutter [3] for a derivation of these boundary conditions.) We should note that our reactor is allowed to operate either adiabatically ($\beta = 0$) or non-adiabatically ($\beta \neq 0$), but the recycle line is assumed adiabatic with no chemical reaction occuring outside the reactor and with the recycle stream passing instantaneously from the exit to the entrance of the reactor. For either problem (namely, with or without recycle) the mathematical formulation is completed with the addition of prescribed initial conditions,

(1.11) $$u(x, 0) = \phi(x), \qquad 0 \leq x \leq 1,$$

(1.12) $$w(x, 0) = \psi(x), \qquad 0 \leq x \leq 1.$$

For ease of presenting the main ideas and in order to avoid lengthy algebraic manipulations, we shall confine all our analyses to the above problems. However, we should point out that the major portion of our methods and results apply equally well to general systems of nonlinear parabolic equations of the form

(1.13) $$\frac{\partial v_i}{\partial t} = L_i v_i + F_i(v_1, \ldots, v_n), \qquad (i = 1, \ldots, N)$$

in some domain D subject to appropriate boundary and initial conditions. Here the L_i represent uniformly elliptic partial differential operators, and the F_i represent nonlinear functions of the v_i which "resemble" the rate functions of chemical kinetics.

For a simple continuous stirred tank reactor, the governing equations are

(1.14) $$\frac{da}{dt} = -a - \beta(a - u_c) + f(a, b),$$

(1.15) $$\frac{db}{dt} = -b + g(a, b),$$

subject to appropriate initial conditions. Here a represents dimensionless temperature, and b represents dimensionless concentration. In Section 3 of Chapter II we shall derive these equations from (1.1), (1.2) by singular perturbation methods; equations (1.14), (1.15) will be shown to govern the first terms in the outer expansion as $P \to 0$ of the solutions of (1.1), (1.2). Thus, as is well-known in the chemical engineering literature (see [3], for example), the continuous stirred tank reactor is the limiting case of the tubular reactor as the Peclet number vanishes. In Chapter IV we apply a different singular perturbation method to (1.1), (1.2) as $P \to \infty$; in this case we must treat the plug flow tubular reactor which is the limiting case of the tubular reactor as the Peclet number increases without bound.

As is so often the case in applied mathematics, much of the work is already contained in the proper formulation of the problem together with useful non-dimensionalization of the governing equations. We have simply presented the equations we shall study. The particular significance of this choice of non-dimensional variables is that the problems are particularly well-suited to singular perturbation methods for both large and small Peclet number P, an approach which has apparently not been used before for these problems. We feel that the results of our analyses, in fact, yield all possible phenomena which can occur, and thus, this particular approach is the most useful one to take in treating these problems. In various numerical checks, A. B. Poore [4] has found that our analysis for large Peclet number ($P \to \infty$) provides good results for P as low as 2 or 3, and our analysis for small Peclet number ($P \to 0$) provides good results for P close to unity.

Much of our analysis will rely upon certain properties of the solutions of (1.14), (1.15) in the special case that

$$(1.16) \qquad f(a,b) = DB(1-b)e^a, \qquad g(a,b) = B(1-b)e^a.$$

This is the situation which represents a first order Arrhenius reaction. In this case, (1.14)-(1.16) constitute an autonomous phase-plane system which can be successfully studied by the classical analytical and topological methods associated with singular points of the system, their stability, and Poincaré-Bendixon type arguments for the investigation of trajectories and orbits. There is a lengthy chemical engineering literature associated with these studies culminating with the paper of V. Hlavacek, M. Kubicek, and J. Jelinek [2]. Reference [2] also contains a good summary of previous work. By far the most complete treatment of the problem (1.14)-(1.16) is contained in the recent work of A. B. Poore [4], [5], who has accounted for all possible situations which can occur. The rest of this first chapter is a summary of some of Poore's results [4] for the system (1.14)-(1.16). We shall simply state the main results and illustrate them by means of several diagrams; all proofs and details can be found in [4].

Perhaps the simplest presentation of Poore's results can be made by means of Figures I.1 to I.13 presented at the end of this chapter. The (β, B)-plane is divided into six distinct regions (I-VI) as shown in Fig. I.1. Now, refer to any of the Figures I.1 to I.13. For example, consider Figure I.12 which depicts the situation for any fixed value of β and B in region V of Fig. I.1. Thus, for fixed β, B in region V, we see that for all $D > 0$ there exists a unique critical point (steady state) which is stable for $D \in (0, D_3) \cup (D_4, \infty)$ and unstable for $D \in (D_3, D_4)$. For $D \in (D_3, D_4)$ there exists a stable periodic orbit surrounding the unstable critical point. As another example, consider Fig. I.6. We see that for fixed β, B in region III of Fig. I.1, there exists a unique critical point (steady state) for $D \in (0, D_2) \cup (D_1, \infty)$, three critical points for $D \in (D_2, D_1)$ and two critical points for $D = D_1$ or $D = D_2$. The middle and upper critical points for $D \in (D_2, D_4)$ are unstable, and at $D = D_4$ there is a stable branch of periodic orbits which branches to the left from the upper branch. All other figures are to be interpreted in the same manner. The analytical statements of these results are contained in the following theorems which we present without proof: All proofs can be found in the above mentioned work of Poore [4].

Theorem I.1. Let

$$m_1 = \tfrac{1}{2} - \tfrac{1}{2}\sqrt{1 - \frac{4(1+\beta)}{B}}\ , \qquad m_2 = \tfrac{1}{2} + \tfrac{1}{2}\sqrt{1 - \frac{4(1+\beta)}{B}}\ ,$$

$$D_i = \frac{m_i}{1-m_i} \exp\left(\frac{-Bm_i}{1+\beta} - \frac{\beta u_c}{1+\beta}\right), \quad (i = 1, 2)\ .$$

Then, (i) when $B \leq 4(1+\beta)$ or when $B > 4(1+\beta)$ and $D \in (0, D_2) \cup (D_1, 1)$, there exists one and only one critical point of the autonomous system (1.14)-(1.16),

(ii) when, $B > 4(1+\beta)$ and $D = D_1$ or $D = D_2$, there exist two critical points of (1.14)-(1.16), and

(iii) when $B > 4(1+\beta)$ and $D \in (D_2, D_1)$ there exist three critical points of (1.14)-(1.16).

Define the regions I through VI of Fig. I.1 as follows:

Region I: $0 \leq B < \min\left[4(1+\beta),\ 3+\beta+2\sqrt{2+\beta}\ \right]$.

Region II: $4(1+\beta) < B < \frac{(1+\beta)^3}{\beta}$ for $0 < \beta \leq \frac{\sqrt{5}-1}{2}$ and

$4(1+\beta) < B < 3+\beta+2\sqrt{2+\beta}$ for $\frac{\sqrt{5}-1}{2} \leq \beta \leq \frac{7}{9}$.

Region III: $\frac{(1+\beta)^3}{\beta} < B$ for $0 \leq \beta < \infty$.

Region IV: $4(1+\beta) < B < \frac{(1+\beta)^3}{\beta}$ for $\beta > 1$.

Region V: $3+\beta+2\sqrt{2+\beta} < B < 4(1+\beta)$ for $\beta > \frac{7}{9}$.

Region VI: $3+\beta+2\sqrt{2+\beta} < B < \frac{(1+\beta)^3}{\beta}$ for $\frac{\sqrt{5}-1}{2} < \beta \leq \frac{7}{9}$

or $4(1+\beta) < B < \frac{(1+\beta)^3}{\beta}$ for $\frac{7}{9} \leq \beta \leq 1$.

Theorem II.2. Let $R = \{(a, b) \mid -\infty < a < \infty,\ 0 < b < 1\}$, and let

$$s_1 = \frac{B+1+\beta}{2} - \frac{1}{2B}\sqrt{(B+1+\beta)^2 - 4B(2+\beta)}\ ,$$

$$s_2 = \frac{B+1+\beta}{2} + \frac{1}{2B}\sqrt{(B+1+\beta)^2 - 4B(2+\beta)}\ .$$

Define D_i $(i+1, \cdots, 4)$ by $D_1 = D(m_1)$, $D_2 = D(m_2)$ for $B \geq 4(1+\beta)$, and $D_3 = D(s_1)$, $D_4 = D(s_2)$ for $B \geq 3+\beta+2\sqrt{2+\beta}$, where

$$D(\tau) = \frac{\tau}{1-\tau} \exp\left(\frac{-\beta\tau}{1+\beta} - \frac{\beta u_c}{1+\beta}\right).$$

Then,

i. (See Fig. I.2.) For β, B in region I, m_1 and m_2 are imaginary while either s_1 or s_2 are imaginary or $s_2 > s_1 > 1$. For each $D \geq 0$ there is a unique critical point of (1.14)-(1.16) which is an asymptotically stable node or spiral. Furthermore, all trajectories in the domain R tend to the critical point.

ii. (See Fig. I.3.) For β, B in region II, we have $0 < m_1 < m_2 < 1$, and for $B < 3+\beta+2\sqrt{2+\beta}$ we have s_1 and s_2 imaginary, but for $B \geq 3+\beta+2\sqrt{2+\beta}$ we have s_1 and s_2 real with $0 < m_1 < s_1 < s_2 < m_2 < 1$. The critical point (a, b) is an asymptotically stable spiral or node for $b > m_2$ or $b < m_1$ and is an unstable saddle point for $m_1 < b < m_2$. For $D \in (D_2, D_1)$ there exist three critical points with the middle one a saddle point. For $D \in (0, D_2) \cup (D_1, \infty)$ there is exactly one critical point.

iii. (See Figs. I.4, I.5, I.6.) For β, B in region III there are three critical points for $D \in (D_2, D_1)$ and one critical point for $D \in (0, D_2) \cup (D_1, \infty)$. We have $0 < m_1 < s_1 < m_2 < s_2 < 1$. The critical point is an asymptotically stable node or spiral for $b \in (0, m_1) \cup (s_2, 1)$ and an unstable node or spiral for $b \in (m_2, s_2)$.

iv. (See Figs. I.7, I.8.) For β, B in region IV there are three critical points for $D \in (D_2, D_1)$ and one for $D \in (0, D_2) \cup (D_1, \infty)$. We have $0 < s_1 < m_1 < m_2 < s_2 < 1$. The critical point is an asymptotically stable node or spiral for $b \in (0, s_1) \cup (s_2, 1)$, a saddle for $b \in (m_1, m_2)$, and an unstable spiral or node for $b \in (s_1, m_1) \cup (m_2, s_2)$.

v. (See Figs. I.12, I.13.) For β, B in region V there is exactly one critical point for all $D > 0$. m_1 and m_2 are imaginary, but s_1 and s_2 are real and $0 < s_1 < s_2 < 1$. For $b \in (0, s_1) \cup (s_2, 1)$ the critical point is a stable node or spiral, and for $b \in (s_1, s_2)$ the critical point is an unstable node or spiral.

vi. (See Figs. I.9, I.10, I.11.) For β, B in region VI there are three critical points for $D \in (D_2, D_1)$ and one for $D \in (0, D_2) \cup (D_1, \infty)$. We have $0 < m_1 < m_2 < s_1 < s_2 < 1$. The critical point is a stable spiral or node for $b \in (0, m_1) \cup (m_2, s_1) \cup (s_2, 1)$, an unstable spiral or node for $b \in (s_1, s_2)$, and a saddle for $b \in (m_1, m_2)$.

The analytical results which support the periodic solutions sketched in the figures will be the subject of Chapter II.

REFERENCES

1. Cohen, D. S., Multiple solutions and periodic oscillations in nonlinear diffusion processes, to appear.

2. Hlavacek, V., Kubicek, M., and Jelinek, J., Modeling of chemical reactors -XVIII: Stability and oscillatory behavior of the CSTR, Chem. Eng. Sci., 25 (1970) 1441-1461.

3. McGowin, C. R., and Perlmutter, D., D., Tubular reactor steady state and stability characteristics, AIChE Journal, 17 (1971) 831-849.

4. Poore, A. B., Stability and bifurcation phenomena in chemical reactor theory, Ph.D. thesis, California Institute of Technology, 1972.

5. Poore, A. B., Multiplicity, stability, and bifurcation of periodic solutions in problems arising from chemical reactor theory, to appear.

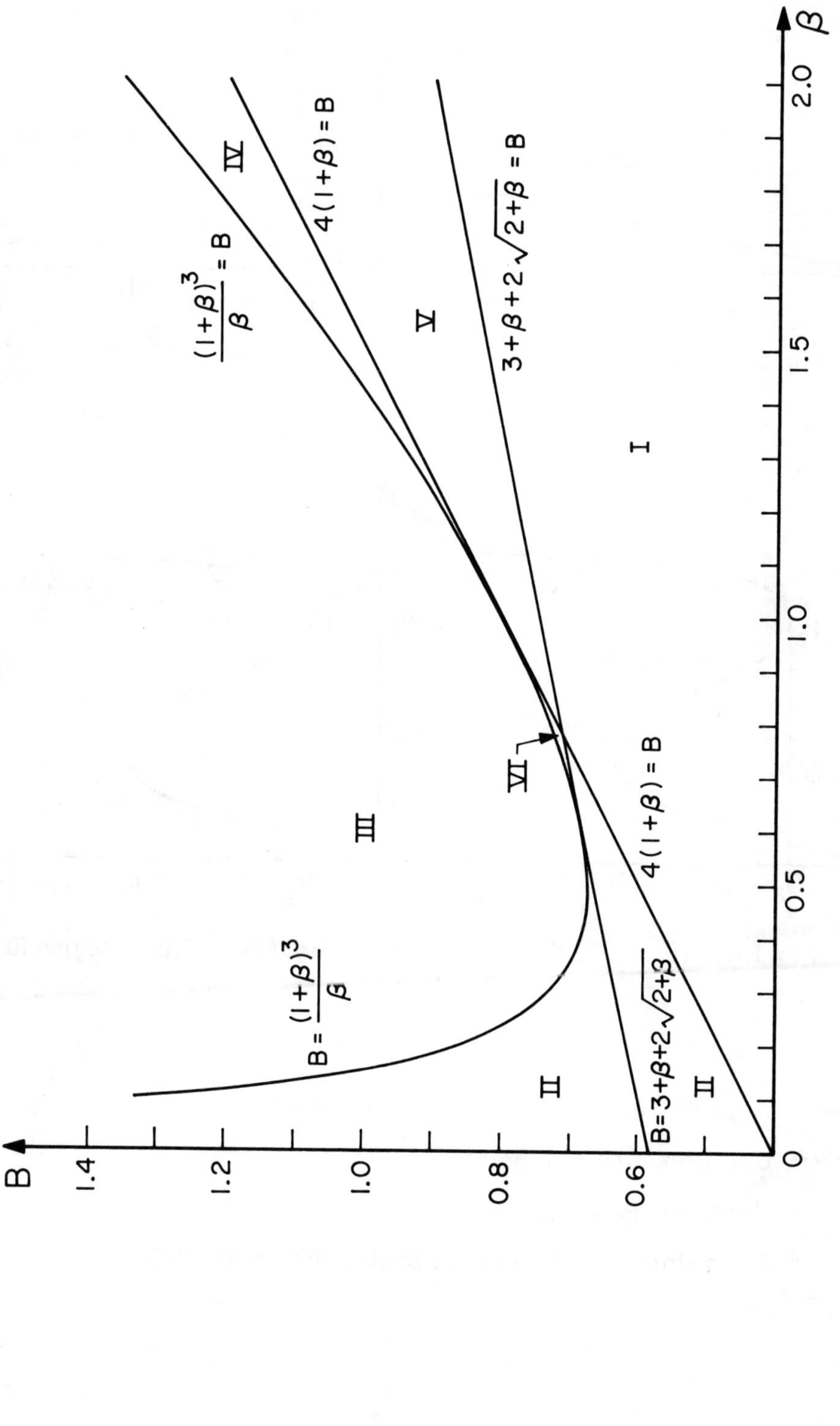
B
1.4
1.2
1.0
0.8
0.6
0
0.5
1.0
1.5
2.0
β
$B = \frac{(1+\beta)^3}{\beta}$
$\frac{(1+\beta)^3}{\beta} = B$
$4(1+\beta) = B$
$4(1+\beta) = B$
$3+\beta+2\sqrt{2+\beta} = B$
$B = 3+\beta+2\sqrt{2+\beta}$
I
II
II
III
IV
V
VI

Fig. I.1

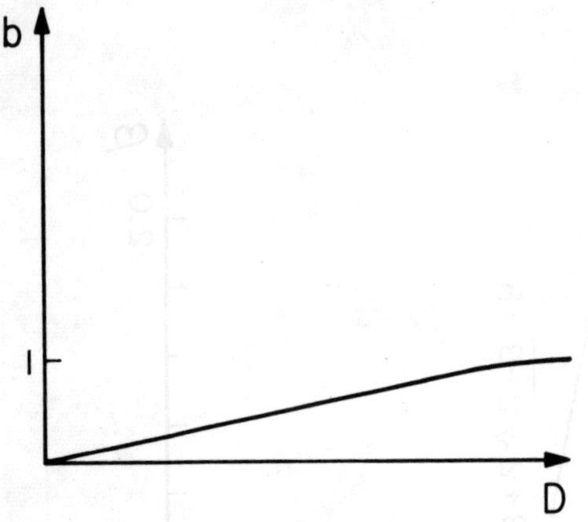

Fig. I.2. $(\beta, B) \in$ Region I

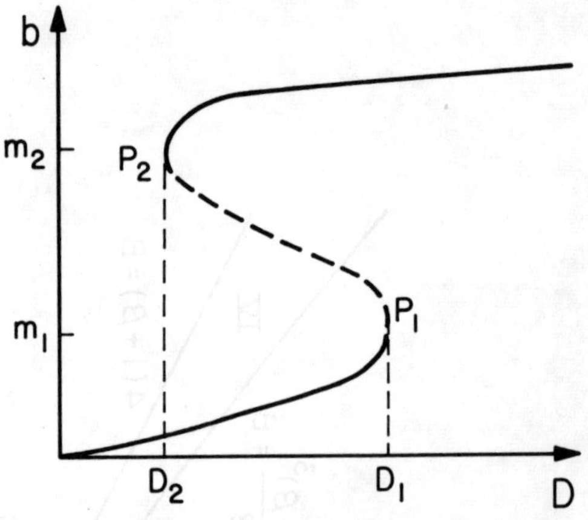

Fig. I.3. $(\beta, B) \in$ Region II

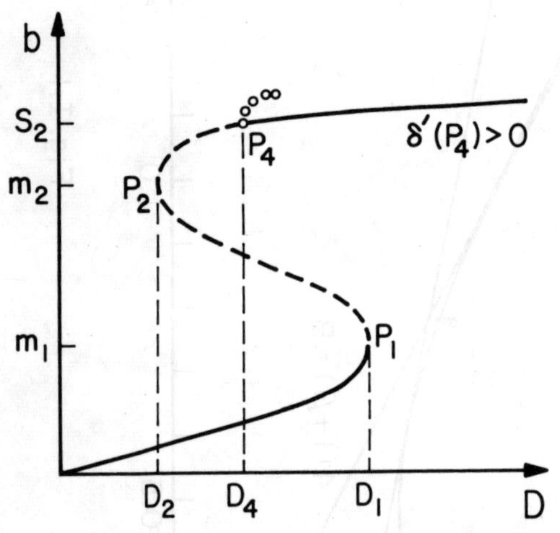

Fig. I.4. $(\beta, B) \in$ Region III

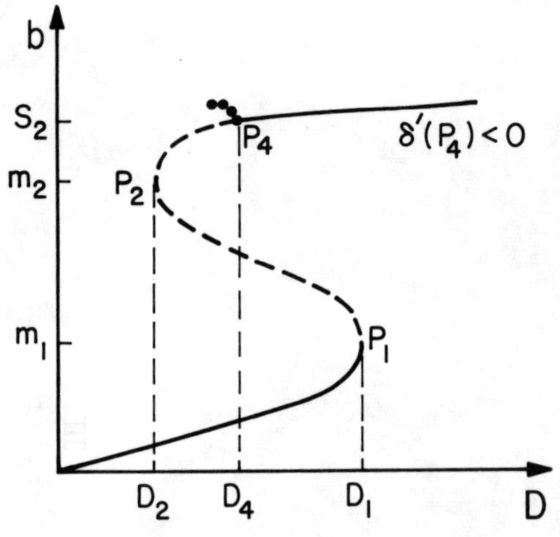

Fig. I.5. $(\beta, B) \in$ Region III

——— Asymptotically Stable Critical Point

– – – – Unstable Critical Point

∘ ∘ ∘ ∘ Unstable Periodic Orbit

• • • • Asymptotically Orbitally Stable Periodic Orbit

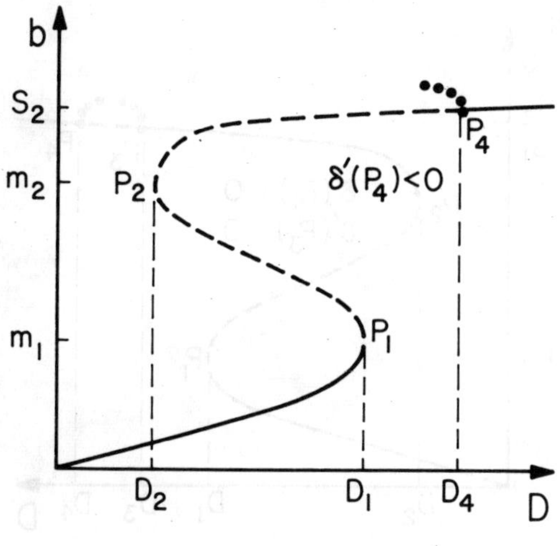

Fig. I.6. $(\beta,B) \in$ Region III

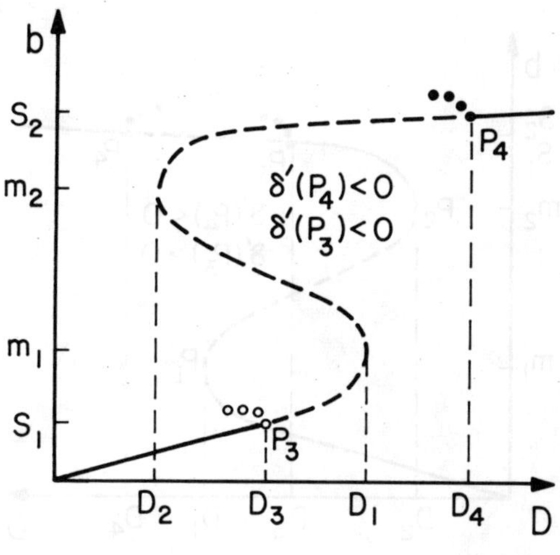

Fig. I.7. $(\beta,B) \in$ Region IV

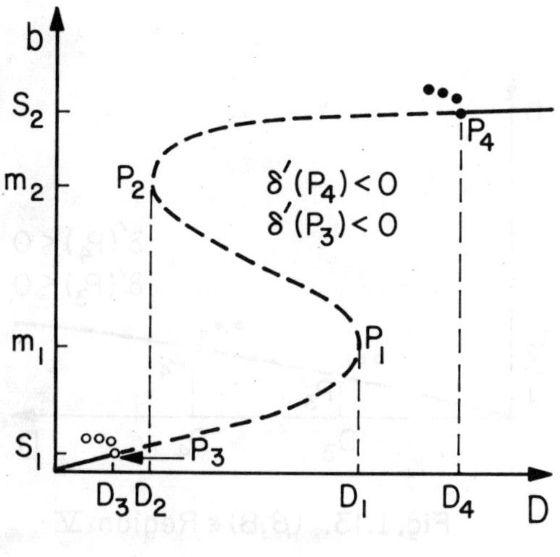

Fig. I.8. $(\beta,B) \in$ Region IV

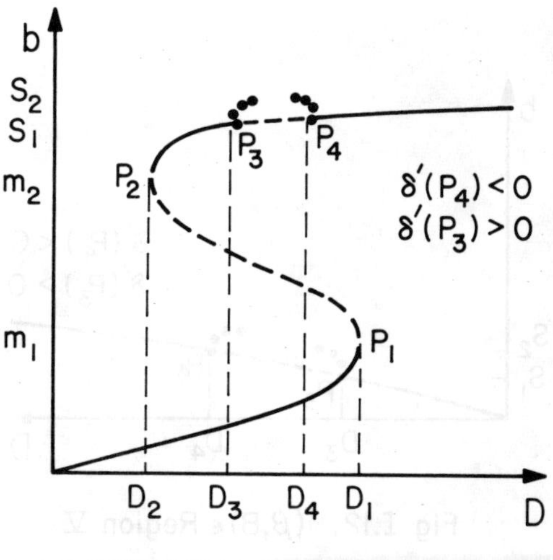

Fig. I.9. $(\beta,B) \in$ Region VI

——— Asymptotically Stable Critical Point

– – – – Unstable Critical Point

○ ○ ○ ○ Unstable Periodic Orbit

● ● ● ● Asymptotically Orbitally Stable Periodic Orbit

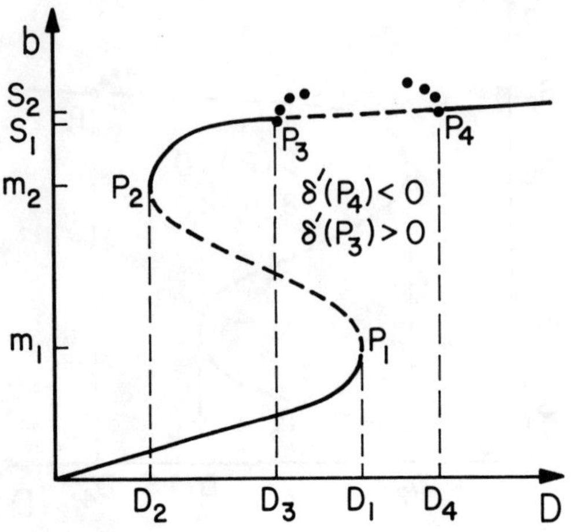

Fig. I.10. $(\beta, B) \in$ Region VI

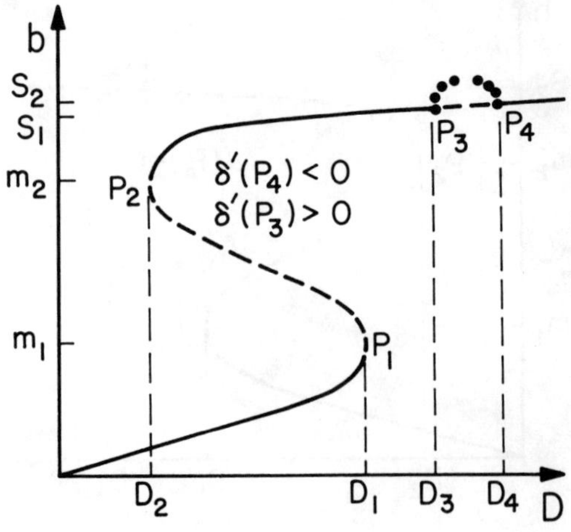

Fig. I.11. $(\beta, B) \in$ Region VI

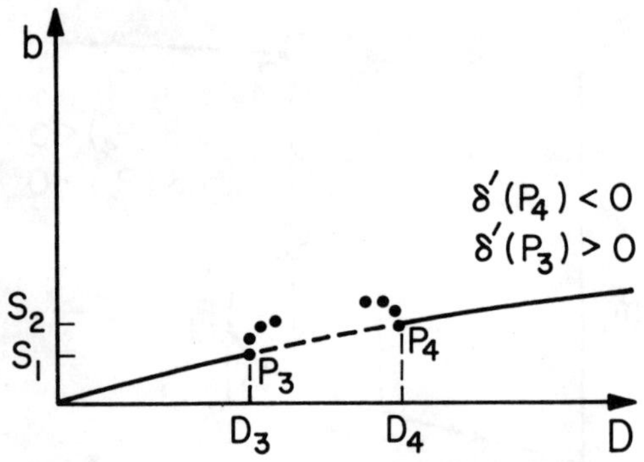

Fig. I.12. $(\beta, B) \in$ Region V

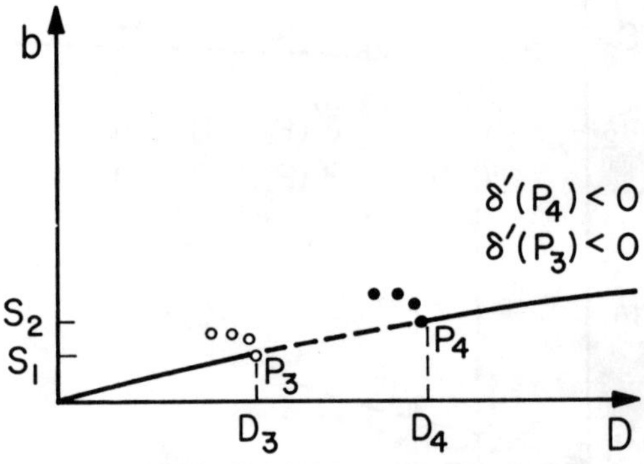

Fig. I.13. $(\beta, B) \in$ Region V

——— Asymptotically Stable Critical Point

– – – – Unstable Critical Point

○ ○ ○ ○ Unstable Periodic Orbit

● ● ● ● Asymptotically Orbitally Stable Periodic Orbit

Chapter II

MULTIPLE PERIODIC OSCILLATIONS IN TUBULAR REACTORS

II.1. Introduction. One of the more interesting recent developments in chemical reactor theory has been the discovery of oscillatory stationary states in the temperature and concentration fields. For the most part these have been observed experimentally [14] or by some machine computation [6], [9], [14]. In the case of a simple continuous stirred tank reactor, Hlavacek, Kubicek, and Jelinek [5] have given analytical plausibility arguments for the appearance of limit cycles. However, there seems to be no treatment of the more difficult non-adiabatic tubular flow reactors. Such a study is given here.

The entire question of oscillatory solutions and their stability is intimately connected with previous considerations of bifurcation, multiplicity, existence, and stability of positive solutions of certain parabolic initial-boundary value problems describing the tubular reactors. These are examined in depth by A. B. Poore [12], [13], who has studied all these questions for the system

$$\left.\begin{aligned} &(1.1) \quad T_t = \frac{1}{P} T_{xx} - T_x - \beta(T - T_c) + DB(1-C)d^{\frac{T}{1+\gamma T}} \\ &(1.2) \quad C_t = \frac{1}{P} C_{xx} - C_x + D(1-C)e^{\frac{T}{1+\gamma T}} \end{aligned}\right\} \quad 0 < x < 1, \quad t > 0 ,$$

$$(1.3) \qquad T_x(0,t) - aT(0,t) = 0 , \qquad t \geq 0 ,$$

$$(1.4) \qquad C_x(0,t) - \alpha C(0,t) = 0 , \qquad t \geq 0 ,$$

$$(1.5) \qquad T_x(1,t) = 0 , \qquad t \geq 0 ,$$

$$(1.6) \qquad C_x(1,t) = 0 , \qquad t \geq 0 ,$$

$$(1.7) \qquad T(x,0) = \Phi(x) , \qquad 0 \leq x \leq 1 ,$$

$$(1.8) \qquad C(x,0) = \Psi(x) , \qquad 0 \leq x \leq 1 .$$

These are the equations for the temperature T and concentration C in a non-adiabatic tubular reactor, and the other quantities are prescribed physical constants.

By appropriate simple transformations the question of oscillatory solutions can be reduced to a study of

$$\left.\begin{aligned} &(1.9) \quad u_t = \frac{1}{\epsilon} u_{xx} - u_x + [\alpha(\lambda)+1] u - \beta(\lambda) w + f(\lambda, u, w)] \\ &(1.10) \quad w_t = \frac{1}{\epsilon} w_{xx} - w_x + \beta(\lambda) u + [\alpha(\lambda)+1)] w + g(\lambda, u, w) \end{aligned}\right\} \quad 0 < x < 1, \quad t > 0 ,$$

$$u_x(0,t) - \epsilon u(0,t) = 0 , \qquad t \geq 0 , \tag{1.11}$$

$$w_x(0,t) - \epsilon w(0,t) = 0 , \qquad t \geq 0 , \tag{1.12}$$

$$u_x(1,t) = 0 , \qquad t \geq 0 , \tag{1.13}$$

$$w_x(1,t) = 0 , \qquad t \geq 0 , \tag{1.14}$$

$$u(x,0) = \phi(x), \qquad 0 \leq x \leq 1 , \tag{1.15}$$

$$w(x,0) = \psi(x), \qquad 0 \leq x \leq 1 . \tag{1.16}$$

The motivation for obtaining this form of the equations is given in [12] and is common in the chemical engineering literature [5], [6]. (We shall briefly indicate how (1.1)-(1.16) result from (1.1)-(1.8) later.) Here u and w, respectively, represent the non-dimensionalized temperature and concentration of equations (1.1)-(1.8). The other quantities are known from the physical situation, and we shall state their properties in Section II.2.

We shall study the system (1.9)-(1.16) in the case that the Peclet number ϵ is small; i.e., $0 < \epsilon << 1$. In Section II.2 we briefly summarize the pertinent previously developed theory for system (1.9)-(1.16), and we shall collect certain mathematical machinery which we shall need in the subsequent analysis of this problem. In Section II.3 by a simple singular perturbation procedure, we shall show that to first order in ϵ the study of (1.9)-(1.16) can be reduced to the consideration of a far more tractable set of nonlinear ordinary differential equations. In Section II.4 we analyze the nonlinear ordinary differential equations specifically for oscillatory solutions. This is accomplished formally by a multi-time scale method (i.e., the so-called "two-timing" method [4], [8].) Not only does this method produce the periodic solutions but the stability of the solutions is also immediately resolved without recourse to further techniques. In Section II.5 we indicate a far more complex situation occurring in the system (1.9)-(1.16). The complete details (including rigorous proofs) for all cases possible are given in references [3], [12] and [13]. We shall describe the results for one case here.

II.2. Summary of previous results and heuristic motivation for new results.

In order to establish the results of the present paper, we shall need certain previously established facts concerning the steady state problem corresponding to (1.1)-(1.8). Rather than rederive these results we shall simply state them and refer the reader to the appropriate mathematical and chemical engineering literature.

It is well-established [1], [2], [7], [11] that there can exist multiple steady (i.e., $\frac{\partial}{\partial t} \equiv 0$) states for the system (1.1)-(1.8). In fact, for all ranges of the physical parameters the numbers of these steady states and their (infinitesimal linearized) stability is known [11], [12]. Two distinct situations will concern us here. These are best described by referring to Figures II.1 and II.2. In both figures we illustrate the behavior of the amplitude of the temperature $\left(\|T\| = \max_{0 \leq x \leq 1} [T(x)]\right)$ as the parameter D (the Damkohler number) varies. Such figures are called response diagrams.

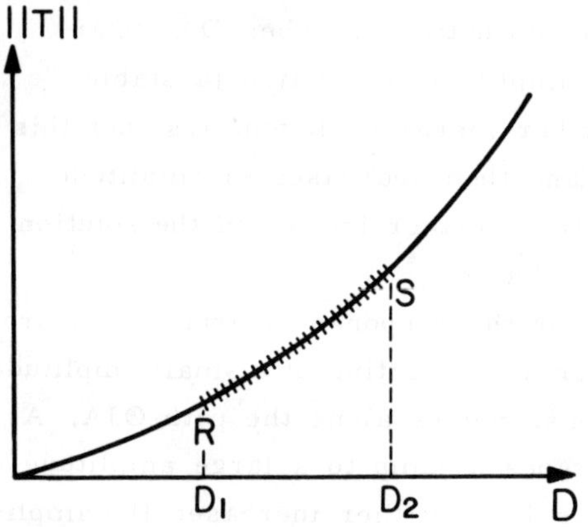

Fig. II.1.

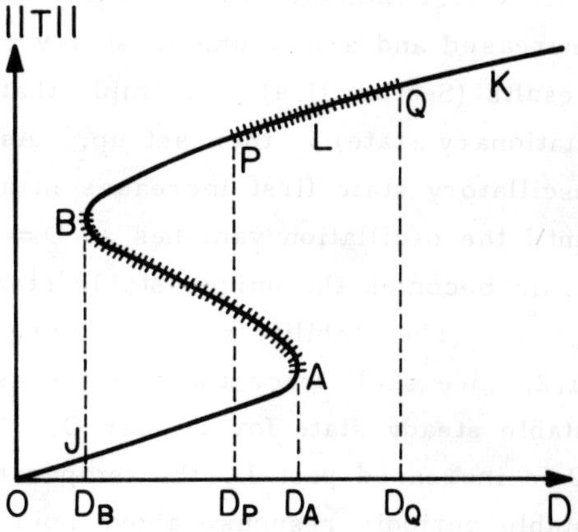

Fig. II.2.

Figure II.1 shows that for values of the physical parameters β, B, P, T_c, λ, a, α in certain ranges (and these ranges are given explicitly by A. B. Poore [12], [13]),there is a unique solution of the steady state problem for all values of the Damkohler number D. The cross-hatching represents instability; that is, based on linearized stability theory for the system (1.1)-(1.8), it is found that for all values of D in $D_1 \leq D \leq D_2$ the unique steady state is unstable. Based on both experiments and machine computation for special values of the parameters, R. A. Schmitz [9] has concluded that in such situations an oscillatory stationary state is set up so that, in fact, no stable steady state does exist but a periodic solution of the time dependent equations (1.1)-(1.8) evolves for D in the range $D_1 \leq D \leq D_2$. Thus, it is reasonable to expect that points R and S are bifurcation points for the time-dependent (parabolic) problem (1.1)-(1.8), and from these points branches of periodic solutions bifurcate. In Section II.4 we shall show (formally) that this can be the case, but in fact, a more complex situation can actually occur, and we discuss this in Section II.5.

Figure II.2 illustrates a steady state response diagram for a different set of values of the parameters [12], [13]. The cross-hatching again represents instability. In this case we shall establish (also formally) that the points P and

Q are bifurcation points for the time dependent problem, and from these points branches of periodic solutions bifurcate.

Our illustrations, Figures II.1 and II.2, indicate two of the six distinct situations which occur for the steady state problem corresponding to equations (1.1)-(1.8). All six cases and the conditions for their occurrence are given by A. B. Poore [12]. We have described two typical situations here. Our analysis applies equally well to all cases, and it resolves the stability considerations of the response diagrams for the non-adiabatic tubular reactor. Thus, for example, Figure II.1 indicates that in the steady state as D is increased, so also is $\|T\|$ increased and a new unique steady state is set up until D reaches D_1. Our results (Section II.4) then imply that a small amplitude oscillation (a stable stationary state) is then set up. As D is further increased it appears that this oscillatory state first increases in amplitude and then decreases in amplitude until the oscillation vanishes at $D = D_2$. As D is further increased the solution again becomes the unique stable steady state solution.

The stability story is more complex for the response diagram of Figure II.2. We shall trace the process as D is increased starting at a small amplitude stable steady state for D near O. The response moves along the path OJA. As D is increased past D_A the temperature undergoes a jump to a large amplitude stable periodic response above point L, and as D is further increased the amplitude of this oscillation decreases until the oscillation vanishes at point Q. The stable steady state is followed up the branch through point K as D increases still further. Now, as D is decreased the process follows the branch KQLPB, the solution being steady from K to Q and from P to B with a stable oscillation of increasing, then decreasing amplitude from Q to P. As D is decreased below D_B, there is an extinction as the response jumps to point J and then follows the path J to O.

The above arguments are plausible and indeed very satisfying within the context of chemical reactor theory. We shall establish their truth in the next sections. The main points are that, for the time dependent nonlinear equations governing non-adiabatic tubular reactors, the points R and S of Figure II.1 and P and Q are of Figure II.2 are bifurcation points and that from these bifurcation points at least two distinct solutions branches emanate. One branch represents a solution not depending explicity on time (but nevertheless still a solution of the time dependent equations), and the other branch at each bifurcation point represents a branch of time periodic solutions. Thus, the bifurcation diagrams for the time dependent system look like those illustrated in Figures II.3 and II.4. A more complex bifurcation diagram corresponding to the situation of Figure II.2 is discussed later in Section II.5.

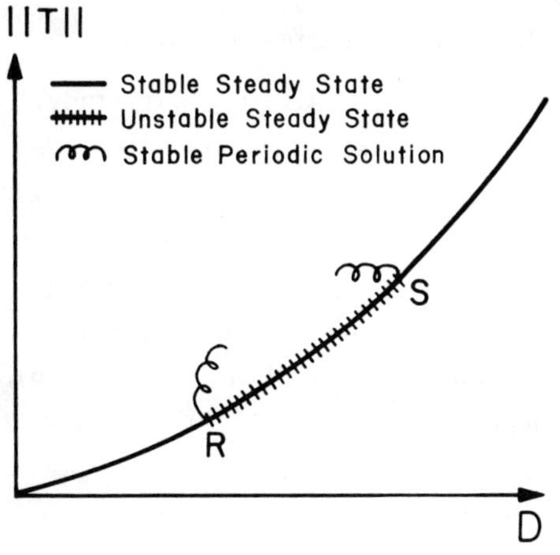

Fig. II.3.

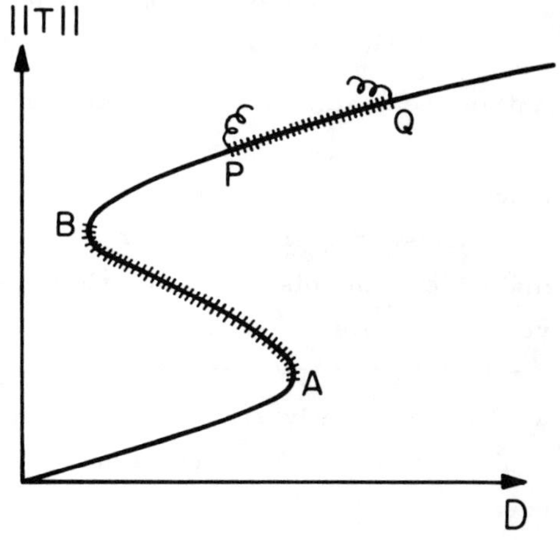

Fig. II.4.

C. R. McGowin and D. D. Perlmutter [11] and A. B. Poore [12], [13] have given complete tabulations of all the points on the steady state response diagrams for (1.1)-(1.8) where changes of (linearized) stability take place. (Poore gives complete analytical characterizations of these points for small $\epsilon > 0$.) Thus, the values and characterizations of T, C and D at the points R, S, P, Q of Figures II.1 and II.2 are known. The present analysis begins at this stage. Suppose, for example, that $T_1(x)$ and $C_1(x)$ represent the steady state solutions at point P of Figure II.2 when $D = D_P$. Then, let $u(x,t) = T(x,t) - T_1(x)$, $w(x,t) = C(x,t) - C_1(x)$, and $\lambda = D - D_P$. Equations (1.1)-(1.8) can then be written in the form (1.9)-(1.16) for the unknowns u and w. Hence, $u \equiv w \equiv 0$ at $\lambda = 0$ are solutions corresponding to the steady state solutions of (1.1)-(1.8), and we shall look for the bifurcation of time periodic solution branches from this point. In fact, equations (1.1)-(1.8) take the form (1.9)-(1.16) when we perform the same change of variables near any of the points R, S, P or Q. Therefore, our arguments of this section will be established by a demonstration that stable oscillatory solutions bifurcate from the point $\lambda = 0$, $\|u\| = \|w\| = 0$.

Motivated by all the preceding discussion we now study the system (1.9) -(1.16) under the following conditions:

H-1: $f(\lambda, u, w)$ and $g(\lambda, u, w)$ are smooth functions of (λ, u, w) satisfying $f(\lambda_0, 0, 0) = g(\lambda_0, 0, 0) = 0$.

H-2: f and g contain no linear terms in u and w near $(\lambda, u, w) = (\lambda_0, 0, 0)$; that is,

$$f(0,0,0) = f_u(0,0,0) = f_w(0,0,0) = 0 ,$$

$$g(0,0,0) = g_u(0,0,0) = g_w(0,0,0) = 0 .$$

H-3: $\alpha(\lambda)$ and $\beta(\lambda)$ are smooth functions of λ satisfying $\alpha(\lambda_0) = 0$ for some $\lambda = \lambda_0$, and $\beta(\lambda) > 0$ for all λ.

H-4: $\epsilon > 0$ satisfies $0 < \epsilon << 1$, and all other quantities are O(1).

Within the framework of the nonlinear diffusion processes motivating our study, the functions f, g, α and β are, in fact, infinitely continuously differentiable functions of all their arguments. Strictly speaking, we do not need such strong smoothness requirements for our analysis; however, for ease of presenting the main ideas in this paper, we shall assume that "smooth" implies this strong requirement of C^∞ functions. The hypothesis H-2 is simply a formulation of the fact that in performing the transformations to take (1.1)-(1.8) into (1.9)-(1.16), we have explicitly subtracted the linear terms near $(\lambda, u, w) = (\lambda_0, 0, 0)$ from the nonlinearities. Thus, the linearized system about $(\lambda. u, w) = (\lambda_0, 0, 0)$ is simply equations (1.9)-(1.16) with the functions f and g deleted. The value $\lambda = \lambda_0$ corresponds to the value of D at the points R, S, P or Q of Figures II.1 and II.2. At such a point it is found that $\alpha(\lambda_0) = 0$. Thus, our problem (1.1)-(1.16) under the conditions H-1 to H-4 contains the problem of the non-adiabatic tubular reactor as a special case and encompasses a much wider class of problems.

<u>II.3. The reduced system</u>. By a singular perturbation procedure we shall now reduce the problem (1.9)-(1.16) to a somewhat simpler system of nonlinear ordinary differential equations which we shall study in Section II.4 using the "two-timing" method.

Assume that $0 < \epsilon << 1$. We now construct the asymptotic expansions of the solutions of (1.9)-(1.16) as $\epsilon \to 0$. In this case we find, by the techniques of singular perturbation theory [4], that there is an initial boundary layer of thickness $O(\epsilon)$ near $t = 0$ for all x in $0 \leqslant x \leqslant 1$. Away from this boundary layer the form of the asymptotic expansion (the outer solution) is given by

$$(3.1) \qquad u(x,t,\epsilon) \sim \sum_{n=0}^{\infty} u_n(x,t)\epsilon^n, \qquad w(x,t,\epsilon) \sim \sum_{n=0}^{\infty} w_n(x,t)\epsilon^n .$$

Inserting (3.1) into equations (1.9)-(1.14), we find by equating like powers of ϵ, that to order ϵ^2 we obtain

$$(3.2) \qquad \begin{cases} \dfrac{\partial^2 u_0}{\partial x^2} = 0 , \\ \dfrac{\partial u_0}{\partial x}(0,t) = 0 , \\ \dfrac{\partial u_0}{\partial x}(1,t) = 0 , \end{cases} \qquad \begin{cases} \dfrac{\partial^2 w_0}{\partial x^2} = 0 , \\ \dfrac{\partial w_0}{\partial x}(0,t) = 0 , \\ \dfrac{\partial w_0}{\partial x}(1,t) = 0 . \end{cases}$$

$$(3.3) \qquad \frac{\partial^2 u_1}{\partial x^2} = \frac{\partial u_0}{\partial t} + \frac{\partial u_0}{\partial x} - [\alpha(\lambda)+1]u_0 + \beta(\lambda)w_0 - f(\lambda, u_0, w_0) ,$$

$$(3.4) \qquad \frac{\partial^2 w_1}{\partial x^2} = \frac{\partial w_0}{\partial t} + \frac{\partial w_0}{\partial x} - \beta(\lambda)u_0 - [\alpha(\lambda)+1]w_0 - g(\lambda, u_0, w_0) ,$$

(3.5) $$\frac{\partial u_1}{\partial x}(0,t) = u_0(0,t) ,$$

(3.6) $$\frac{\partial w_1}{\partial x}(0,t) = w_0(0,t) ,$$

(3.7) $$\frac{\partial u_1}{\partial x}(1,t) = 0 ,$$

(3.8) $$\frac{\partial w_1}{\partial x}(1,t) = 0 .$$

Equations (3.2) imply that

(3.9) $$u_0(x,t) = a(t) , \qquad w_0(x,t = b(t) ,$$

where a(t) and b(t) are, at this stage, arbitrary functions of t. In order to determine them we must proceed to the next step in the perturbation procedure.

Using (3.9), we can write the differential equations (3.3) and (3.4) as

(3.10) $$\frac{\partial^2 u_1}{\partial x^2} = \frac{da}{dt} - [\alpha(\lambda)+1]a+\beta(\lambda)b-f(\lambda,a,b) \equiv A(t) ,$$

(3.11) $$\frac{\partial^2 w_1}{\partial x^2} = \frac{db}{dt} -\beta(\lambda)a-[\alpha(\lambda)+1]b-g(\lambda,a,b) \equiv B(t) .$$

Thus, $u_1(x,t) = \frac{1}{2}A(t)x^2 + c_1(t)x + c_2(t)$, and $w_1(x,t) = \frac{1}{2}B(t)x^2 + c_3(t)x + c_4(t)$, where the $c_i(t)$, $i=1,\cdots 4$, are arbitrary functions. The boundary conditions (3.5) and (3.6) imply that $c_1(t) = a(t)$ and $c_3(t) = b(t)$. Finally, to satisfy the boundary conditions (3.7) and (3.8), we find that we must have $A(t) + a(t) = 0$ and $B(t) + b(t) = 0$; that is, we must have

(3.12) $$\frac{da}{dt} = \alpha(\lambda)a-\beta(\lambda)b+f(\lambda,a,b) ,$$

(3.13) $$\frac{db}{dt} = \beta(\lambda)a+\alpha(\lambda)b+g(\lambda,a,b) .$$

Thus, the functions a(t) and b(t) are determined from compatibility conditions (3.12) and (3.13) which they must satisfy in order to generate a consistent perturbation procedure.

We now obtain the appropriate initial conditions for a(t) and b(t) by a standard matching procedure with the asymptotic form in the initial boundary layer. Let $\tilde{t} = t/\epsilon$, and let $u(x,t) = u(x,\epsilon\tilde{t}) \equiv U(x,\tilde{t})$, $w(x,t) = w(x,\epsilon\tilde{t}) \equiv W(x,\tilde{t})$. With this change of variables we find that, to first order in ϵ, the equations (1.9) -(1.16) becomes

$$(3.14)\quad \begin{cases} U_{\tilde{t}} = U_{xx}, & W_{\tilde{t}} = W_{xx}, \\ U_x(0,\tilde{t}) = 0, & W_x(0,\tilde{t}) = 0, \\ U_x(1,\tilde{t}) = 0, & W_x(1,\tilde{t}) = 0, \\ U(x,0) = \phi(x), & W(x,0) = \psi(x). \end{cases}$$

Therefore,

$$(3.15)\quad U(x,\tilde{t}) = A_0 + \sum_{n=1}^{\infty} A_n e^{-n^2\pi^2\tilde{t}} \cos n\pi x,$$

$$(3.16)\quad W(x,\tilde{t}) = B_0 + \sum_{n=1}^{\infty} B_n e^{-n^2\pi^2\tilde{t}} \cos n\pi x,$$

where

$$(3.17)\quad A_0 = \int_0^1 \phi(\xi)d\xi, \qquad A_n = 2\int_0^1 \phi(\xi)\cos n\pi\xi \, d\xi,$$

$$(3.18)\quad B_0 = \int_0^1 \psi(\xi)d\xi, \qquad B_n = 2\int_0^1 \psi(\xi)\cos n\pi\xi \, d\xi.$$

Equations (3.15) and (3.16) give the first term (i.e., the zero order term) in the asymptotic expansion of the solution (the inner solution) of (1.9)-(1.16) in the boundary layer. The standard matching of inner and outer solutions now requires that

$$(3.19)\quad \lim_{t\to 0} [u_0(x,t)] = \lim_{\tilde{t}\to\infty} [U(x,\tilde{t})],$$

$$(3.20)\quad \lim_{t\to 0} [w_0(x,t)] = \lim_{\tilde{t}\to 0} [W(x,\tilde{t})].$$

Therefore, $a(0) = A_0$ and $b(0) = B_0$, and the equations governing $a(t)$ and $b(t)$ are

$$(3.21)\quad \frac{da}{dt} = \alpha(\lambda)a - \beta(\lambda)b + f(\lambda,a,b),$$

$$(3.22)\quad \frac{db}{dt} = \beta(\lambda)a + \alpha(\lambda)b + g(\lambda,a,b),$$

$$(3.23)\quad a(0) = A_0 = \int_0^1 \phi(\xi)d\xi,$$

$$(3.24)\quad b(0) = B_0 = \int_0^1 \psi(\xi)d\xi.$$

The equations (3.14)-(3.18) show clearly that, to first order in ϵ, the mechanism governing the solution of (1.9)-(1.16) is initially the standard linear diffusion process. After a time of the order $O(\epsilon)$, the process, to first order in ϵ, is governed by the system (3.21)-(3.24); we shall now study this system.

II.4. Periodic solutions. The equations (3.21), (3.22) constitute a standard autonomous system normally treated by classical phase-plane techniques, and in fact, much of what we shall establish here can be derived using a considerable amount of well-established literature. However, we shall not follow such an approach for two reasons: (i) System (1.1)-(1.8) describes the simplest type of non-adiabatic tubular reactor. For large Peclet numbers (i.e., $P >> 1$) our methods apply equally well to considerably more complicated problems in reactor theory [6], [11]. In these more difficult problems the equations analogous to (3.21), (3.22) which arise are not standard autonomous phase-plane systems, but much more difficult systems. Nevertheless, they can be handled by the techniques we shall develop here. (ii) The "two-timing" method which we shall use produces formulas for the solutions which are immediately interpretable physically and from which the stability of the solutions is also immediately resolved without recourse to further analysis. Thus, this method is simply much easier to use even on the standard phase-plane systems (and it applies equally well to the more difficult problems). When it works the technique is spectacularly successful as has so often been the case where it has been applied in other problems (see [4] and [8] for examples).

The motivation for a two-timing approach as well as the proper scaling for the asymptotic analysis comes from the following reasoning: Consider, for example, the upper branch of the response diagram of Figure II.2 when we increase D from slightly below D_P to slightly above D_P so that we pass through the point P where the steady state changes from stable to unstable. As discussed in Section II.2, we have concluded that the steady states to the right of point P are unstable based upon a linearized perturbation theory which implies that perturbations from this steady state will initially grow exponentially in time. This (linearized) exponentially growing function cannot represent the solution for very long because clearly the nonlinear terms must then become important. If, in fact, this exponentially growing function tends to a stable oscillatory solution, as we conjectured in Section II.2, then growth on another time scale must come into play so that in some sense the perturbation from the unstable steady state should exhibit a more or less typical multi-time scale representation; namely, we expect a representation of the form $u(x,t) = A(\tau)P(t^*)$ where $P(t^*)$ represents a periodic oscillation on a so-called "fast-time" t^* and $A(\tau)$ represents "slow-time" modulation which perhaps approaches a constant value as time $t \to \infty$. Change of stability in a somewhat different type of heat conduction problem has also been studied in this way with a two-timing method by B. J. Matkowsky [10].

For ease of presentation here and in order not to obscure the basic method with lengthy algebraic calculations, we shall now perform our investigation on the system (3.21)-(3.24) for the special case that

$$f(\lambda, a, b) = -\lambda a^3, \qquad g(\lambda, a, b) = -\mu\lambda a b^2,$$

(4.1)

$$\beta(\lambda) = \text{constant}\ \beta \qquad \alpha(\lambda_0) = \alpha'(\lambda_0) = 0, \quad \alpha''(\lambda_0) \neq 0.$$

In fact, such a choice is an excellent model for a simple first order reaction for the system (1.1)-(1.8), where the β here corresponds to the β in equation (1.1) and μ corresponds to the transformed B of equation (1.1), see [5]. We would like to point out that our entire analysis can be carried out for the general system (1.9)-(1.16) under conditions H-1 to H-4 as well as for the general reactor problems involving higher order chemical kinetics for the rate functions f and g. This general treatment necessarily requires considerably more complicated algebraic manipulation, and in fact, for the general reactor kinetics, some of the algebraic equations require the numerical specification of certain physical contents and numerical procedures for solving involved algebraic expressions. Such specific information would be of use only in specific chemical processes, but, in fact, for certain interesting commonly occurring chemical constants, A. B. Poore [12] has carried out these calculations.

The only necessary tool which we shall need in carrying out the two-timing formalism is an elementary fact from ordinary differential equations which we shall state in the form of an easily referenced lemma.

Lemma. The general solution of

$$\frac{dx}{dt} + y = m\sin t + n\cos t,$$

$$\frac{dy}{dt} - x = p\sin t + q\cos t$$

is

$$x(t) = A\sin t + B\cos t + \left(\frac{m-q}{2}\right)t\sin t + \left(\frac{n+p}{2}\right)t\cos t + \left(\frac{n-p}{2}\right)\sin t,$$

$$y(t) = -A\cos t + B\sin t + \left(\frac{n+p}{2}\right)t\sin t - \left(\frac{m-q}{2}\right)t\cos t + \left(\frac{m+q}{2}\right)\sin t.$$

Thus, in order to suppress secular terms (i.e., in order to have solutions bounded for all $t \geq 0$) it is sufficient to require $m-q = 0$ and $n+p = 0$.

We define δ by the relationship $\delta = \lambda - \lambda_0$, and we assume that

(4.2) $$a \equiv a(\overset{*}{t}, \tau) = \delta a_1(\overset{*}{t}, \tau) + \delta^2 a_2(\overset{*}{t}, \tau) + \cdots,$$

(4.3) $$b \equiv b(\overset{*}{t}, \tau) = \delta b_1(\overset{*}{t}, \tau) + \delta^2 b_2(\overset{*}{t}, \tau) \cdots,$$

where the "slow time" τ and "fast time" $\overset{*}{t}$ are defined by

(4.4) $$\tau = \delta^2 t \ ,$$

(4.5) $$\overset{*}{t} = (1+\delta\omega_1 + \delta^2\omega_2 + \cdots)t \ .$$

Much of the technique has already been employed with judicious choice of the forms (4.2-(4.5). The further manipulation (see Cole [4] or Kevorkian [8] for its exposition) requires that the ω_i and the other unknowns which will occur will be chosen according to the principle that we suppress secular terms in such a way that we generate a self-consistent procedure for determining bounded functions $a_i(\overset{*}{t},\tau)$ and $b_i(\overset{*}{t},\tau)$ with modulation only on the slow time scale τ. We shall now carry out this procedure.

With the definitions (4.4) and (4.5) we find that $\frac{d}{dt} = (1+\delta\omega_1+\delta^2\omega_2+\cdots)\frac{\partial}{\partial \overset{*}{t}} + \delta^2\frac{\partial}{\partial\tau}$. Thus, substituting (4.2-(4.5) into (3.21), (3.22) under the requirements (4.1) and equating the coefficients of like powers of δ, we obtain

(4.6) $$\begin{cases} \dfrac{\partial a_1}{\partial \overset{*}{t}} = -\beta b_1 \ , \\ \dfrac{\partial b_1}{\partial \overset{*}{t}} = \beta a_1 \ , \end{cases}$$

(4.7) $$\begin{cases} \dfrac{\partial a_2}{\partial \overset{*}{t}} + \beta b_2 = -\omega_1 \dfrac{\partial a_1}{\partial \overset{*}{t}} \ , \\ \dfrac{\partial b_2}{\partial \overset{*}{t}} - \beta a_2 = -\omega_1 \dfrac{\partial b_1}{\partial \overset{*}{t}} \ , \end{cases}$$

(4.8) $$\begin{cases} \dfrac{\partial a_3}{\partial \overset{*}{t}} + \beta b_3 = -\dfrac{\partial a_1}{\partial\tau} - \omega_1 \dfrac{\partial a_2}{\partial \overset{*}{t}} - \omega_2 \dfrac{\partial a_1}{\partial \overset{*}{t}} + \dfrac{\alpha''(\lambda_0)}{2} a_1 - \lambda_0 a_1^3 \\ \dfrac{\partial b_3}{\partial \overset{*}{t}} - \beta a_3 = -\dfrac{\partial b_1}{\partial\tau} - \omega_1 \dfrac{\partial b_2}{\partial \overset{*}{t}} - \omega_2 \dfrac{\partial b_1}{\partial \overset{*}{t}} + \dfrac{\alpha''(\lambda_0)}{2} b_1 - \mu\lambda_0 a_1 b_1^2 \ . \end{cases}$$

The solution of (4.6) is given by

(4.9) $$\begin{aligned} a_1(\overset{*}{t},\tau) &= A(\tau)\sin\beta\overset{*}{t} + B(\tau)\cos\beta\overset{*}{t} \ , \\ b_1(\overset{*}{t},\tau) &= -A(\tau)\cos\beta\overset{*}{t} + B(\tau)\sin\beta\overset{*}{t} \ , \end{aligned}$$

where the unknown functions $A(\tau)$ and $B(\tau)$ will be determined at a later stage of the perturbation procedure. With these formulas for $a_1(\overset{*}{t},\tau)$ and $b_1(\overset{*}{t},\tau)$ used on the right-hand side of (4.7), we find that (4.7) becomes

$$\frac{\partial a_2}{\partial t^*} + \beta b_2 = -\omega_1 \beta A(\tau)\cos\beta t^* + \omega_1 \beta B(\tau)\sin\beta t^* ,$$

(4.10)

$$\frac{\partial b_2}{\partial t^*} - \beta a_2 = \omega_1 \beta A(\tau)\sin\beta t^* - \omega_1 \beta B(\tau)\cos\beta t^* .$$

Using our lemma to suppress secular terms, we see immediately that we must require that $\omega_1 = 0$. Thus,

$$a_2 (t^*, \tau) = C(\tau)\sin\beta t^* + D(\tau)\cos\beta t^* ,$$

(4.11)

$$b_2 (t^*, \tau) = -C(\tau)\cos\beta t^* + D(\tau)\sin\beta t^* ,$$

where the $C(\tau)$ and $D(\tau)$ are to be determined at a later stage in our perturbation scheme. The system (4.8) now becomes

(4.12)

$$\left\{\begin{aligned}
\frac{\partial a_3}{\partial t^*} + \beta b_3 = &- \frac{dA}{d\tau}\sin\beta t^* - \frac{dB}{d\tau}\cos\beta t^* - \omega_2 \beta A\cos\beta t^* + \omega_2 \beta B\sin\beta t^* \\
&+ \frac{\alpha''(\lambda_0)}{2} A\sin\beta t^* + \frac{\alpha''(\lambda_0)}{2} B\cos\beta t^* - \lambda_0 A^3 \sin^3 \beta t^* \\
&- 3\lambda_0 A^2 B\sin^2 \beta t^* \cos\beta t^* - 3\lambda_0 AB^2 \sin\beta t^* \cos^2 \beta t^* - \lambda_0 B^3 \cos^3 \beta t^* , \\
\frac{\partial b_3}{\partial t^*} - \beta a_3 = &\frac{dA}{d\tau}\cos\beta t^* - \frac{dB}{d\tau}\sin\beta t^* - \omega_2 \beta A\sin\beta t^* - \omega_2 \beta B\cos\beta t^* \\
&\frac{-\alpha''(\lambda_0)}{2} A\cos\beta t^* + \frac{\alpha''(\lambda_0)}{2} B\sin\beta t^* - \mu\lambda_0 AB^2 \sin^3 \beta t^* \\
&- \mu\lambda_0 (B^3 - 2A^2B)\sin^2\beta t^* \cos\beta t^* - \mu\lambda_0(A^3 - 2AB^2)\sin\beta t^* \cos\beta t^* - \mu\lambda_0 A^2 B\cos^3\beta t^* .
\end{aligned}\right.$$

By employing certain well-known trigonometric identities (or equivalently, by developing the right-hand sides in a Fourier series), we can write (4.12) as

(4.13)

$$\left\{\begin{aligned}
\frac{\partial a}{\partial t^*} + \beta b_3 = &\left[- \frac{dA}{d\tau} + \omega_2 \beta B + \frac{\alpha''(\lambda_0)}{2} A - \frac{3}{4}\lambda_0 A^3 - \frac{3}{4}\lambda_0 AB^2\right]\sin\beta t^* \\
&+ \left[- \frac{dA}{d\tau} - \omega_2 \beta A + \frac{\alpha''(\lambda_0)}{2} B - \frac{3}{4}\lambda_0 A^2 B - \frac{3}{4}\lambda_0 B^3\right]\cos\beta t^* \\
&+ \text{(higher harmonics)}, \\
\frac{\partial b_3}{\partial t^*} - \beta a_3 = &\left[- \frac{dB}{d\tau} - \omega_2 \beta A + \frac{\alpha''(\lambda_0)}{2} B - \frac{3}{4}\mu\lambda_0 AB^2 - \frac{1}{4}\mu\lambda_0(A^3 - 2AB^2)\right]\sin\beta t^* \\
&+ \left[\frac{dA}{d\tau} - \omega_2 \beta B - \frac{\alpha''(\lambda_0)}{2} A - \frac{1}{4}\mu\lambda_0(B^3 - 2A^2B) - \frac{3}{4}\mu\lambda_0 A^2 B\right]\cos\beta t^* \\
&+ \text{(higher harmonics)}.
\end{aligned}\right.$$

Now, suppression of secular terms according to our lemma implies that we require

$$
(4.14)\qquad
\begin{aligned}
&-2\frac{dA}{d\tau} + 2\omega_2\beta B + \alpha'' A - \frac{3}{4}\lambda_0 A^3 - \frac{3}{4}\lambda_0 AB^2 + \frac{1}{4}\mu\lambda_0(B^3 - 2A^2B) + \frac{3}{4}\mu\lambda_0 A^2B = 0 ,\\
&-2\frac{dB}{d\tau} - 2\omega_2\beta A + \alpha'' B - \frac{3}{4}\lambda_0 A^2B - \frac{3}{4}\lambda_0 B^3 - \frac{3}{4}\mu\lambda_0 AB^2 - \frac{1}{4}\mu\lambda_0(A^3 - 2AB^2) = 0
\end{aligned}
$$

or equivalently,

$$
(4.15)\qquad
\begin{aligned}
&-2\frac{dA}{d\tau} + 2\omega_2\beta B + \alpha''(\lambda_0)A - \frac{3}{4}\lambda_0 A(A^2+B^2) + \frac{1}{4}\mu\lambda_0 B(A^2+B^2) = 0 ,\\
&-2\frac{dB}{d\tau} - 2\omega_2\beta A + \alpha''(\lambda_0)B - \frac{3}{4}\lambda_0 B(A^2+B^2) - \frac{1}{4}\mu\lambda_0 A(A^2+B^2) = 0 .
\end{aligned}
$$

The periodic nature of the solutions of (3.21), (3.22) for small $\delta > 0$ can now be established by employing a device due to J. D. Cole [4]. Multiply the first of equations (4.15) by A and the second by B and then add the equations to obtain

$$
(4.16)\qquad -\frac{dR}{d\tau} + \alpha''(\lambda_0)R - \frac{3}{4}\lambda_0 R^2 = 0 ,
$$

where

$$
(4.17)\qquad R = A^2 + B^2 .
$$

The solution of (4.16) is easily found to be

$$
(4.18)\qquad R(\tau) = \frac{4\alpha''(\lambda_0)}{3\lambda_0}\,\frac{1}{1+e^{-\alpha''(\lambda_0)\tau}} .
$$

Thus, equations (4.2), (4.3), (4.17), (4.18) clearly show an approach to a limit cycle of amplitude

$$
(4.19)\qquad 2\left[\frac{\alpha''(\lambda_0)}{3\lambda_0}\right]^{1/2}\delta^{1/2} .
$$

Thus, as claimed in Section II.2, we have established that periodic solution branches of the form (4.19) to $O(\delta)$ bifurcate from the points R and S of Figure II.1 and from the points P and Q of Figure II.2. Note that the stability of these oscillatory branches follows immediately by writing equation (4.16) in the form

$$
\frac{dR}{d\tau} = \frac{3\lambda_0}{4}\left[\frac{4\alpha''(\lambda_0)}{3\lambda_0} - R\right]
$$

from which we can immediately conclude that, for $R(\tau)$ less (greater) than its steady state value $4\alpha''(\lambda_0)/3\lambda_0$, we have $\frac{dR}{d\tau}$ greater (less) than zero implying motion toward the steady state (i.e., stability).

II.5 Multiple stable and unstable periodic solutions. For values of the physical parameters $\beta, B, P, T_c, \lambda, a, \alpha$ in certain ranges (see [12]), the situation illustrated in Figure II.2 remains the same with regard to steady states and their stability, but the bifurcation of periodic solutions is very different from that illustrated in Figure II.4. The situation which occurs is that illustrated in Figure II.5, and we shall now discuss this case.

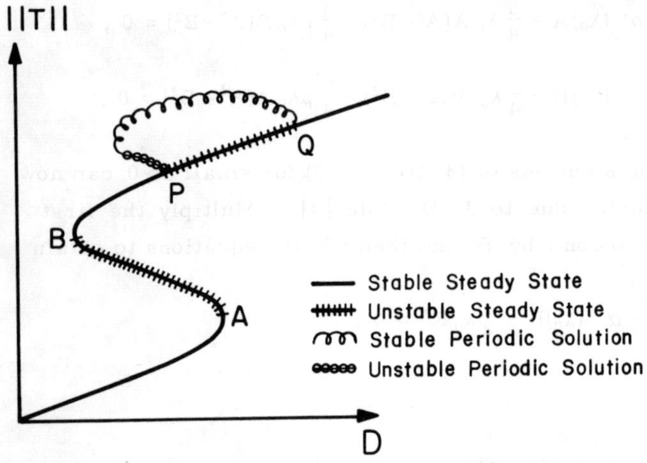

Fig. II.5

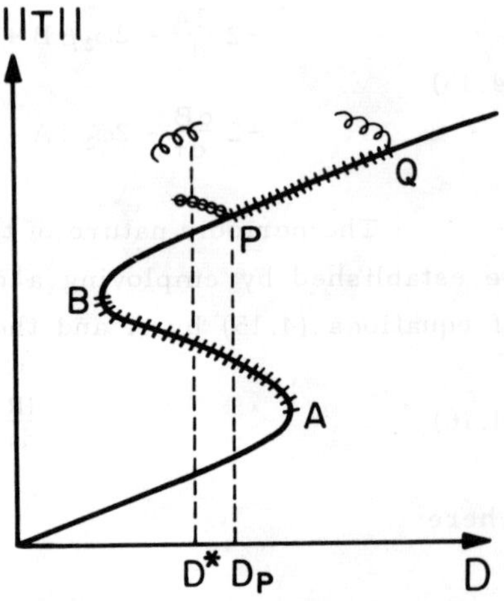

Fig. II.6

For certain values of the parameters yielding the situation of Figure II.2, we find that the "two-timing" procedure of Section II.4 is applicable at point Q but will not work at point P. By suitable changes of variables, however, it can be shown that the autonomous system (3.21)-(3.24) can be re-cast into a form to which a modified version of a theory of K. O. Friedrichs can be applied.(See [3], [12] or [13] for reference to Friedrichs' work. The theory necessary for the present analysis is given in detail in [12].) The analysis based on Friedrichs' theory does not yield formulas as nice as those found with the multi-time scale methods. However, Friedrichs' theory is rigorous, and our modifications enable us to rigorously establish the results shown in Figure II.6. That is, we can prove that an unstable branch of periodic solutions bifurcates to the left at point P. Furthermore, for sufficiently small $D_P - D^*$, we can establish the existence of a stable periodic solution surrounding the unstable one. The existence of a stable branch of periodic solutions bifurcating to the left from point Q is also proved. (All this is carried out in [3] and [12].) Therefore, Figure II.6 contains the information proved.

Highly heuristic and formal perturbation techniques together with the computational investigations of Hlavacek, Kubicek, and Jelinek [5] force us to conclude that the diagram of Figure II. 6 is completed as shown in Figure II. 5.

The figures of Chapter I (i.e., Figs. I. 2 to I.13) contain all possible situations which can occur. The supporting analysis is given in [3] and [12]. At each point where there is a change of stability in the steady state, the algebraic sign of a certain quantity (labeled $\delta'(P)$ in the figures) indicates the direction of branching of a periodic solution branch. The derivation of $\delta'(P)$ together with the stability results for all solution branches is given by A. P. Poore [12] and D. S. Cohen and A. B. Poore [3].

REFERENCES

1. Cohen, D.S., Multiple stable solutions of nonlinear boundary value problems arising in chemical reactor theory, SIAM J. Appl. Math., 20 (1971) 1-13.
2. Cohen, D.S., Multiple solutions of singular perturbation problems, SIAM J. Math. Anal., 3 (1972) 72-82.
3. Cohen, D.S. and Poore, A.B., Multiple stable and unstable periodic solutions of the nonlinear equations of chemical reactor theory, to appear.
4. Cole, J.D., Perturbation Methods in Applied Mathematics, Blaisdell Publishing Co., Waltham, Mass. 1968.
5. Hlavacek, V., Kubicek, M., and Jelinek, J., Modeling of chemical reactors -XVII: Stability and oscillatory behavior of the CSTR, Chem. Eng. Sci., 25 (1970) 1441-1461.
6. Hlavacek, V., and Hofmann, H., Modeling of chemical reactors-XIX: Transient axial heat and mass transfer in tubular reactors. The stability considerations-I. Chem. Eng. Sci., 25 (1970) 1517-1526.
7. Hlavacek, V., Hofmann, H., and Kubicek, M., Modeling of chemical reactors -XXIV: Transient axial heat and mass transfer in tubular reactors. The stability considerations-II. Chem. Eng. Sci., 26 (1971) 1629-1634.
8. Kevorkian, J., The two variable expansion procedure for the approximate solution of certain nonlinear differential equations. Lecture in Appl. Math., No. 7, Space Math., Part III, Amer. Math. Soc., 1966. Also in Proc. Yale Univ. Summer Institute on Dynamical Astronomy.
9. Lindburg, R. C., and Schmitz, R.A., Dynamics of heterogeneous reaction at a stagnation point: Numerical study of nonlinear transient effects, Int. J. Heat Mass Transfer, 14 (1971) 718-721.
10. Matkowsky, B.J., A simple nonlinear dynamic stability problem, Bulletin Amer. Math. Soc., 76 (1970) 620-625.
11. McGowin, C.R., and Perlmutter, D.D., Tubular reactor steady state and stability characteristics, AIChE Journal, 17 (1971) 831-849.
12. Poore, A.B., Stability and bifurcation phenomena in chemical reactor theory, Ph.D. thesis, California Institute of Technology, 1972.
13. Poore, A.B., Multiplicity, stability and bifurcation of periodic solutions in problems arising from chemical reactor theory, to appear.
14. Winegardner, D.K., and Schmitz, R.A., Dynamics of heterogeneous reaction at a stagnation point, AIAA Journal, 5 (1967) 1589-1595.

Chapter III

THE STEADY ADIABATIC REACTOR WITH RECYCLE

III.1. Introduction. Experimental and computer results (see [2] for references to these works) clearly show that multiple stable and unstable steady states exist. The governing mechanism, however, is more complicated than that in the corresponding steady reactor without recycle. Presumably, analytical treatment has previously been unsuccessful because iteration schemes relying on maximum principles, certain shooting techniques, and certain perturbation methods are not applicable due to the boundary conditions (Chapter I, equations (1.8), (1.9)) which incorporate data from both ends of the interval. We shall show how such a problem can be treated by a singular perturbation method.

We shall consider the case where the Peclet number P is very large, that is, $0 < \frac{1}{P} = \epsilon << 1$. In the steady state $(\frac{\partial}{\partial t} \equiv 0)$ where the reactor is operating adiabatically $(\beta = 0)$, the appropriate equations of Chapter I reduce to

(1.1) $$\epsilon u'' - u' + f(u,w) = 0 , \qquad 0 < x < 1 ,$$

(1.2) $$\epsilon w'' - w' + g(u,w) = 0 , \qquad 0 < x < 1 ,$$

(1.3) $$\epsilon u'(0) - u(0) + Ru(1) = -(1-R)u ,$$

(1.4) $$\epsilon w'(0) - w(0) + Rw(1) = -(1-R) ,$$

(1.5) $$u'(1) = 0 ,$$

(1.6) $$w'(1) = 0 .$$

We shall consider simple first order Arrhenius kinetics so that

(1.7) $$f(u,w) = -g(u,w) = Dwe^{\frac{-k}{u}} .$$

This problem is sufficiently simple so that the equations for u and w can be uncoupled in the following way: Add equation (1.2) to equation (1.1) to obtain

(1.8) $$\epsilon v'' - v' = 0, \qquad 0 < x < 1 ,$$

where

(1.9) $$v(x) = u(x) + w(x) .$$

Similarly, adding equation (1.4) to equation (1.3) yields

(1.10) $$\epsilon v'(0) - v(0) + Rv(1) = -(1-R)(1+u_0) .$$

Similarly, (1.5) and (1.6) imply that

(1.11) $$v'(1) = 0 .$$

The solution of (1.8), (1.10), (1.11) is

(1.12) $$v(x) \equiv 1 + u_0 \ .$$

Eliminating v(x) from (1.9) and (1.12) yields

(1.13) $$w(x) = 1 + u_0 - u(x) \ .$$

Therefore,

(1.14) $$f(u,w) = Dwe^{\frac{-k}{u}} = D(1+u_0-u)e^{\frac{-k}{u}} \ .$$

Hence, our problem for u reduces to

(1.18) $$\epsilon u'' - u' + F(u) = 0 \ , \qquad 0 < x < 1 \ ,$$

(1.19) $$\epsilon u'(0) - u(0) + Ru(1) = -(1-R)u_0 \ ,$$

(1.20) $$u'(1) = 0 \ ,$$

where

(1.21) $$F(u) = D(1+u_0-u)e^{\frac{-k}{u}} \ .$$

A sketch of F(u) is given in Figure III.1

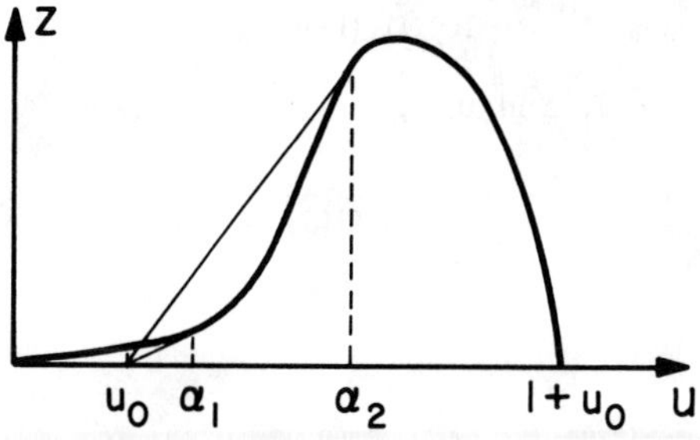

Fig. III.1

III.2. <u>The perturbation method</u>. We now construct asymptotic expansions of the solutions of (1.18)-(1.21) as $\epsilon \to 0$ by standard techniques [4] of singular perturbation theory. In all cases $F(u) = O(1)$, and there is a boundary layer of thickness $O(\epsilon)$ near $x = 1$. In order to cope with the mixed boundary condition (1.19), suppose temporarily that we have found the first term, call it y(x), of the expansion (the outer expansion) away from this boundary layer. Then, in the boundary layer we introduce a new length $\tilde{x} = \frac{1-x}{\epsilon}$, and let $u(x) = u(1-\epsilon\tilde{x}) \equiv U(\tilde{x})$. Then, the first term of the expansion (the inner expansion) near $x=1$ is given by

(2.1) $$U'' + U' = 0 ,$$

(2.2) $$U'(0) = 0 ,$$

(2.3) $$U(\infty) = y(1) .$$

The boundary condition (2.3) expresses the proper condition for matching the inner and outer expansions. The solution of (2.1)-(2.3) is $U(\tilde{x}) \equiv y(1)$, and thus, the first term $y(x)$ of the outer expansion is a uniformly valid expansion to order $O(\epsilon)$ on the entire interval $0 \leqslant x \leqslant 1$. Hence, the term $u(1)$ in boundary condition (1.19) can be retained as it stands to order $O(\epsilon)$ in determining $y(x)$. Thus, $y(x)$ is given by

(2.4) $$-y' + F(y) = 0 ,$$

(2.5) $$y(0) = Ry(1) + (1-R)u_0 .$$

Now, equation (2.4) implies

(2.6) $$\int_{y(0)}^{y(1)} \frac{dy}{F(y)} = \int_0^x dx = 1 ,$$

and using (2.5) we obtain

(2.7) $$\int_{Ry(1)+(1-R)u_0}^{y(1)} \frac{dy}{F(y)} = 1 ,$$

or, equivalently, for fixed R and u_0 ,

(2.8) $$G(\xi) = 1 ,$$

where

(2.9) $$G(\xi) \equiv \int_{R\xi+(1-R)u_0}^{\xi} \qquad \text{and } \xi = y(1) .$$

Therefore, to order $O(\epsilon)$, the original problem (1.18)-(1.21) has as many solutions as there are roots of the equation (2.8). Hence, our study now reduces to an investigation of the equation (2.8). A detailed study of this equation together with rigorous proofs of the validity of our asymptotic approximations of the multiple solutions of (1.18)-(1.21) is given in the paper of D. S. Cohen and H. B. Keller [1]. In these notes we shall simply now give the results and a brief outline of the pertinent steps in the analysis of (2.8), (2.9).

III. 3. Multiple solutions. It is obvious from physical considerations and it can be shown analytically [3] that all solutions $u(x)$ of (1.18)-(1.21) are such that $u_0 \leq u(x) \leq 1+u_0$. The solutions ξ of (2.8) also possess this property as we shall now see. The feed stream temperature u_0 is less than the exit temperature $u(1) \sim y(1)$, and $u_0 < y(1)$ implies $Ry(1) + (1-R)u_0 < y(1)$. Thus, the lower limit of the integral in (2.7) is always less than the upper limit. Now, $G(u_0) = 0$, $G'(u_0) = (1-R)/F(u_0) > 0$, and since the integrand in (2.9) possesses a non-integrable singularity at $\xi = 1+u_0$, we have $G(1+u_0) = \infty$. Therefore, the graph of the function $G(\xi)$ in the interval $u_0 \leq \xi \leq 1+u_0$ shows that G increases from zero at $\xi = u_0$ and it goes off to infinity at $\xi = 1+u_0$. We shall now show that, in some connected interval somewhere in $u_0 < \xi < 1+u_0$, we have $G'(\xi) < 0$. Now,

(3.1) $$G'(\xi) = \frac{1}{F(\xi)} - \frac{R}{F(R\xi+(1-R)u_0)} .$$

In order to establish conditions under which $G'(\xi) < 0$, we shall need the following construction (see Figure III.1): Plot the curve $z = F(u)$. From the point $(u, z) = (u_0, 0)$ draw tangent lines to the curve $z = F(u)$. Denote the values of u at the points of tangency by α_1 and α_2. (For some values of k and D in the non-linearity $F(u)$, the points may not exist.)

Lemma III.1. (See Figure III.1) If the points α_1 and α_2 exist, then $G'(\xi) < 0$ for all ξ such that $\alpha_1 < \xi < \alpha_2$ and $\alpha_1 < R\xi + (1-R)u_0 < \alpha_2$.

Proof. Let $t = u - u_0$, and let $F(u) = F(t+u_0) \equiv h(t)$. The tangent to the curve $z = h(t)$ at the point $(s, h(s))$ is given by

$$z = h'(s)t + [h(s) - h'(s)s] .$$

Thus, $h(s) - h'(s)s < 0$ implies that the tangent at any point must intersect the negative z-axis. Now, note that

$$\frac{d}{dt}\left[\frac{h(t)}{t}\right] = \frac{h'(t)t - h(t)}{t^2} .$$

Therefore, $h(t) - h'(t)t < 0$ implies that $\frac{d}{dt}\left[\frac{h(t)}{t}\right] > 0$ which implies that $\frac{h(t)}{t}$ is an increasing function of t. Thus, since $R\xi + (1-R)u_0 - u_0 < \xi - u_0$, then

$$\frac{h(R\xi+(1-R)u_0-u_0)}{R\xi+(1-R)u_0-u_0} < \frac{h(\xi-u_0)}{\xi-u_0} ,$$

or equivalently,

$$\frac{F(R\xi+(1-R)u_0)}{R\xi+(1-R)u_0-u_0} < \frac{F(\xi)}{\xi-u_0} .$$

Therefore,

$$F(R\xi+(1-R)u_0) < RF(\xi) ,$$

from which, according to (3.1), it immediately follows that $G'(\xi) < 0$. Q.E.D.

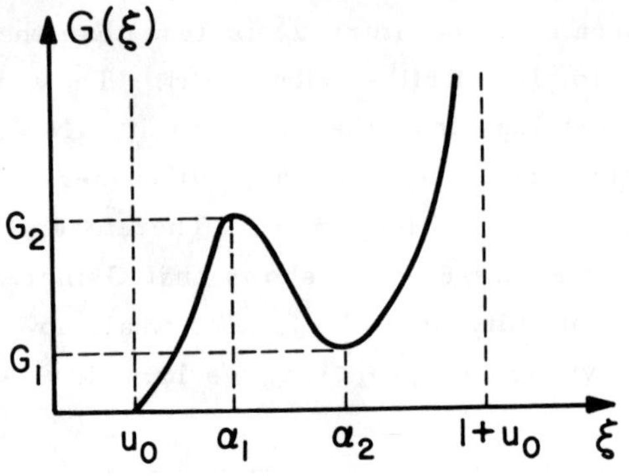

Fig. III.2.

We have established that if the points α_1 and α_2 exist, there exists an interval in $u_0 < \xi < 1+u_0$ where $G'(\xi) < 0$. Thus, the graph of $G(\xi)$ is roughly as illustrated in Figure III.2. The mechanism controlling the multiplicity is now obvious, and to order $O(\epsilon)$ we have the

Theorem III.2. (i) If $G_1 < 1 < G_2$, then there exist three solutions. (ii). If $G_1 = 1$ or if $G_2 = 1$, then there exist two solutions. (iii) If $0 \leq 1 \leq G_1$ or if $G_2 < 1$, then there exists a unique solution. Furthermore, if the numbers α_1 and α_2 do not exist, then $G(\xi)$ is a continuous monotone increasing function of ξ, and there exists a unique solution.

It can be shown [1] that for various values of the parameters R, u_0, D and k, all three cases of the above theorem occur.

We shall not pursue the problem with recycle further here. Stability questions and the more difficult recycle problems with time dependence and non-adiabatic ($\beta \neq 0$) reactors have been treated by D.S. Cohen and H.B. Keller [1] together with proofs of the validity of the singular perturbation approximations of the solutions.

REFERENCES

1. Cohen, D.S., and Keller, H.B., Multiple solutions of the nonlinear boundary value problems of tubular chemical reactors with recycle, to appear.

2. Cohen, D.S., Multiple solutions and periodic oscillations in nonlinear diffusion processes, to appear.

3. Cohen, D.S., Multiple stable solutions of nonlinear boundary value problems arising in chemical reactor theory, SIAM J. Appl. Math., 20 (1971) 1-13.

4. Cole, J.D., Perturbation Methods in Applied Mathematics, Blaisdell Publishing Co., Waltham, Mass., 1968.

Chapter IV

THE ADIABATIC TUBULAR REACTOR WITH LARGE PECLET NUMBER

IV.1. Introduction. We shall be concerned with the following nonlinear parabolic boundary value problem:

(1.1) $$u_t = \epsilon u_{xx} - u_x + F(\lambda, u), \qquad 0 < x < 1, \quad t > 0,$$

(1.2) $$u_x(0,t) - au(0,t) = 0, \qquad t \geq 0,$$

(1.3) $$u_x(1,t) = 0, \qquad t \geq 0,$$

(1.4) $$u(x,0) = h(x), \qquad 0 \leq x \leq 1.$$

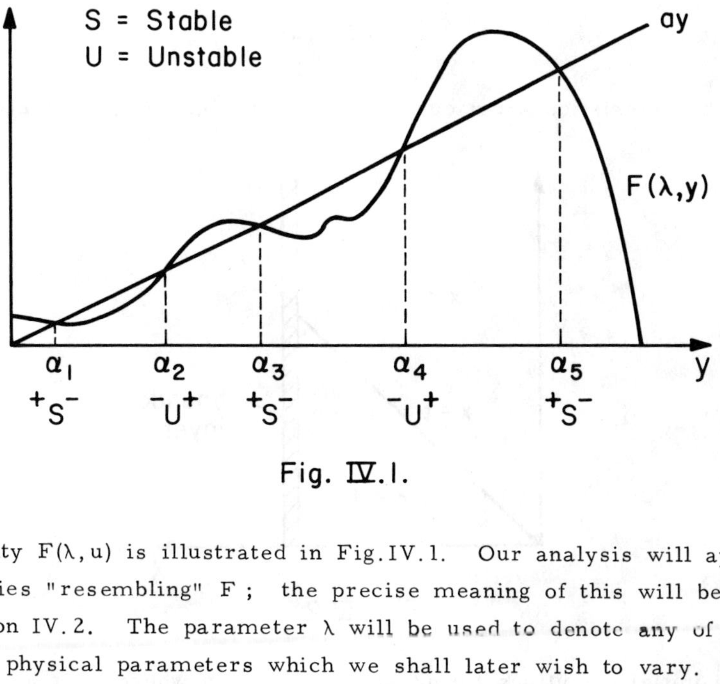

Fig. IV.1.

The nonlinearity $F(\lambda, u)$ is illustrated in Fig.IV.1. Our analysis will apply to all nonlinearities "resembling" F ; the precise meaning of this will be made clear in Section IV.2. The parameter λ will be used to denote any of the dimensionless physical parameters which we shall later wish to vary. Throughout this chapter we shall assume that $0 < \epsilon << 1$ (i.e., large Peclet number) and $F(\lambda, u) = O(1)$.

Our goal is to investigate the multiplicity and stability of solutions of (1.1)-(1.4). We wish to point out that whereas the equations of Chapters I-III are exact in the sense that they are derivable from first principles of chemical reactor theory, the equations (1.1)-(1.4) are not. The present problem is a commonly used model for the simple adiabatic tubular reactor with large Peclet number. It is not difficult to derive the exact equations. However, their study involves a considerable amount of algebraic complexity in conceptually simple situations, and it is felt that the present formulation (1.1)-(1.4) retains all essential features, including the important correct description of multiplicity and stability.

IV.2. Multiple solutions. We now construct asymptotic expansions of the solutions of (1.1)-(1.4) as $\epsilon \to 0$ by standard techniques [3]. We find that for all time $t \geq 0$ there is a boundary layer of thickness $O(\epsilon)$ near $x=1$. Away from this boundary layer, the first term of the expansion (the outer expansion) is governed by

$$u_t + u_x = F(\lambda, u) , \qquad 0 < x < 1, \quad t > 0, \tag{2.1}$$

subject to appropriate conditions. From the standard theory for first order partial differential equations [4], equation (2.1) is equivalent to solving the ordinary differential equations

$$\frac{du}{ds} = F(\lambda, u) \tag{2.2}$$

along the characteristics x-t = constant. From Fig. IV.2 we see that the

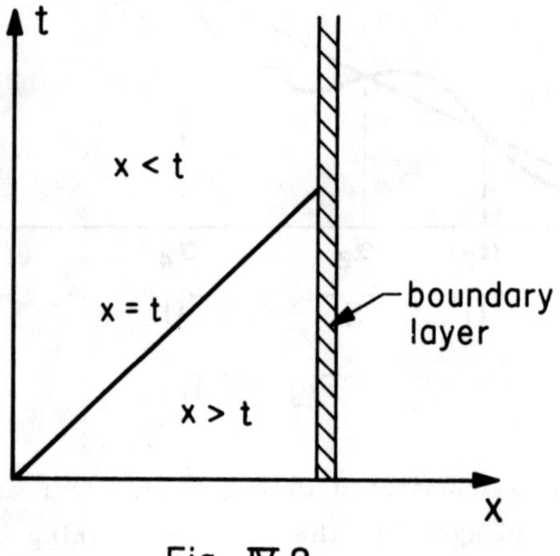

Fig. IV.2

appropriate initial conditions for (2.2) in the region where $x > t$ are given by equation (1.4); that is, equation (1.4) provides the initial values of u along the axis $t=0$ from which the equation (2.2) can be solved along the lines x-t = constant. Similarly, we see that to solve (2.2) along the lines x-t = constant in the region $x < t$, we shall need initial values for (2.2) along the axis $x=0$. To determine these initial values, we evaluate (2.1) at $x=0$ and note that (2.1) and (1.2) together imply that

$$u_t(0,t) + au(0,t) = F(\lambda, u(0,t)) , \tag{2.3}$$

which according to (1.4) must be solved subject to

$$(2.4) \qquad u(0,0) = h(0).$$

Equivalently,

$$(2.5) \qquad \frac{dy}{dt} = F(\lambda, y) - ay ,$$

$$(2.6) \qquad y(0) = h(0) ,$$

where $y(t) \equiv u(0,t)$. Thus, the solutions of (2.5), (2.6) are the proper initial conditions necessary to complete the asymptotic approximation for small ϵ in the region $x < t$.

To summarize, the first term in the outer expansion of the solutions of (1.1)-(1.4) as $\epsilon \to 0$ is given by the solution of (2.1) subject to the condition (1.4) along $t = 0$ in the region $x > t$ and subject to the solutions of (2.5), (2.6) along $x = 0$ in the region $x < t$.

The multiplicity of solutions is now controlled by (2.5), (2.6). That we have multiple solutions can be seen as follows: First, consider steady states (i.e., $\frac{\partial}{\partial t} \equiv 0$). There are as many steady states as there are roots of the equation $ay = F(\lambda, y)$. These are illustrated in Fig.IV.1 and are denoted by $\alpha_i (i = 1, \cdots, 5)$. Thus, it seems reasonable to suppose that for the nonlinearity, illustrated equations (2.5), (2.6) have five distinct solutions which in the limit as $t \to \infty$ approach the $\alpha_i (i = 1, \cdots, 5)$.

The following heuristic arguments lead us to conclude that the steady states alternate in stability as illustrated in Fig.IV.1: The algebraic signs a little to each side of the steady states α_i denote the sign of $\frac{dy}{dt}$ in (2.5) at these points. Thus, we see that the solutions of (2.5) tend to run away from α_2 and α_4 and tend to run to α_1, α_3, or α_5. (Note that solutions can not oscillate because a first order autonomous ordinary differential equation $y' = f(y)$ can not possess periodic solutions unless $f(y)$ is multi-valued.) Since y is the value of u at $x = 0$, our arguments apply only at $x = 0$. The goal of the next section is to support these formal arguments in general.

<u>IV.3. Stability of multiple solutions</u>. Denote a solution of (2.3) by $u(0,t) \equiv \alpha(t)$. We have seen in Section IV.2 that as $t \to \infty$, if $\alpha_i < y(0) = u(0,0) = h(0) < \alpha_{i+1}$, then

$$(3.1) \qquad u(0,t) \equiv \alpha(t) \to \begin{cases} \alpha_i & \text{if } i \text{ odd} , \\ \alpha_{i+1} & \text{if } i \text{ even} . \end{cases}$$

Now, the solution $u(x,t)$ of

(3.2) $$u_t + u_x = F(\lambda, u) ,$$

(3.3) $$u(0,t) = \alpha(t)$$

is given implicitly by

(3.4) $$x = G(u(x,t)) - G(\alpha(t-x)) ,$$

where

(3.5) $$G(u) = \int_0^u \frac{du}{F(\lambda, u)} .$$

Thus, as $t \to \infty$, we find that

(3.6) $$x = G(v(x)) - G(\alpha_i) ,$$

which is the solution of the steady state problem (to lowest order in ϵ)

(3.7) $$\frac{dv}{dx} = F(\lambda. v) ,$$

(3.8) $$v(0) = \alpha_i .$$

Therefore, we can conclude that the steady states $v_i(x)$ for i odd are stable and the steady states $v_i(x)$ for i even are unstable. Furthermore, to first order in ε, our procedure allows us to trace the approach to the steady states from arbitrary initial data.

The effect of any physical parameters (which we denote by λ) is now clear. For example, we can see from Fig. IV. 1 that varying λ may introduce new roots α_i or delete old roots α_i, depending on how we shift or distort the graph of $F(\lambda, y)$. Such a study would, in fact, have meaning within the context of the practical chemical engineering literature; it is clear that our analysis applies to all cases, and we conclude that for large Peclet number reactors the multiple steady states alternate in stability. For the steady states rigorous existence proofs supporting our singular perturbation constructions have been given by D. S. Cohen [2] and H. B. Keller [5], and the steady state problem corresponding to (1.1)-(1.4) with a simple first order Arrhenius rate function has been investigated by D. S. Cohen [1].

REFERENCES

1. Cohen, D.S., Multiple stable solutions of nonlinear boundary value problems arising in chemical reactor theory, SIAM J. Appl. Math., 20 (1971) 1-13.

2. Cohen, D.S., Multiple solutions of singular perturbation problems, SIAM J. Math. Anal., 3 (1972) 72-82.

3. Cole, J.D., Perturbation Methods in Applied Mathematics, Blaisdell, Waltham, Mass., 1968.

4. Courant, R., and Hilbert, D., Methods of Mathematical Physics, Vol.II, Interscience, New York, 1966.

5. Keller, H.B., Existence theory for multiple solutions of a singular perturbation problem, SIAM J. Math. Anal., 3 (1972) 86-92.

Chapter V

BRANCHING OF SOLUTIONS OF NONLINEAR ELLIPTIC EQUATIONS

V.1. Introduction. The results of this chapter are based on the doctoral dissertion J. P. Keener [4]. We shall simply formulate the problems and state the main results; all proofs and details together with several studies we shall not pursue here can be found in [4]. Large portions of Keener's thesis are used here unchanged.

We shall study branching phenomena for nonlinear elliptic boundary value problems of the form

$$(1.1)\qquad \begin{aligned} Lu + f(\lambda, \tau, u) &= 0\,, \qquad x \in D,\\ Bu &= 0\,, \qquad x \in \partial D\,. \end{aligned}$$

The nonlinearity $f(\lambda, \tau, u)$ satisfies

(i) $f(\lambda, 0, 0) = 0$ for all real λ (the unforced case),

(ii) $f(\lambda, \tau, 0) \neq 0$ if $\tau \neq 0$ (the forced case).

Our main interest is to compare the problems (i.e., the forced and unforced cases) in the limit as the forcing τ vanishes and to illustrate the difference in phenomena which can occur when τ is small but not zero as compared with the better known and more frequently studied case when $\tau = 0$. Even in the case that $\tau = 0$, our results remove critical restrictions usually placed on the problem in previous analyses [12]. (Note that the essential difference between cases (i) and (ii) is that in case (i), $u \equiv 0$ is a solution of (1.1) whereas in case (ii) $u \equiv 0$ is not a solution of (1.1).)

In (1.1) $x = (x_1, \cdots, x_n)$ and L is the uniformly elliptic second order operator defined on D by

$$(1.2)\qquad Lu \equiv \sum_{i,j=1}^{n} a_{ij}(x) \frac{\partial^2 u}{\partial x_i \partial x_j} + \sum_{j=1}^{n} a_j(x) \frac{\partial u}{\partial x_j} - a_0(x)u\,.$$

The boundary operator B is defined on ∂D by

$$(1.3)\qquad Bu = b_0(x)u + b_1(x) \sum_{j=1}^{n} \beta_j(x) \frac{\partial u}{\partial x_j}\,,$$

where for notational purposes we will denote

$$\frac{\partial u}{\partial \beta} = \sum_{j=1}^{n} \beta_j(x) \frac{\partial u}{\partial x_j}\,.$$

We denote by $C^{k+\alpha}(\Omega)$ the space of real valued functions which are k times continuously differentiable on a point set Ω and have Hölder continuous k^{th} derivatives on Ω with Hölder exponent α. We assume that D is a bounded domain in $\mathbb{R}^n$ with boundary ∂D of class $C^{2+\alpha}$. The coefficients $a_{ij}(x)$, $a_j(x)$, $a_0(x) > 0$ are assumed to be $C^{2+\alpha}(\bar{D})$, $C^{1+\alpha}(\bar{D})$ and $C^{\alpha}(\bar{D})$ respectively, while $b_0(x)$, $b_1(x)$, $\beta_i(x)$ are in $C^{1+\alpha}(\partial D)$ for some $\alpha \epsilon (0,1)$. The uniform ellipticity of L implies that for all unit vectors $y = (y_1, \cdots, y_n)$

(1.4) -i) $$\sum_{i,j=1}^{n} a_{ij}(x)\, y_i y_j \geq a > 0\,, \qquad x \epsilon D\,.$$

Taking $n_i(x)$ to be the components of the unit outward normal at $x \epsilon \partial D$, we assume that the coefficients of the boundary operator B satisfy

(1.4) -ii) $$\sum_{j=1}^{n} \beta_j(x)\, n_j(x) > 0\,, \qquad \sum_{j=1}^{n} \beta_j^2(x) = 1$$

and that ∂D can be decomposed into $\partial D = \partial D_1 \cup \partial D_2$, where

(1.4) -iii) $$b_0(x) > 0\,, \qquad b_1(x) \equiv 0\,, \qquad x \epsilon \partial D_1\,,$$

(1.4) -iv) $$b_0(x) \geq 0, \qquad b_1(x) > 0\,, \qquad x \epsilon \partial D_2\,.$$

The assumed smoothness assumptions on L and B are sufficient to assure us that, for $F(x) \epsilon C^{\alpha}(\bar{D})$, the linear problem

(1.5) $$\begin{aligned} Lu(x) &= F(x)\,, && x \epsilon D\,, \\ Bu(x) &= 0\,, && x \epsilon \partial D\,, \end{aligned}$$

has a unique solution, $u(x) \epsilon C^{2+\alpha}(\bar{D})$ ([9], pp. 134-136). These assumptions further imply that L and B satisfy the strong maximum principle [10] which leads to

Proposition (1). If $\phi(x) \epsilon C'(\bar{D}) \cap C^2(\bar{D})$, then

i - $L\phi \leq$ on D, $\quad B\phi \geq 0$ on $\partial D \Rightarrow \phi(x) \geq 0$ on $\bar{D}$,

ii - $L\phi <$ on D, $\quad B\phi \geq 0$ on $\partial D \Rightarrow \phi(x) > 0$ on D.

Furthermore, if $\phi(x) = 0$ for some $x \epsilon \partial D$, then

$$\frac{\partial \phi}{\partial \alpha} < 0, \qquad x \epsilon \partial D,$$

where $\frac{\partial}{\partial \alpha}$ is the directional derivative taken in any outward direction.

We will assume that $f(\lambda, \tau, u) \epsilon C^{\alpha}(D)$ whenever $u \epsilon C^{2+\alpha}(D)$ and the partial derivative at λ, τ, u satisfies $f_u(\lambda, \tau, u) \epsilon C^{\alpha}(D)$ when $u \epsilon C^{2+\alpha}(D)$. All other derivatives up to and including third order are assumed to be continuous on D if $u \epsilon C^{2+\alpha}(D)$. Although $f(\lambda, \tau, u)$ is allowed to depend on x, this dependence will not be explicitly shown. For convenience in simplifying certain algebraic manipulations, we shall assume that $f_\lambda(\lambda, 0, 0) = f_{\lambda\lambda}(\lambda, 0,) = f_{\lambda\lambda\lambda}(\lambda, 0, 0) = \cdots : 0$. (Note that this assumption does not imply that f is linear in λ. For example, we could have $f(\lambda, \tau, u) = \lambda u + \lambda^2 u$.)

The standard bifurcation problem with $\tau = 0$ has been treated in numerous places [2], [6], [7], [8], [11], [14]. One result of these studies [8] is that branching can occur at a point $(\lambda, u) = (\lambda_0, 0)$ only if there are nontrivial solutions of the problem

$$L\phi + f_u(\lambda_0, 0, 0)\phi = 0, \qquad x \in D,$$

(1.6)

$$B\phi = 0, \qquad x \in \partial D.$$

For the forced case with $\tau_0 \neq 0$, a point (λ_0, u_0) can be a branching point of (1.1) only if there are nontrivial solutions to the problem resulting from linearization of (1.1) about the known solution at $\lambda = \lambda_0$. That is, there must be nontrivial solutions of

$$L\psi + f_u(\lambda_0, \tau_0, u_0)\psi = 0, \qquad x \in D,$$

(1.7)

$$B\psi = 0, \qquad x \in \partial D,$$

where (λ_0, τ_0, u_0) satisfy (1.1). Solutions (u, ψ, λ, τ) satisfying both (1.1) and (1.7) will be referred to as non-isolated solutions of (1.1) corresponding to the point λ.

To provide a starting point for our investigation, we will assume that there is a number λ_0 and a nontrivial function $\phi_0(x) \in C^{2+\alpha}(\overline{D})$ which satisfy (1.6). <u>The quadruple</u> $(u, \psi, \lambda, \tau) = (0, \phi_0, \lambda_0, 0)$ <u>will be referred to as a trivial nonisolated solution of</u> (1.1). We will also assume that all solutions of (1.6) are multiples of $\phi_0(x)$.

By defining the inner product

(1.10) $$\langle u, v \rangle = \int_D u(x)\, v(x)dx,$$

we can define adjoint operators L^* and B^* to be those operators satisfying

(1.11) $$\langle v, Lu \rangle - \langle u, L^* v \rangle = 0,$$

whenever $u, v, \in C^{2+\alpha}(\overline{D})$ and $Bu = 0$, $B^* v = 0$. The operators which result from this definition are given by [1]

(1.12) $$L^* v = \sum_{i,j=1}^{n} \frac{\partial^2 (a_{ij}(x)v)}{\partial x_i \partial x_j} - \sum_{j=1}^{n} \frac{\partial (a_j(x)v)}{\partial x_j} - a_0(x)v, \quad x \in D.$$

$B^*v = 0$ is defined by requiring

$$(1.13)\qquad P[u,v] = \sum_{i,j=1}^{n}\left[\frac{\partial u}{\partial x_i} a_{ij} v - \frac{\partial}{\partial x_j}(a_{ij} v)u\right] + \sum_{i=0}^{n} a_i iv = 0, \qquad x \in \partial D,$$

when $Bu = 0$. For $\rho(x) \in C^{\alpha}(\overline{D})$, whenever

$$(1.14)\qquad \begin{aligned} L\phi + \rho(x)\phi &= 0, & x \in D, \\ B\phi &= 0, & s \in \partial D, \end{aligned}$$

has a nontrivial solution, we know, from the study of spectral theory for compact operations [3], that the associated adjoint problem

$$(1.15)\qquad \begin{aligned} L^*\phi^* + \rho(x)\phi^* &= 0, \\ B^*\phi^* &= 0, \end{aligned}$$

also has a nontrivial solution, and the null space of equation (1.14) is of the same dimension as the null space of (1.15). The Fredholm Alternative Theorem [1] holds for solutions of

$$(1.16)\qquad Lv + \rho(x)v = g(x).$$

Specifically, this asserts that (1.16) has a solution $v(x) \in C^{+\alpha}(\overline{D})$ provided $g(x) \in C^{\alpha}(\overline{D})$ and

$$(1.17)\qquad \langle g(x), \phi_0^* \rangle = 0,$$

where ϕ_0^* is a solution of (1.15). Let $\Lambda(x) \in C(\overline{D})$ be a "weight function" such that $\langle \phi_0(x), \phi_0^*(x)\,\Lambda(x)\rangle \neq 0$. We make the stronger assumption that if the solution $v(x)$ in (1.16) is made unique by requiring the orthogonality condition

$$(1.18)\qquad \langle v(x), \phi_0^*(x)\,\Lambda(x)\rangle = 0,$$

then there exists a constant $G > 0$ such that

$$(1.19)\qquad \|v\|_{\infty} \leq G\|g\|_{\infty}.$$

The notation has been chosen with an eye toward generalizations [5]. If we wanted L to be an operator in a real Hilbert space H, then the inner product (1.10) could be chosen appropriately. The inequality (1.19) could be assumed to hold in the induced norm of H, and many of the results that follow would be true with only a slight change of wording.

V.2. Perturbation theory for nonisolated solutions. In this section we will develop a formal perturbation scheme which indicates the form of nontrivial nonisolated solutions of (1.1). We will show that this scheme is well-defined and can be carried out to arbitrary order provided the nonlinearity $f(\lambda, \tau, u)$ is sufficiently differentiable in each of the arguments λ, τ and u. It will be our task in later sections to show the validity of this perturbation scheme.

Suppose that the quadruple $(u, \phi, \lambda, \tau) = (0, \phi_0, \lambda_0, 0)$ is a known nonisolated solution of (1.1). Our hope is that this solution is an element of a branch of nonisolated solutions and that this branch can be represented parametrically with some parameter ϵ. If this parametric representation is also sufficiently differentiable at the known solution $(u, \psi, \lambda, \tau) = (0, \phi_0, \lambda_0, 0)$, then we can expand the parametric representation in a Taylor series about a known solution We choose the parameter ϵ so that $\big(u(x,\epsilon), \psi(x,\epsilon), \lambda(\epsilon), \tau(\epsilon)\big) = (0, \phi_0, \lambda_0, 0)$ when $\epsilon = 0$.

The first n+1 terms of this Taylor expansion will be referred to as the nth perturbation expansion for nonisolated solutions of (1.1) and will be of the form

$$
(2.1) \quad \begin{cases}
\tilde{u}^n(x,\epsilon) = \epsilon(u_0 + \epsilon u_1 + \cdots + \epsilon^n u_n), \\
\tilde{\psi}^n(x,\epsilon) = \phi_0 + \epsilon\psi_1 + \epsilon^2\psi_2 + \cdots + \epsilon^n \psi_n, \\
\tilde{\lambda}^n(\epsilon) = \lambda_0 + \epsilon\lambda_1 + \cdots + \epsilon^n \lambda_n, \\
\tilde{\tau}^n(\epsilon) = \epsilon(\tau_0 + \epsilon\tau_1 + \cdots + \epsilon^n \tau_n) .
\end{cases}
$$

Substituting (2.1) directly into (1.1) and (1.7), expanding the nonlinear terms in powers of ϵ, and then equating coefficients of like powers of ϵ, we obtain

$$
(2.2) \quad \begin{cases}
Lu_0 + f_u(\lambda_0, 0, 0)u_0 = -f_\tau)\lambda_0, 0, 0)\tau_0, & x \in D, \\
Bu_0 = 0, & x \in \partial D,
\end{cases}
$$

$$
(2.3) \quad \begin{cases}
Lu_1 + f_u(\lambda_0, 0, 0)u_1 = -\Big\{ f_{\lambda u}(\lambda_0, 0, 0)\lambda_1 u_0 + f_{\lambda\tau}(\lambda_0, 0, 0)\lambda_1\tau_0 \\
\quad + f_{\tau u}(\lambda_0, 0, 0)u_0\tau_0 + \tfrac{1}{2} f_{uu}(\lambda_0, 0, 0)u_0^2 \\
\quad + \tfrac{1}{2} f_{\tau\tau}(\lambda_0, 0, 0)\tau_0^2 + f_\tau(\lambda_0, 0, 0)\tau_1 \Big\}, & x \in D \\
Bu_1 = 0, & x \in \partial D,
\end{cases}
$$

$$(2.4)\quad \begin{cases} L\phi_0 + f_u(\lambda_0, 0, 0)\phi_0 = 0, & x \in D, \\ B\phi_0 = 0, & x \in \partial D, \end{cases}$$

$$(2.5)\quad \begin{cases} L\psi_1 + f_u(\lambda_0, 0, 0)\psi_1 = -\Big\{ f_{u\lambda}(\lambda_0, 0, 0)\lambda_1 \phi_0 + f_{uu}(\lambda_0, 0, 0)\phi_0 u_0 \\ \qquad + f_{u\tau}(\lambda_0, 0, 0)\tau_0 \phi_0 \Big\}, & x \in D, \\ B\psi_1 = 0, & x \in \partial D. \end{cases}$$

Since the operator $L^* + f_u(\lambda_0, 0, 0)$ has a null space spanned by ϕ_0^*, we know by the Fredholm alternative theorem that equations (2.2)-(2.5) can be solved if and only if the right-hand side of each equation is orthogonal to ϕ_0^* as in (1.17). This condition determines the constants λ_1, τ_0 and τ_1 in (2.2)-(2.5). Furthermore these solutions will not be unique, since we may add any multiple of ϕ_0 to the solution. To make the solutions unique, we require that

$$(2.6)\quad \langle \psi(x), \phi_0^*(x) f_{\lambda u}(\lambda_0, 0, 0) \rangle = 1, \qquad \langle u(x), \phi_0^*(x) f_{\lambda u}(\lambda_0, 0, 0) \rangle = \epsilon.$$

This places a restriction on the terms of the perturbation expansion (2.1), requiring that

$$(2.7)\quad \langle \phi_0, \phi_0^* f_{\lambda u}(\lambda_0, 0, 0) \rangle = 1, \qquad \langle u_0, \phi_0^* f_{\lambda u}(\lambda_0, 0, 0) \rangle = 1,$$

$$(2.8)\quad \langle \psi_i, \phi_0^* f_{\lambda u}(\lambda_0, 0, 0) \rangle = 0, \qquad \langle u_i, \phi_0^* f_{\lambda u}(\lambda_0, 0, 0) \rangle = 0$$

for $i = 1, 2, 3 \ldots$.

In order to solve (2.2), the Fredholm alternative theorem requires that

$$(2.9)\quad \tau_0 \langle f_\tau(\lambda_0, 0, 0), \phi_0^* \rangle = 0 .$$

Assuming that $\langle f_\tau(\lambda_0, 0, 0), \phi_0^* \rangle \neq 0$, we must have $\tau_0 = 0$. With $\tau_0 = 0$, equations (2.2) and (2.4) are identical so that, applying (2.7), we obtain

$$(2.10)\quad u_0(x) = \phi_0(x).$$

Using this information, equation (2.3) becomes

$$(2.11)\quad \begin{cases} Lu_1 + f_u(\lambda_0, 0, 0)u_1 = -\{f_{\lambda u}(\lambda_0, 0, 0)\lambda_1 \phi_0 + \frac{1}{2} f_{uu}(\lambda_0, 0, 0)\phi_0^2 + f_\tau(\lambda_0, 0, 0)\tau_1\}, & x \in D, \\ Bu_1 = 0, & x \in \partial D. \end{cases}$$

Applying the Fredholm alternative theorem to (2.11), we have

(2.12) $$\lambda_1 \langle f_{\lambda u}(\lambda_0,0,0)\phi_0, \phi_0^* \rangle + \tau_1 \langle f_\tau(\lambda_0,0,0), \phi_0^* \rangle = -\tfrac{1}{2}\langle f_{uu}(\lambda_0,0,0)\phi_0^2, \phi_0^* \rangle .$$

Similarly, from equation (2.5) we get

(2.13) $$\lambda_1 \langle f_{\lambda u}(\lambda_0, 0, 0)\phi_0, \phi_0^* \rangle = -\langle f_{uu}(\lambda_0, 0, 0)\phi_0^2, \phi_0^* \rangle .$$

Equations (2.12) and (2.13) are two linear, simultaneous equations for λ_1 and τ_1. The determinant of this system is

(2.14) $$D = \langle f_\tau(\lambda_0, 0, 0)\phi_0^* \rangle \cdot \langle f_{\lambda u}(\lambda_0, 0, 0)\phi_0, \phi_0^* \rangle$$

so that these equations can be solved provided $D \neq 0$. If $D \neq 0$, the solution of (2.12)-(2.13) is

(2.15) $$\lambda_1 = -\frac{\langle f_{uu}(\lambda_0,0,0), \phi_0^2, \phi_0^* \rangle}{\langle f_{\lambda u}(\lambda_0,0,0)\phi_0, \phi_0^* \rangle}, \qquad \tau_1 = \frac{1}{2}\frac{\langle f_{uu}(\lambda_0,0,0)\phi_0^2, \phi_0^* \rangle}{\langle f_\tau(\lambda_0,0,0), \phi_0^* \rangle} .$$

Of interest in many applications is the relationship between λ, the "buckling load," and τ, the "imperfection amplitude." According to (2.1)

$$\tilde{\tau} = \epsilon^2 \tau_1 + O(\epsilon^3) \quad ;$$

so if $\tau_1 \neq 0$, we can find $\lambda = \lambda(\tau)$ approximately. In particular,

(2.16) $$\tilde{\tau} = \frac{\epsilon^2}{2} \frac{\langle f_{uu}(\lambda_0, 0, 0)\phi_0^2, \phi_0^* \rangle}{\langle f_\tau(\lambda_0, 0, 0), \phi_0^* \rangle} + O(\epsilon^3)$$

and

(2.17) $$\tilde{\lambda} = \lambda_0 - \epsilon \frac{\langle f_{uu}(\lambda_0, 0, 0)\phi_0^2, \phi_0^* \rangle}{\langle f_{\lambda u}(\lambda_0, 0, 0)\phi_0, \phi_0^* \rangle} + O(\epsilon^2)$$

can be combined to give

(2.18) $$\tilde{\lambda} = \lambda_0 \pm \tau^{1/2} \frac{[2\langle f_\tau(\lambda_0,0,0), \phi_0^* \rangle \cdot \langle f_{uu}(\lambda_0,0,0)\phi_0^* \rangle]^{1/2}}{\langle f_{\lambda u}(\lambda_0,0,0)\phi_0, \phi_0^* \rangle} + O(\tau),$$

where τ must be restricted so that λ is real.

In many applications, $f_{uu}(\lambda_0,0,0) \equiv 0$ so that (2.18) is not valid. Suppose there is an integer p such that

(2.19) $$\frac{\partial^k f(\lambda_0, 0, 0)}{\partial u^k} = 0, \quad 2 \leq k \leq p, \qquad \langle \frac{\partial^{p+1} f(\lambda_0, 0, \ )}{\partial u^{p+1}} \phi_0^{p+1}, \phi_0^* \rangle \neq 0.$$

Then the perturbation equations can be shown to reduce to

(2.20) $$\begin{cases} Lu_k + f_u(\lambda_0, 0, 0)u_k = 0, & x \in D, \ k = 0, 1, 2, \cdots, p-1, \\ Bu_k = 0, & x \in \partial D, \end{cases}$$

(2.21) $$\begin{cases} Lu_p + f_u(\lambda_0, 0, 0)u_p = -\Big[\lambda_p f_{\lambda u}(\lambda_0, 0, 0)u_0 + \dfrac{\partial^{p+1} f(\lambda_0, 0, 0)}{\partial u^{p+1}} \dfrac{u_0^{\,p+1}}{(p+1)!} \\ \qquad + f_\tau(\lambda_0, 0, 0)\tau_p\Big], & x \in D, \\ Bu_p = 0 & x \in \partial D, \end{cases}$$

and

(2.22) $$\begin{cases} L\psi_k + f_u(\lambda_0, 0, 0)\psi_k = 0, & x \in D, \quad k = 0, 1, 2, \cdots, p-1, \\ B\psi_k = 0, & x \in \partial D, \end{cases}$$

(2.23) $$\begin{cases} L\psi_k + f_u(\lambda_0, 0, 0)\psi_p = -\Big[\lambda_p f_{\lambda u}(\lambda_0, 0, 0)\phi_0 + \dfrac{\partial^{p+1} f(\lambda_0, 0, 0)}{\partial u^{p+1}} \dfrac{\phi_0^{\,p+1}}{p!}\Big], & x \in D, \\ B\psi_p = 0, & x \in \partial D, \end{cases}$$

and the conditions (2. 7) and (2. 8) are required to hold.

According to equations (2. 20) and (2. 21), we have

(2. 24) $$u_0(x) = \phi_0(x), \quad u_k(x) = \psi_k(x) = 0, \qquad k = 1, 2, \cdots, p-1 .$$

The calculations used in deriving (2. 20)-(2. 23) show that

(2. 25) $$\lambda_k = \tau_k = 0, \qquad k = 1, 2, \cdots, p-1 .$$

Thus, using (2. 24) and (2. 25) in (2. 1), we find that the solution reduces to

(2. 26) $$\begin{cases} \tilde{u}^p = \epsilon(\phi_0 + \epsilon^p u_p) + O(\epsilon^{p+2}), \\ \tilde{\psi}^p = \phi_0 + \epsilon^p \psi_p + O(\epsilon^{p+1}), \\ \tilde{\lambda}^p = \lambda_0 + \epsilon^p \lambda_p + O(\epsilon^{p+1}), \\ \tilde{\tau}^p = \epsilon^{p+1} \tau_p + O(\epsilon^{p+2}). \end{cases}$$

Invoking the Fredholm alternative theorem in (2.21) and (2.23), we can find λ_p and τ_p. Specifically,

$$\lambda_p\langle f_{\lambda u}(\lambda_0,0,0)\phi_0,\phi_0^*\rangle + \tau_p\langle f_\tau(\lambda_0,0,0),\phi_0^*\rangle = -\frac{1}{(p+1)!}\langle\frac{\partial^{p+1}f(\lambda_0,0,0)}{\partial u^{p+1}}\phi_0^{p+1},\phi_0^*\rangle,$$

(2.27)

$$\lambda_p\langle f_{\lambda u}(\lambda_0,0,0)\phi_0,\phi_0^*\rangle = -\frac{1}{p!}\langle\frac{\partial^{p+1}f(\lambda_0,0,0)}{\partial u^{p+1}}\phi_0^{p+1},\phi_0^*\rangle,$$

so that

$$\tau_p = \frac{p}{(p+1)!}\frac{\langle\frac{\partial^{p+1}f(\lambda_0,0,0)}{\partial u^{p+1}}\phi_0^{p+1},\phi_0^*\rangle}{\langle f_\tau(\lambda_0,0,0),\phi_0^*\rangle}. \tag{2.28}$$

At the outset, we assumed conditions (2.19) that assured us that $\tau_p \neq 0$. Now we can solve for $\lambda = \lambda(\tau)$ approximately. Doing so, we get

$$\lambda = \lambda_0 - \frac{\tau^{\frac{p}{p+1}}}{p!}\left(\frac{(p+1)!}{p}\langle f_\tau(\lambda_0,0,0),\phi_0^*\rangle\right)^{\frac{p}{p+1}}\frac{\left(\langle\frac{\partial^{p+1}f(\lambda_0,0,0)}{\partial u^{p+1}}\phi_0^{p+1},\phi_0^*\rangle\right)^{\frac{1}{p+1}}}{\langle f_{\lambda u}(\lambda_0,0,0)\phi_0,\phi_0^*\rangle} + O(\tau). \tag{2.29}$$

Thus, the buckling load λ is altered by imperfections in the order of $\tau^{\frac{p}{p+1}}$ for τ sufficiently small.

We shall analyze the implications of (2.29) later. We wish to point out here, however, the following facts: If p is even, then there is one nonisolated solution for $\tau > 0$ and one for $\tau < 0$. However, if p is odd, then τ must be restricted so that λ is real which implies that there will be two nonisolated solutions for τ of one sign and no nonisolated solutions for τ of the other sign.

It is easy to show that the perturbation scheme given by (2.1) is well-defined and that the kth terms of the expansion are determined as solutions of linear equations involving previously determined k terms.

V.3. Existence of nonisolated solutions. In Section V.2, we developed a formal perturbation scheme which gave rise to expressions which we hope are approximate solutions of (1.1), (1.7). At this stage, however, we do not even know that (1.1), (1.7) have "nontrivial" solutions. In order to show that such solutions exist, we look for solutions of (1.1), (1.7) in a form suggested by the perturbation method, namely,

$$\begin{cases} u(x,\epsilon) = \epsilon\phi_0 + \epsilon^2 v(x,\epsilon), \\ \psi(x,\epsilon) = \phi_0 + \epsilon\chi(x,\epsilon), \\ \lambda(\epsilon) = \lambda_0 + \epsilon\mu(\epsilon), \\ \tau(\epsilon) = \epsilon^2\eta(\epsilon), \end{cases} \tag{3.1}$$

where $\phi_0(x)$ satisfies (1.6). In addition, we require that

(3.2) $$\langle v(x,\epsilon),\ \phi_0^*(x) f_{\lambda u}(\lambda\tau, 0, 0)\rangle = 0, \qquad \langle \chi(x,\epsilon),\ \phi_0^*(x) f_{\lambda u}(\lambda_0, 0, 0)\rangle = 0 .$$

We must show that for some nontrivial range of the parameter ϵ, $0 \leq |\epsilon| \leq \epsilon_0$, the functions $v(x,\epsilon)$, $\chi(x,\epsilon)$, $\mu(\epsilon)$, $\eta(\epsilon)$ exist and are bounded uniformly in ϵ. If this can be shown, then as ϵ approaches zero, the solutions (3.1) approach the trivial solution $(u, \psi, \lambda, \tau) = (0, \phi_0, \lambda_0, 0)$ continuously. Furthermore, the solutions (3.1) constitute a family of nonisolated solutions of (1.1) depending continuously on the parameter ϵ. Upon substituting (3.1) into (1.1) and (1.7), we obtain

(3.3) $$\begin{cases} Lv + f_u(\lambda_0,0,0)v = -\dfrac{1}{\epsilon^2}\Big[f(\lambda,\tau,u) - f_u(\lambda_0,0,0)u\Big] \\ \quad = -\Big[\eta(\epsilon)\displaystyle\int_0^1 f_\tau(\lambda, s\tau, u)ds + \mu(\phi_0+\epsilon v)\int_0^1\!\!\int_0^1 f_{\lambda u}(\lambda_0 + s\epsilon\mu, 0, tu)ds\,dt \\ \quad + (\phi_0+\epsilon v)^2 \displaystyle\int_0^1\!\!\int_0^1 f_{uu}(\lambda_0, 0, stu)s\,dt\,ds\Big], \qquad x \in D, \\ \quad \equiv P(v, \mu, \eta, \epsilon; x), \\ Bv = 0, \qquad x \in \partial D, \end{cases}$$

(3.4) $$\begin{cases} L\chi + f_u(\lambda_0, 0, 0)\chi = -\dfrac{1}{\epsilon}\Big[f_u(\lambda,\tau,u) - f_u(\lambda_0, 0, 0)\Big]\psi \\ \quad = -\Big[\mu\displaystyle\int_0^1 f_{\lambda u}(\lambda_0 + s\epsilon\mu, 0, u)ds + \epsilon\eta\int_0^1 f_{\tau u}(\lambda, s\tau, u)ds \\ \quad + (\phi_0+\epsilon v)\displaystyle\int_0^1 f_{uu}(\lambda_0, 0, su)ds\Big](\phi_0 + \epsilon\chi), \\ \quad \equiv Q(v, \chi, \mu, \epsilon; x), \qquad x \in D, \\ B\chi = 0, \qquad x \in \partial D. \end{cases}$$

In deriving (3.3) and (3.4) we have found it convenient to use the identity

(3.5) $$g(a) - g(b) = (a-b)\int_0^1 \frac{dg}{dx}(sa + (1-s)b)ds .$$

Equations (3.3) and (3.4) are of the form (1.16) and can be solved for v and χ only if the orthogonality conditions

(3.6) $$\langle P(v, \mu, \eta, \epsilon; x), \phi_0^*\rangle = 0, \qquad \langle Q(v, \chi, \mu, \eta, \epsilon; x),\ \phi_0^*\rangle = 0$$

hold. These solutions, if they exist, are only determined to within an additive multiple of ϕ_0, unless the conditions (3.2) are satisfied.

The orthogonality condition (3.6) provides the method by which we intend to solve (3.4) and (3.5). We will solve them iteratively, first by choosing values of η and μ so that (3.6) holds and then solving (3.3) and (3.4) for the functions v and χ. With the new functions v and χ, we must choose new values of η and μ so that (3.6) again holds, and the process continues indefinitely. If we can show that this process converges, then we will have found a solution of (3.3) and (3.4). This iteration scheme is a modification of the standard technique of Lyapunov and Schmidt [14] suggested by the treatment in [6] of the bifurcation problem (1.1) with $\tau = 0$.

To formulate the contraction mapping, we introduce the sets

(3.7) $$\mathcal{B}_K = \{y(x) \mid y(x) \in C^{2+\alpha}(D), \|y\| \leq K, \langle y(x), \phi_0^*(x) f_{\lambda u}(\lambda_0, 0, 0)\rangle = 0\},$$

(3.8) $$\mathcal{I}_K = \{\eta \mid |\eta| \leq K\},$$

(3.9) $$S_1(\rho, \Gamma) = \{(\lambda, \tau, u; x) \mid \lambda = \lambda_0 + \epsilon\mu, \tau = \epsilon^2\eta, u = \epsilon\phi_0 + \epsilon^2 v, \; x \in \overline{D},$$
$$v \in \mathcal{B}_\Gamma;\ \mu, \eta \in \mathcal{I}_\Gamma;\ 0 \leq |\epsilon| \leq \rho\}.$$

Notice that $S_1(\rho, \Gamma)$ depends on ρ and $\rho\Gamma$ but not on Γ alone. For each v(x), $\chi(x) \in \mathcal{B}_K$ and $\eta, \mu \in \mathcal{I}_K$, a transformation T_ϵ is defined for each ϵ in $0 \leq |\epsilon| \leq \epsilon_1$ by

$$T_\epsilon[v, \chi, \mu, \eta] = [\tilde{v}, \tilde{\chi}, \tilde{\mu}, \tilde{\eta}],$$

where

(3.10) $$\begin{cases} \tilde{\mu}\langle(\phi_0 + \epsilon\chi)\int_0^1 f_{\lambda u}(\lambda_0 + s\epsilon\mu, 0, u)ds, \phi_0^*\rangle \\ \qquad = -\langle(\phi_0 + \epsilon v)(\phi_0 + \epsilon\chi)\int_0^1 f_{uu}(\lambda_0, 0, su)ds, \phi_0^*\rangle \\ \qquad - \epsilon\eta\langle(\phi_0 + \epsilon\chi)\int_0^1 f_{\tau u}(\lambda, s\tau, u)ds, \phi_0^*\rangle, \end{cases}$$

(3.11) $$\begin{cases} \tilde{\eta}\langle\int_0^1 f_\tau(\lambda, s\tau, u)ds, \phi_0^*\rangle = -\tilde{\mu}\langle(\phi_0 + \epsilon v)\int_0^1\int_0^1 f_{\lambda u}(\lambda_0 + s\epsilon\mu, 0, tu)ds\,dt, \phi_0^*\rangle \\ \qquad - \langle(\phi_0 + \epsilon v)^2\int_0^1\int_0^1 f_{uu}(\lambda_0, 0, stu)s\,dt\,ds, \phi_0^*\rangle, \end{cases}$$

$$(3.12)\quad \begin{cases} L\tilde{v} + f_u(\lambda_0, 0, 0)\tilde{v} = -\Big[\tilde{\eta}\int_0^1 f_\tau(\lambda, s\tau, u)ds \\ \qquad + \tilde{\mu}(\phi_0+\epsilon v)\int_0^1\int_0^1 f_{\lambda u}(\lambda_0+s\epsilon\mu, 0, tu)dsdt \\ \qquad + (\phi_0+\epsilon v)^2\int_0^1\int_0^1 f_{uu}(\lambda_0, 0, stu)sdtds\Big], \qquad x\in D, \\ B\tilde{v} = 0, \qquad x\in\alpha D, \\ \langle \tilde{v}, \phi_0^* f_{\lambda u}(\lambda_0, 0, 0)\rangle = 0, \end{cases}$$

$$(3.13)\quad \begin{cases} L\tilde{\chi} + f_u(\lambda_0, 0, 0)\tilde{\chi} = -\Big[\tilde{\mu}(\phi_0+\epsilon\chi)\int_0^1 f_{\lambda u}(\lambda_0+s\epsilon\mu, 0, u)ds \\ \qquad + \epsilon\eta(\phi_0+\epsilon\chi)\int_0^1 f_{\tau u}(\lambda, s\tau, u)ds \\ \qquad + (\phi_0+\epsilon v)(\phi_0+\epsilon\chi)\int_0^1 f_{uu}(\lambda_0, 0, su)ds\Big], \qquad x\in D, \\ B\tilde{\chi} = 0, \qquad x\in\partial D, \\ \langle\tilde{\chi}, \phi_0^* f_{\lambda u}(\lambda_0, 0, 0)\rangle = 0. \end{cases}$$

This definition of T_ϵ induces an iteration procedure in a natural way. Suppose we let an initial iterate be $(v^0(\epsilon, x), \chi^0(\epsilon, x), \mu^0(\epsilon), \eta^0(\epsilon))$. Then we define the sequence of iterates $\{(v^\nu(\epsilon, x), \chi^\nu(\epsilon, x), \mu^\nu(\epsilon), \eta^\nu(\epsilon))\}$ by

$$(3.14)\qquad \left[v^{\nu+1}, \chi^{\nu+1}, \mu^{\nu+1}, \eta^{\nu+1}\right] = T_\epsilon\left[v^\nu, \chi^\nu, \mu^\nu, \eta^\nu\right].$$

Theorem 3.1. Let $S_1 = S_1(\rho, \Gamma)$ for some fixed $\rho \leq 1$, $\rho \leq 1$. Suppose that

$$(3.15)\qquad \begin{aligned} &f(\lambda, \tau, u)\in C^\alpha(S_1),\ f_u(\lambda, \tau, u)\in C^\alpha(S_1), \\ &f_\tau(\lambda, \tau, u), f_\lambda, f_{\lambda u}, f_{uu}, f_{\tau u}, f_{uuu}, f_{\lambda uu}, f_{\tau uu}, f_{\tau\tau u}, f_{\lambda\lambda u}, f_{\lambda\tau u}, f_{\tau\tau}\in C(S_1). \end{aligned}$$

and that $\langle f_\tau(\lambda_0, 0, 0), \phi_0^*\rangle \cdot \langle\phi_0 f_{\lambda u}(\lambda_0, 0, 0), \phi_0^*\rangle \neq 0$. Then, there are real positive constants ϵ_0 and K, $\epsilon_0 \leq \rho$, $\epsilon_0 K \leq \rho\Gamma$ such that the mapping T_ϵ given by (3.10)-(3.13) maps $U_K = (\mathcal{B}_K \times \mathcal{B}_K \times \mathcal{I}_K \times \mathcal{I}_K)$ into U_K, and T_ϵ is a contraction on U_K for all ϵ, $0 \leq |\epsilon| \leq \epsilon_0$. Furthermore, the problem (1.1), (1.7) has a nontrivial solution of the form (3.1) where $v(x,\epsilon)$, $\chi(x,\epsilon)$, $\mu(\epsilon)$, $\eta(\epsilon)$ satisfy (3.3), (3.4), (3.6) and are the limits of the iteration scheme generated by T_ϵ for any initial iterates in U_K.

The proof is given in the previously mentioned doctoral thesis of J. P. Keener [4].

Theorem 3.1 assures us that nonisolated solutions of (1.1) are of the form (3.1), where $v(\epsilon, x)$, $\chi(\epsilon, x)$, $\mu(\epsilon)$, and $\eta(\epsilon)$ are uniformly bounded by K for $|\epsilon| \leq \epsilon_0$. To know more about the quantitative behavior of the solution, we would like to know more about $\mu(\epsilon)$ and $\eta(\epsilon)$. We know that $\mu(\epsilon)$ and $\eta(\epsilon)$ are fixed points of (3.10) and (3.11), respectively. Suppose that there is an integer p such that

$$\frac{\partial^k f(\lambda_0, 0, 0)}{\partial u^k} = 0, \qquad 2 \leq k \leq p,$$

(3.16)

$$\left\langle \frac{\partial^{p+1} f(\lambda_0, 0, 0)}{\partial u^{p+1}} \phi_0^{p+1}, \phi_0^* \right\rangle \neq 0,$$

and assume that all third derivatives of $f(\lambda, \tau, u)$ exist and are continuous. Then, applying the identity (3.5) to (3.10) and (3.11), we find

$$\mu\Big(\langle \phi_0 f_{\lambda u}(\lambda_0,0,0), \phi_0^* \rangle + O(\epsilon)\Big) + \eta\, O(\epsilon) = -\frac{\epsilon^{p-1}}{p!} \left\langle \frac{\partial^{p+1} f(\lambda_0,0,0)}{\partial u^{p+1}} \phi_0^{p+1}, \phi_0^* \right\rangle + O(\epsilon^p), \tag{3.17}$$

$$\eta\Big(\langle f_\tau(\lambda_0,0,0), \phi_0^* \rangle + O(\epsilon)\Big) + \mu\Big(\langle \phi_0 f_{\lambda u}(\lambda_0,0,0), \phi_0^* \rangle + O(\epsilon)\Big) \tag{3.18}$$

$$= -\frac{\epsilon^{p-1}}{(p+1)!} \left\langle \phi_0^{p+1} \frac{\partial^{p+1} f(\lambda_0,0,0)}{\partial u^{p+1}}, \phi_0^* \right\rangle + O(\epsilon^p).$$

Although (3.17) and (3.18) include implicit dependence on $\mu(\epsilon)$ and $\eta(\epsilon)$ in the $O(\epsilon)$ and $O(\epsilon^p)$ terms, we know that $|\mu(\epsilon)| \leq K$, $|\eta(\epsilon)| \leq K$ for $|\epsilon| \leq \epsilon_0$, and this permits the determination of the asymptotic form of $\mu(\epsilon)$ and $\eta(\epsilon)$ as $|\epsilon| \to 0$.

The system (3.17), (3.18) can be solved for ϵ sufficiently small, to give

$$\mu(\epsilon) = -\frac{\epsilon^{p-1}}{p!} \frac{\left\langle \frac{\partial^{p+1} f(\lambda_0,0,0)}{\partial u^{p+1}} \phi_0^{p+1}, \phi_0^* \right\rangle}{\langle \phi_0 f_{\lambda u}(\lambda_0,0,0), \phi_0^* \rangle} + O(\epsilon^p), \tag{3.19}$$

$$\eta(\epsilon) = \epsilon^{p-1} \frac{p}{(p+1)!} \frac{\left\langle \frac{\partial^{p+1} f(\lambda_0,0,0)}{\partial u^{p+1}} \phi_0^{p+1}, \phi_0^* \right\rangle}{\langle f_\tau(\lambda_0,0,0), \phi_0^* \rangle} + O(\epsilon^p). \tag{3.20}$$

Coupling (3.19) and (3.20) with the form of the solution (3.1), we see that the perturbation solution (2.26)-(2.28) is asymptotic to the solution (3.1) as $\epsilon \to 0$. In Section V.4, we will show that this is true for the perturbation scheme with any number of terms.

V.4. Comparison of iteration scheme and perturbation procedure. In Section V.3 we exhibited a mapping T_ϵ whose fixed point gave rise to solutions of (1.1), (1.7) for each ϵ, $0 \leq |\epsilon| \leq \epsilon_0$. The iterations generated by T_ϵ were found to converge to the fixed point for all initial iterates in U_K.

In this section, we will examine the iterations generated by the initial iterate

(4.1) $$w^0 = \left(v^0(\epsilon, x), \chi^0(\epsilon, x), \mu^0(\epsilon), \eta^0(\epsilon)\right) = (0,0,0,0).$$

It is relatively simple to show [4] that there exists a constant M such that

(4.2) $$\|w^{k+1} - w^k\| \leq (|\epsilon| M)\|w^k - w^{k-1}\|,$$

where ϵ^0 is chosen so that $\epsilon_0 M < 1$. Applying (4.2) recursively, we find that

(4.3) $$\|w^{k+1} - w^k\| \leq (|\epsilon| M)^k \|w^1 - w^0\| \leq (|\epsilon| M)^k K.$$

A simple application of the triangle inequality implies

$$\|w^{k+m} - w^k\| \leq K(|\epsilon| M)^k\left[(|\epsilon| M)^{m-1} + (|\epsilon| M)^{m-2} + \cdots + (|\epsilon| M) + 1\right]$$

$$= K(|\epsilon| M)^k \frac{1-(|\epsilon| M)^m}{1-|\epsilon| M},$$

and passing to the limit as $m \to \infty$, we get

(4.5) $$\|w - w^k\| \leq K \frac{(|\epsilon| M)^k}{1-|\epsilon| M},$$

where $w = \left(v(\epsilon, x), \chi(\epsilon, x), \mu(\epsilon), \eta(\epsilon)\right)$ is a solution of (3.3), (3.4), (3.6). Thus, we have

(4.6) $$\|w - w^k\| = O(|\epsilon|^k).$$

We now interpret this information in terms of the solutions of (1.1), (1.7) in the form (3.1). The sequence $\{w^k\}$ corresponds for ϵ fixed to finding a sequence $\left(u^k(\epsilon, x), \psi^k(\epsilon, x), \mu^k(\epsilon), \eta^k(\epsilon)\right)$ where

$$(4.7)\qquad \begin{cases} u^k(\epsilon, x) = \epsilon\phi_0 + \epsilon^2 v^k(\epsilon, x), \\ \psi^k(\epsilon, x) = \phi_0 + \epsilon\chi^k(\epsilon, x), \\ \lambda^k(\epsilon) = \lambda_0 + \epsilon\mu^k(\epsilon), \\ \tau^k(\epsilon) = \epsilon^2\eta^k(\epsilon), \end{cases}$$

with initial iterate

$$(4.8)\qquad (u^0, \psi^0, \lambda^0, \tau^0) = (\epsilon\phi_0, \phi_0, \lambda_0, 0).$$

Furthermore, (4.6) tells us that

$$(4.9)\qquad \begin{cases} \|u-u^k\| = O(|\epsilon|^{k+2}), \\ \|\psi-\psi^k\| = O(|\epsilon|^{k+1}), \\ \|\lambda-\lambda^k\| = O(|\epsilon|^{k+1}), \\ \|\tau-\tau^k\| = O(|\epsilon|^{k+2}). \end{cases}$$

We now have the following theorem (see Keener [4] for proof):

Theorem 4.1. Let the hypotheses of Theorem 3.1 hold, and let (2.30) and (2.31) be satisfied for all ϵ, $|\epsilon| \leq \epsilon_0$. Let $(\tilde{u}^n, \tilde{\psi}^n, \tilde{\lambda}^n, \tilde{\tau}^n)$ be of the form (2.1) with $u_i(x)$, $\psi_i(x)$, bounded on $\overline{D}$ for $i = 1, 2, \cdots n$. Then, the iterates $(u^n, \psi^n, \lambda^n, \tau^n)$ of (4.7) and the perturbation expansions $(\tilde{u}^n, \tilde{\psi}^n, \tilde{\lambda}^n, \tilde{\tau}^n)$ of (2.1) and (2.34)-(2.37) satisfy

$$(4.10)\qquad \begin{cases} \|u^n(\epsilon, x)-\tilde{u}^n(\epsilon, x)\| = O(|\epsilon|^{n+2}), \\ \|\psi^n(\epsilon, x)-\tilde{\psi}^n(\epsilon, x)\| = O(|\epsilon|^{n+1}), \\ |\lambda^n(\epsilon)-\tilde{\lambda}^n(\epsilon)| = O(|\epsilon|^{n+1}), \\ |\tau^n(\epsilon)-\tilde{\tau}^n(\epsilon)| = O(|\epsilon|^{n+2}). \end{cases}$$

Note that an application of the triangle inequality to (4.9), (4.10) immediately implies that the perturbation method is asymptotic to the known solution as $\epsilon \to 0$.

V.5. Extension of a solution branch from a nonisolated solution. In Sections V.2-V.4 we showed that there exist non-trivial nonisolated solutions of (1.1) depending continuously on a parameter ϵ for $|\epsilon| \leq \epsilon_0$. In this section we shall investigate conditions under which a nonisolated solution of (1.1) is an element of a non-trivial solution branch with τ fixed. To do so we shall construct the solution branch of (1.1) which contains a given nonisolated solution.

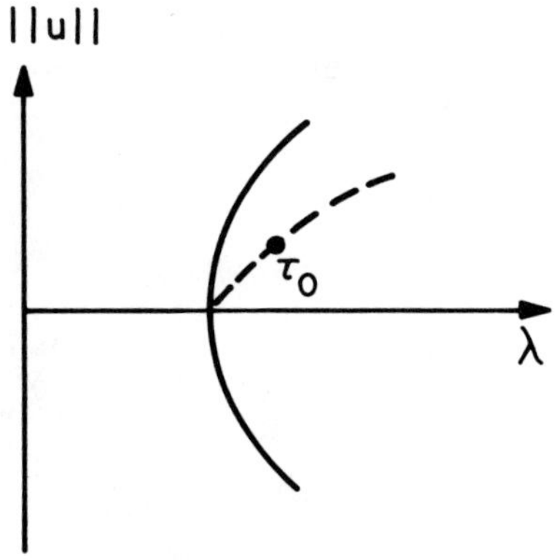

Fig. V.1.

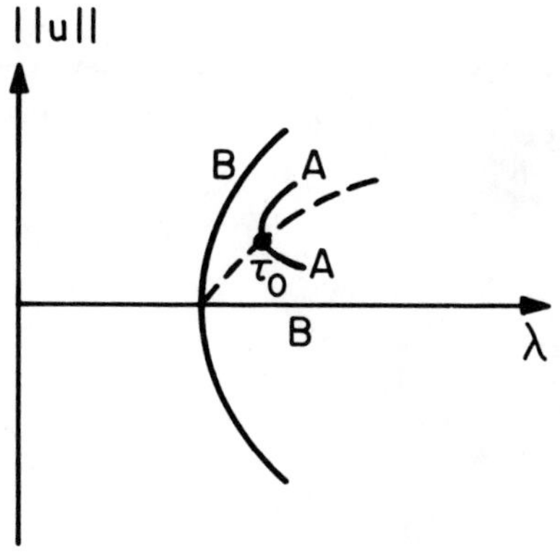

Fig. V.2

Perhaps these statements can be made clearer with the help of Figures V.1 and V.2. Suppose, for example, that for $\tau = 0$ our bifurcation diagram is given by the solid lines of Fig. V.1. The non-trivial nonisolated solutions (for τ small but not zero) of Sections V.2 - V.4 are given by the dashed curve of Figure V.1. (In a manner of speaking, this dashed curve represents the locus of "double zeroes" of the problem (1.1); that is, the operator $Lu + f(\lambda, \tau, u)$ and its first derivative (the Frechet derivative) both vanish on this curve. Thus, all points on the dashed curve are possible branching points for a non-trivial solution branch.) The goal of the present section is to fix τ (say $\tau = \tau_0 \neq 0$) and show that the specific non-trivial nonisolated solution at τ_0 (i.e., the "double zero" at τ_0) is a branching point of a non-trivial solution branch. Thus, we expect to complete Fig. V.1 as illustrated in Fig. V.2. The content of our work would then be that the solution branches A exist for the forced problem (for $\tau = \tau_0 \neq 0$), and as $\tau \to 0$ these branches approach the solution branches B for the unforced problem. All possible situations are accounted for; these are illustrated in Figures V.4 to V.6 (presented at the end of this chapter) which we shall fully explain in Section 6.

To support the claims and conjectures of the preceding paragraph, we proceed as follows: Suppose $\tau = \tau_0 \neq 0$ is fixed arbitrarily, and let

$$g(\lambda, u) \equiv f(\lambda, \tau_0, u). \tag{5.1}$$

Then, (1.1) becomes

$$(5.2) \quad \begin{cases} Lu + g(\lambda, u) = 0, & x \in D, \\ Bu = 0, & x \in \partial D, \end{cases}$$

where $g(\lambda, 0) \neq 0$. Suppose that $u \equiv w_0(x)$ is a nonisolated solution of (5.2) for $\lambda = \mu_0$. Accordingly, there exist functions $\psi_0(x)$ and $\psi_0^*(x)$ which satisfy

$$(5.3) \quad \begin{cases} L\psi + g_u(\mu_0, w_0(x))\psi = 0, & x \in D, \\ B\psi = 0, & x \in \partial D, \end{cases}$$

and

$$(5.4) \quad \begin{cases} L^*\psi^* + g_u(\mu_0, w_0(x))\psi^* = 0, & x \in D, \\ B^*\psi^* = 0, & x \in \partial D. \end{cases}$$

We want to find solution sets $(\mu, w(x))$ of (5.2), if they exist, such that $\mu - \mu_0$ and $w(x) - w_0(x)$ are small. A natural way to proceed is to use the perturbation method to suggest the form of such solutions and then to construct a contraction mapping which shows that the suggested form leads to solutions. Suppose we assume an expansion of (μ, w) in powers of δ which has the form

$$(5.5) \quad \begin{cases} \widetilde{w}(x, \delta) = w_0(x) + \delta w_1(x) + \delta^2 w_2(x) + \cdots, \\ \widetilde{\mu}(\delta) = \mu_0 + \delta\mu_1 + \delta^2 \mu_2 + \cdots. \end{cases}$$

Substituting (5.5) into (5.2), expanding $g(\widetilde{\mu}, \widetilde{w})$ in powers of δ, and equating the coefficients of like powers of δ leads to the equations

$$(5.6) \quad \begin{cases} Lw_0 + g(\mu_0, w_0) = 0, & x \in D, \\ Bw_0 = 0, & x \in \partial D, \end{cases}$$

$$(5.7) \quad \begin{cases} Lw_1 + g_u(\mu_0, w_0)w_1 = -g_\lambda(\mu_0, w_0)\mu_1, & x \in D, \\ Bw_1 = 0, & x \in \partial D, \end{cases}$$

and

$$Lw_2 + g_u(\mu_0, w_0)w_2 = -\Big[g_\lambda(\mu_0, w_0)\mu_0 + \tfrac{1}{2} g_{uu}(\mu_0, w_0)w_1^2$$

(5.8) $$+ g_{\lambda u}(\mu_0, w_0)\mu_1 w_1 + \tfrac{1}{2} g_{\lambda\lambda}(\mu_0, w_0)\mu_1^2\Big], \qquad x \in D,$$

$$Bw_2 = 0, \qquad x \in \partial D,$$

provided the derivatives g_λ, $g_{\lambda\lambda}$, $g_{\lambda u}$ and g_{uu} exist and are continuous. In order that w_2 be uniquely determined we require that

$$\langle w_2, \overset{*}{\psi_0} g_{\lambda u}(\mu_0, w_0)\rangle = 0.$$

Equation (5.6) is automatically satisfied by our definition of μ_0 and $w_0(x)$. Because ψ_0 satisfies (5.3), the Fredholm alternative theorem implies that (5.7) can be solved only if

(5.9) $$\mu_1\langle g_\lambda(\mu_0, w_0), \psi_0^* \rangle = 0 .$$

If we assume that $\langle g_\lambda(\mu_0, w_0), \psi_0^* \rangle \neq 0$, then (5.9) implies that

(5.10) $$\mu_1 = 0, \quad w_1(x) = \psi_0(x).$$

Finally, the Fredholm alternative theorem applied to (5.8) implies

(5.11) $$\mu_2\langle g_\lambda(\mu_0, w_0), \psi_0^* \rangle = -\tfrac{1}{2}\langle g_{uu}(\mu_0, w_0)\psi_0^2, \psi_0^* \rangle .$$

Thus, the perturbation method indicates that solutions of (5.2) are of the form

(5.12) $$\begin{cases} \tilde{w}(x, \delta) = w_0(x) + \delta\psi_0(x) + O(\delta^2), \\ \tilde{\mu}(\delta) = \mu_0 + \delta^2\mu_2 + O(\delta^3), \\ \text{where } \mu_2 = -\tfrac{1}{2} \dfrac{\langle g_{uu}(\mu_0, w_0)\psi_0^2, \psi_0^* \rangle}{\langle g_\lambda(\mu_0, w_0), \psi_0^* \rangle} . \end{cases}$$

Motivated by the results of the perturbation method (5.12), we propose to look for solutions of (5.2) of the form

(5.13) $$\begin{cases} w(x, \delta) = w_0(x) + \delta\psi_0(x) + \delta^2 y(x, \delta), \\ \mu(\delta) = \mu_0 + \delta^2 \nu(\delta), \\ \text{where } \langle y(x, \delta), \psi_0^*(x) g_{\lambda u}(\mu_0, w_0)\rangle = 0 . \end{cases}$$

On substituting (5.13) into (5.2) we get

$$(5.14)\quad \begin{cases} Ly + g_u(\mu_0, w_0)y = -\frac{1}{\delta^2}\Big[g(\mu, w) - g(\mu_0, w_0) - g_u(\mu_0, w_0)(w - w_0)\Big] \\ \qquad = -\Big[\nu \int_0^1 g_\lambda(\mu_0 + \delta^2 s\nu_1 w)ds \\ \qquad\quad + (\psi_0 + \delta y)^2 \int_0^1\int_0^1 g_{uu}\Big(\mu_0, w_0 + \delta st(\psi_0 + \delta y)\Big) s\,dt\,ds\Big] \\ \qquad = P(y, \nu, \delta; x), \qquad x \in D, \\ By = 0, \qquad x \in \partial D, \\ \langle y, \psi_0^* g_{\lambda u}(\mu_0, w_0)\rangle = 0. \end{cases}$$

Equation (5.14) is of the form (1.16) and can be solved for y only if $\langle P(y, \nu, \delta; s), \psi_0^* \rangle = 0$.

As before, we expect that we shall be able to find a solution of (5.14) for δ sufficiently small by employing an iteration procedure. To set up such a procedure, we introduce the sets

$$(5.15)\quad \mathcal{B}_k = \{y(x) \mid y(x) \in C^{2+\alpha}(D), \|y\|_\infty \leq K, \langle y(x), \psi_0^*(x) g_{\lambda u}(\mu_0, w_0)\rangle = 0\},$$

$$(5.16)\quad \mathcal{I}_K = \{\eta \,\big|\, |\eta| \leq K\},$$

$$(5.17)\quad \begin{aligned} S_2(\rho, \Gamma) = \{(\mu, w; x) \mid \mu = \mu_0 + \delta^2 \nu, w = w + \delta\psi_0 + \delta^2 y,\ x \in \overline{D}, \\ 0 \leq |\delta| \leq \rho,\ \nu \in \mathcal{I}_\Gamma, w \in \mathcal{B}_\Gamma\}. \end{aligned}$$

For each y(x) in $\mathcal{B}_K$ and $\nu \in \mathcal{I}_K$, we define the mapping T_δ for each δ in $0 \leq |\delta| \leq \delta_1$ by

$$T_\delta(y, \nu) = (\tilde{y}, \tilde{\nu}),$$

where

$$(5.18)\quad \tilde{\nu}\langle \int_0^1 g_\lambda(\mu_0 + \delta^2 s\nu, w)ds, \psi_0^* \rangle = -\langle (\psi_0 + \delta y)^2 \int_0^1\int_0^1 g_{uu}\Big(\mu_0, w_0 + \delta st(\psi_0 + \delta y)\Big) s\,dt\,ds, \psi_0^* \rangle,$$

$$(5.19)\quad \begin{cases} L\tilde{y}+g_u(\mu_0,w_0)\tilde{y} = -\Big[\tilde{\nu}\int_0^1 g_\lambda(\mu_0+\delta^2 s\nu, w)ds \\ \qquad + (\psi_0+\delta y)^2 \int_0^1\int_0^1 g_{uu}\big(\mu_0, w_0+\delta st(\psi_0+\delta y)\big)\, sdtds\Big], \quad x\in D, \\ B\tilde{y} = 0, \qquad x\in \partial D, \\ \langle \tilde{y}, \psi_0^* g_{\lambda u}(\mu_0, w_0)\rangle = 0\ . \end{cases}$$

Then, after picking some initial iterate (y^0, ν^0), a sequence of iterates $\{y^k, \nu^k\}$ will be generated by

$$(5.20)\qquad \left(y^{k+1},\ \nu^{k+1}\right) = T_\delta\left(y^k,\ \nu^k\right)\ .$$

We now have the following result (see Keener [4] for proof):

Theorem 5.1. Let $S_2 = S_2(\rho,\Gamma)$ for some fixed $\rho \leq 1$, $\rho\Gamma \leq 1$. Suppose that

$$(5.21)\qquad g(\lambda,u), g_u(\lambda,u)\in C^\alpha(S_2),\ g_\lambda, g_{uu}, g_{\lambda u}, g_{\lambda\lambda}, g_{uuu} \in C(S_2)$$

and that $\langle g_\lambda(\mu_0, w_0), \psi_0^*\rangle \neq 0$. Then there exist real positive constants δ_0 and M, $\delta_0 \leq \rho$, $\delta_0 M \leq \rho\Gamma$ such that the mapping T_δ given by (5.18), (5.19) maps $W_M = (\mathcal{B}_M \times \mathcal{I}_M)$ into W_M, and T_δ is a contraction on W_M for all δ, $0 \leq |\delta| \leq \delta_0$. Furthermore, the problem (5.2) has a solution of the form (5.13) for all δ $0 \leq |\delta| \leq \delta_0$, where $y(x,\delta)$ and $\nu(\delta)$ are the limits of the iterates $\{y^k, \nu^k\}$ generated by (5.20) for any initial iterate in W_M.

Just as in Section V.4, we could compare the iteration procedure (5.20) with the perturbation scheme (5.12). Once again we would find that the perturbation scheme is asymptotic to the iteration scheme and that the iteration scheme is asymptotic to the solution as $\delta \to 0$. It will be instructive to examine the asymptotic expansion of the solution $(y(\delta,x), \nu(\delta))$, which is a fixed point of (5.18), (5.19). From (5.18) it is easy to see that

$$(5.22)\qquad \nu(\delta) = \frac{-\langle g_{uu}(\mu_0, w_0)\,\psi_0^2,\ \psi_0^*\rangle}{2\langle g_\lambda(\mu_0, w_0),\ \psi_0^*\rangle} + O(\delta)\ .$$

Substituting (5.22) into (5.13) we see that, to the order which we have taken the solution the exact solution (5.13) and the perturbation solution (5.12) agree asymptotically as $\delta \to 0$.

Knowing the form of the solution (5.13) gives us information about those parameter values μ for which solutions of (5.2) exist. Since $\mu = \mu_0+\delta^2\nu(\delta)$, if $\nu(0) > 0$, then solutions of (5.2) exist in the neighborhood of (w_0, μ_0) for which $\mu > \mu_0$. If $\nu(0) < 0$, then solutions of (5.2) exist in the neighborhood of (w_0,μ_0)

for which $\mu < \mu_0$. In either case, the point $\mu = \mu_0$ is a branching point where the number of solutions of (1.1) changes from zero to two or from two to zero as μ changes from $\mu < \mu_0$ to $\mu > \mu_0$, in the respective cases, $\nu(0) > 0$ and $\nu(0) < 0$. Figure V.3 gives plots of $\mu(\delta)$ versus δ when $\langle g_\lambda(\mu_0, w_0), \psi_0^* \rangle > 0$.

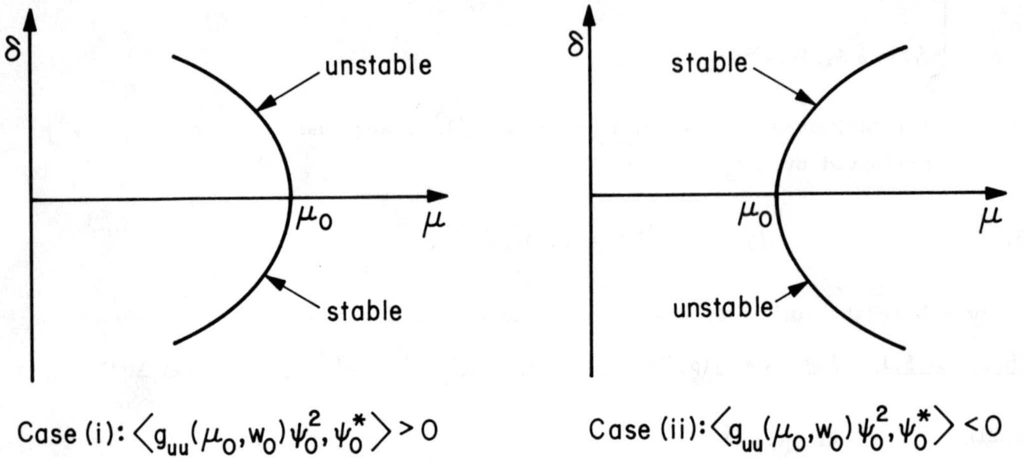

Fig. V.3. Plot of $\mu(\delta)$ in (5.13) for $\langle g_\lambda(\mu_0, w_0), \psi_0^* \rangle > 0$.

We have now shown circumstances under which a nonisolated solution of (1.1) is an element of a solution branch of (1.1) for τ fixed. Since in Section V.3 we were able to show that nontrivial nonisolated solutions of (1.1) do exist, it is natural to ask how Theorem 5.1 applies to the results of Section V.3.

If ϵ_0, found in Section V.3, is sufficiently small, then the resulting eigenvalue $\lambda = \lambda_0 + \epsilon\mu(\epsilon)$ in (3.1) remains isolated so that the null space of (1.7) remains one-dimensional for $|\epsilon| \leq \epsilon_0$. The hypotheses of Theorem 3.1 are sufficient to insure that the hypotheses of Theorem 5.1 hold in $S_1 = S_1(\rho, \Gamma)$ for certain nonzero ρ, Γ, for some fixed ϵ, $|\epsilon| \leq \epsilon_0$. Applying Theorem 5.1, we substitute into (5.13), for $w_0(x)$, $\psi_0(x)$ and μ_0, the nonisolated solutions of (1.1) found in Theorem 3.1 and given in the form (3.1). The resulting solutions of (1.1) are then

$$(5.23)\quad \begin{cases} u = (\epsilon + \delta)\,\phi_0(x) + \epsilon^2 v(x, \epsilon) + \epsilon\delta\,\chi(x, \epsilon) + \delta^2 y(x, \epsilon, \delta), \\ \lambda = \lambda_0 + \epsilon\mu(\epsilon) + \delta^2 \nu(\epsilon, \delta), \\ \tau = \epsilon^2 \eta(\epsilon), \end{cases}$$

where ϵ is fixed, $|\epsilon| \leq \epsilon_0$.

The solution of (1.1) given by (5.23) is valid only if $|\delta| < \delta_0$. However, the number δ_0 is not independent of the number ϵ. In the proof of Theorem 5.1 (see [4]) it is shown that δ_0 is chosen so that

(5.24) $$\delta_0 < \frac{1}{C} = \frac{\gamma}{\tilde{C}} ,$$

where $\gamma = | \langle g_\lambda(\mu_0, w_0), \psi_0^* \rangle |$. Since (μ_0, w_0) are related to ϵ by

(5.25) $$w_0(x) = u(x,\epsilon), \qquad \mu_0 = \lambda(\epsilon),$$

where $\big(u(x,\epsilon), \lambda(\epsilon)\big)$ are of the form (3.1), we can find the dependence of γ on ϵ. Specifically

(5.26) $$\langle g_\lambda(\mu_0,w_0), \psi_0^* \rangle = \langle f_\lambda(\lambda, \tau, u), \psi_0^* \rangle = \epsilon \langle f_{\lambda u}(\lambda_0,0,0)\phi_0, \phi_0^* \rangle + O(\epsilon^2).$$

Note that in (5.26) we have used $\psi_0^* = \phi_0^* + O(\epsilon)$. This can be shown to be true in the same way that it was shown in Section V.3, that $\psi_0 = \phi_0 + O(\epsilon)$, using the fact that $(\phi_0^*)^* = \phi_0$.

Since the constant $\tilde{C}$ in (5.24) is bounded away from zero when $|\epsilon| \leq \epsilon_0$, (5.24) coupled with (5.26) imply that

(5.27) $$\delta_0 = O(\epsilon).$$

Clearly, as ϵ approaches zero, the range of validity of (5.23) decreases. This decrease in the range of validity is not unexpected. As seen in Figures V.4 to V.6 (at the end of this chapter) for $\tau = 0$, the bifurcation solution has a sharp "corner" at $\lambda = \lambda_0$. As $\epsilon \to 0$, the solution branch (5.23) with $\tau \neq 0$ approaches this "corner." But since (5.23) is a smooth function of δ, it cannot have a "corner" when $\epsilon = 0$ so that $\delta_0(\epsilon)$ must approach zero as $\epsilon \to 0$.

We would like to be able to further understand the nature of the solution given by (5.23). Thus, we examine the expression for $\lambda = \lambda(\epsilon, \delta)$. Recall that $\lambda = \lambda_0 + \epsilon\mu(\epsilon) + \delta^2 \nu(\epsilon, \delta)$ where $\nu(\epsilon, \delta)$ is given by (5.22) Since we know the form of μ_0, w_0 and ψ_0 as functions of ϵ, we can rewrite (5.22) as

(5.28) $$\nu(\epsilon, \delta) = - \frac{\epsilon^{p-2}}{2(p-1)!} \frac{\left\langle \dfrac{\partial^{p+1} f(\lambda_0,0,0)}{\partial u^{p+1}} \phi_0^{p+1}, \phi_0^* \right\rangle}{\langle f_{\lambda u}(\lambda_0,0,0)\phi_0, \phi_0^* \rangle} + O(\epsilon^{p-1}) + O(\delta),$$

where the integer p is defined in (3.16). We now see that the nonisolated solution (3.1) corresponds to a branching point for ϵ sufficiently small, since according to (3.16), $\langle \partial^{p+1} f(\lambda_0,0,0) / \partial u^{p+1} \, \phi_0^{p+1}, \phi_0^* \rangle \neq 0$ and hence $\nu(\epsilon,0) \neq 0$.

V. 6. Other solution branches. In previous sections we have shown the existence of branches of (1.1) for τ small. All of the branches contained elements which were nonisolated solutions. Depending on the properties of $f(\lambda, \tau, u)$, for a fixed τ, a given problem may have two, one or possibly no branches with nonisolated solutions. By examining (2.29), we see that if p, defined in (2.19), is even, then there is one nonisolated solution for $\tau > 0$ and one for $\tau < 0$. However, if p is odd, τ must be restricted so that λ is real, which means that there will be two nonisolated solutions for τ of one sign and no nonisolated solutions for τ of the other sign. Work by Simpson and Cohen [13] suggests that this is not the complete story. They find solution branches which have no nonisolated solutions, but points on the branch approach the solution pair $(u, \lambda) = (0, \lambda_0)$ where λ_0 is the principal eigenvalue of (1.6), as $\tau \to 0$. In this section we will show that for all values of τ sufficiently small, (1.1) has at least two distinct solution branches with values of λ in a neighborhood of an eigenvalue λ_0 given by (1.6).

In previous sections we have suggested that the perturbation theory provides a method which will always lead to the desired answer. In this situation, such is not the case. Although we will prove our result by use of a contraction mapping, the mapping is one that is not motivated by a perturbation procedure.

We seek solutions of (1.1) of the form

$$(6.1) \qquad \begin{aligned} u(x,\epsilon) &= \epsilon\,\phi_0 + \epsilon^2 w(x,\epsilon)\ , \\ \lambda(\epsilon) &= \lambda_0 + \epsilon\nu(\epsilon)\ , \end{aligned}$$

for τ fixed, $0 \leq |\tau| \leq \tau_1$. To make $w(x,\epsilon)$ unique we will require

$$(6.2) \qquad \langle w(x,\epsilon),\ \phi_0^* f_{\lambda u}(\lambda_0, 0, 0)\rangle = 0\ .$$

Notice that when $\tau \neq 0$, setting $\epsilon = 0$ in (6.1) does not give a solution of (1.1). This leads us to suspect that (6.1) is valid for $0 < \epsilon_1 \leq |\epsilon| \leq \epsilon_2$ where ϵ_1 and ϵ_2 are related to τ in some way to be determined.

If we substitute (6.1) into (1.1) we find

$$(6.3) \quad \left\{ \begin{aligned} Lw + f_u(\lambda_0, 0, 0)w &= -\frac{1}{\epsilon^2}\{f(\lambda, \tau, u) - f_u(\lambda_0, 0, 0)u\} \\ &= -\Big[\frac{\tau}{\epsilon^2}\int_0^1 f_\tau(\lambda, s\tau, u)ds + \nu(\phi_0+\epsilon w)\int_0^1 f_{\lambda u}(\lambda_0+\epsilon s\nu, 0, tu)ds\,dt \\ &\qquad + (\phi_0+\epsilon w)^2\int_0^1\int_0^1 f_{uu}(\lambda_0, 0, stu)s\,dt\,ds\Big] = R(\nu, \tau, w; \epsilon), \\ Bw &= 0, \\ \langle w, \phi_0 f_{\lambda u}(\lambda_0, 0, 0)\rangle &= 0. \end{aligned} \right.$$

Equation (6.3) is again of the form (1.12) and can be solved only if

(6.4) $$\langle R(\nu, \tau, w; \epsilon), \phi_0^* \rangle = 0 .$$

To formulate the contraction mapping, we again use the set of functions $\mathcal{B}_K$ of (3.7) and the real interval $\mathcal{J}_K$ of (3.8). We also introduce the set

(6.5) $$S_3(\rho, \Gamma) = \{(\lambda, u; x) \mid \lambda = \lambda_0 + \epsilon\nu,\ u = \epsilon\phi_0 + \epsilon^2 w, \qquad x \in \overline{D}$$

$$0 \leq |\epsilon| \leq \rho, \qquad \nu \in \mathcal{J}_\Gamma, \quad w \in \mathcal{B}_\Gamma\} .$$

We define the mapping T_ϵ in the natural way suggested by (6.3), (6.4). That is, for each ϵ in $\epsilon_1 \leq |\epsilon| \leq \epsilon_2$, define $T_\epsilon[w, \nu] = [\tilde{w}, \tilde{\nu}]$ by

(6.6)
$$\tilde{\nu}\langle(\phi_0 + \epsilon w)\int_0^1 f_{\lambda u}(\lambda_0 + s\epsilon\nu, 0, tu)ds\, dt, \phi_0^*\rangle = -\frac{\tau}{\epsilon^2}\langle\int_0^1 f_\tau(\lambda, s\tau, u)ds, \phi_0^*\rangle$$
$$+ \langle(\phi_0 + \epsilon w)^2 \int_0^1\int_0^1 f_{uu}(\lambda_0, 0, stu)s\, dt\, ds, \phi_0^*\rangle ,$$

(6.7)
$$\begin{cases} L\tilde{w} + f_u(\lambda_0, 0, 0)\tilde{w} = -\Big[\tilde{\nu}(\phi_0 + \epsilon w)\int_0^1 f_{\lambda u}(\lambda_0 + \epsilon s\nu, 0, tu)ds\, dt + \frac{\tau}{\epsilon^2}\int_0^1 f_\tau(\lambda, s\tau, u)ds \\ \qquad + (\phi_0 + \epsilon w)^2 \int_0^1\int_0^1 f_{uu}(\lambda_0, 0, stu)s\, dt\, ds\Big], \qquad x \in D, \\ B\tilde{w} = 0, \\ \langle\tilde{w}, \phi_0^* f_{\lambda u}(\lambda_0, 0, 0)\rangle = 0. \end{cases}$$

The mapping T_ϵ induces a natural iteration procedure, which for some initial iterate (w^0, ν^0) is given by

(6.8) $$\left[w^{k+1}, \nu^{k+1}\right] = T_\epsilon\left[w^k, \nu^k\right] \qquad k = 0, 1, 2, \cdots .$$

We now have (see Keener [4] for proof)

Theorem 6.1. Let $S_3 = S_3(\rho, \Gamma)$ for some fixed $\rho \leq 1$, $\rho\Gamma \leq 1$. Suppose that the smoothness assumption (3.15) on $f(\lambda, \tau, u)$ hold on S_3 and that $\langle\phi_0 f_{\lambda u}(\lambda_0, 0, 0), \phi_0^*\rangle \neq 0$. Then, there exist real positive constants ϵ_1, ϵ_2, K, where $\epsilon_1 < \epsilon_2 \leq \rho$, $\epsilon_2 K \leq \rho\Gamma$, such that the mapping T_ϵ, given by (6.6), (6.7), maps $W_K = (\mathcal{B}_K \times \mathcal{J}_K)$ into W_K, and T_ϵ is a contraction on W_K for all ϵ, $0 < \epsilon_1 \leq |\epsilon| \leq \epsilon_2$. Furthermore, the problem (1.1) has nontrivial solutions of the form (6.1), where $w(x, \epsilon)$, $\nu(\epsilon)$ satisfy (6.3), (6.4) and are the limits of the sequence generated by (6.8) for any initial iterate in W_K.

Note that the form of the solutions branches (6.1) implies that two branches result as ϵ takes on positive and negative values, $0 < \epsilon_1 \leq |\epsilon| \leq \epsilon_2$.

When a solution branch has a nonisolated solution, that solution can be found using the technique of Section V.3. J. P. Keener [4] shows that the nonisolated solution found in Section V.3 is an element of a solution branch found in the above Theorem 6.1.

Suppose we represent the nonisolated solution (3.1) by

$$
\begin{aligned}
u_s(x,\epsilon) &= \epsilon\phi_0 + \epsilon^2 v_s(x,\epsilon), \\
\lambda_s &= \lambda_0 + \epsilon\mu_s(\epsilon), \\
\tau_s &= \epsilon^2\eta_s(\epsilon), \\
\psi_s(x,\epsilon) &= \phi_0 + \epsilon\chi_s(x,\epsilon).
\end{aligned} \tag{6.9}
$$

Then $y_s = (u_s, \lambda_s)$ is a nonisolated solution of (1.1) for $\tau = \tau_s$. To be precise, Keener [4] proves the

Theorem 6.2. Let the hypotheses of Theorem 3.1 and Theorem 6.1 hold. Then, the nonisolated solution $y_s = (u_s, \lambda_s)$ of (6.9) lies on the solution branch (6.1) of (1.1), for $\tau = \tau_s$ whenever $|\epsilon| \leq \min\{\epsilon_s, \epsilon_2\}$.

The form of the solution branch (6.1) can be put into the form of (5.23) (the solution branch extended from a nonisolated solution) whenever Theorem 6.2 holds. Since solution branches of the form (5.23) were shown to be unique, the solution branch found here and the solution branch found in Section V.5 must be segments of the same branch whenever Theorem 6.2 holds. If there are values τ which do not give rise to a nonisolated solution of (1.1) as in the case when p is odd, then it seems reasonable to suspect that the solution branch has no nonisolated solutions.

When the nonlinearity $f(\lambda, \tau, u)$ is a positive monotone increasing, concave function for $\tau > 0$, the results of Simpson and Cohen [13] are applicable. In particular, if $\phi_0(x)$ is the positive eigenfunction of (1.6) and λ_0 the corresponding eigenvalue, then the solutions (6.1) are positive if τ_1 and ϵ_2 are sufficiently small and $\epsilon > 0$. However, in this situation (see [13]) positive solutions are unique for each λ so that the solution branch (6.1) must be a segment of the branch of positive solutions found by Simpson and Cohen [13] on which no solutions are nonisolated solutions.

We summarize the results of the foregoing investigation in Figures V.4 to V.6 by showing some possible solution branches which may occur. (See the key on page 61 for explanation of the various curves.)

In addition to the results presented in this chapter Keener [4] has investigated the stability of the various solution branches and the minimal positive and maximal negative solution branches of (1.1). The results of his studies are illustrated in Figures V.3 to V.6 although we shall not attempt to elucidate further here. All results appear in [4] together with applications to imperfection sensitivity in the dynamic buckling of columns and arches.

REFERENCES

1. Courant, R., and Hilbert, D., Methods of Mathematical Physics, Vols. I, II, Interscience, N.Y., 1966.

2. Crandall, M. G. and Rabinowitz, P.H., Bifurcation from simple eigenvalues, J. Funct. Anal., 8 (1971) 321-340.

3. Dunford, N., and Schwartz, J., Linear Operators, Part I, Interscience, N.Y., 1958.

4. Keener, J.P., Some modified bifurcation problems with application to imperfection sensitivity in buckling, Ph.D. thesis. California Institute of Technology, 1972.

5. Keener, J.P., and Keller, H.B., Perturbed Bifurcation Theory, Archive Rat. Mech. Anal., to appear.

6. Keller, H.B., Nonlinear bifurcation, J. Diff. Eq., 7 (1970).

7. Keller, J.B., and Antman, S., Bifurcation Theory and Nonlinear Eigenvalue Problems, Benjamin, N.Y., 1969.

8. Krasnosel'ski, M.A., Topological Methods in the Theory of Nonlinear Integral Equations, MacMillan Co., 1964.

9. Ladyzenskaja, O.A., and Ural'tseva, N.N., Linear and Quasilinear Elliptic Equations, Academic Press, 1968.

10. Protter, M.H., and Weinberger, H.F., Maximum Principles in Differential Equations, Prentice-Hall, Englewood Cliffs, N.J., 1967.

11. Sather, D., Branching of solutions of an equation in Hilbert space, Arch. Rat. Mech. Anal., 36 (1970) 47-64.

12. Sattinger, D.H., Topics in Stability and Bifurcation Theory, University of Minnesota lecture notes, 1972. Also, Sattinger's lectures in the present volume.

13. Simpson, R.B., and Cohen, D.S., Positive solutions of nonlinear elliptic eigenvalue problems, J. Math. Mech. 19 (1970) 895-910.

KEY for FIGURES V.4 to V.6

······· Bifurcation solution when $\tau = 0$.

– – – – Locus of nonisolated solutions – Theorem 3-1.

———— Extension from nonisolated solution for τ fixed – Theorem 5-1.

oooooo Minimal positive and maximal negative solutions (included only when $\phi_0(x) > 0$ for $x \in D$).

===== Solution branches with no nonisolated solutions – Theorem 6-1.

$U = \langle u(x), f_{\lambda u}(\lambda_0, 0, 0)\, \phi_0^*(x) \rangle$

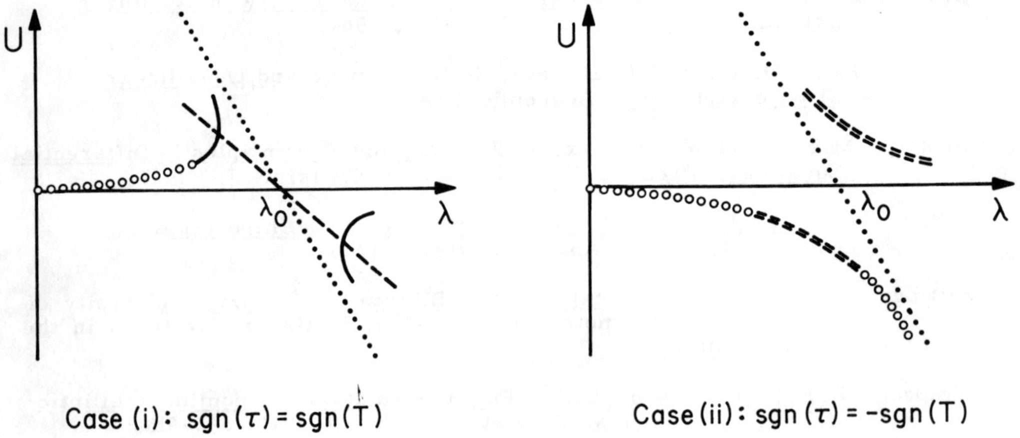

Fig. V.4. Solution branches of (II.1.1) for τ sufficiently small when $\langle f_{uu}(\lambda_0,0,0)\phi_0^2, \phi_0^* \rangle \cdot \langle f_{\lambda u}(\lambda_0,0,0)\phi_0, \phi_0^* \rangle > 0$.

$T = \langle f_{uu}(\lambda_0,0,0)\phi_0^2, \phi_0^* \rangle \cdot \langle f_\tau(\lambda_0,0,0), \phi_0^* \rangle$.

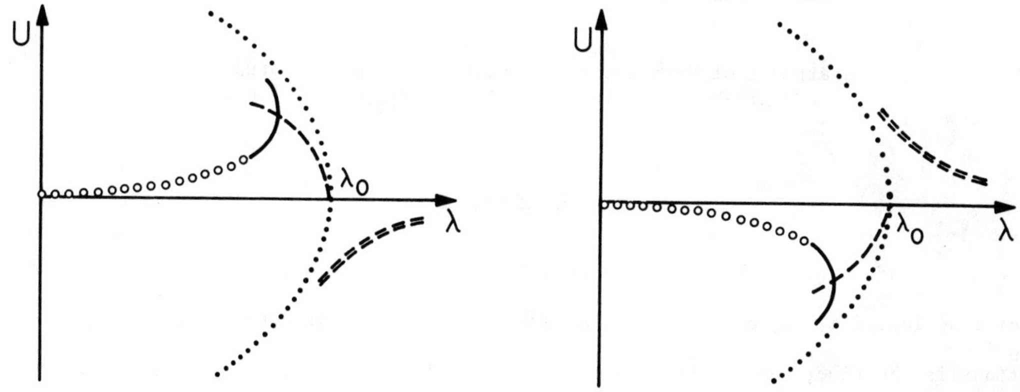

Case (i): sgn(τ) = sgn(T)   Case (ii): sgn(τ) = − sgn(T)

Fig. V.5. Solution branches of (II.1.1) when $f_{uu}(\lambda_0,0,0) = 0$ and $\langle f_{uuu}(\lambda_0,0,0)\phi_0^3, \phi_0^*\rangle \cdot \langle f_{\lambda u}(\lambda_0,0,0)\phi_0, \phi_0^*\rangle > 0$ for τ sufficiently small. $T = \langle f_{uuu}(\lambda_0,0,0)\phi_0^3, \phi_0^*\rangle \cdot \langle f_\tau(\lambda_0,0,0), \phi_0^*\rangle$.

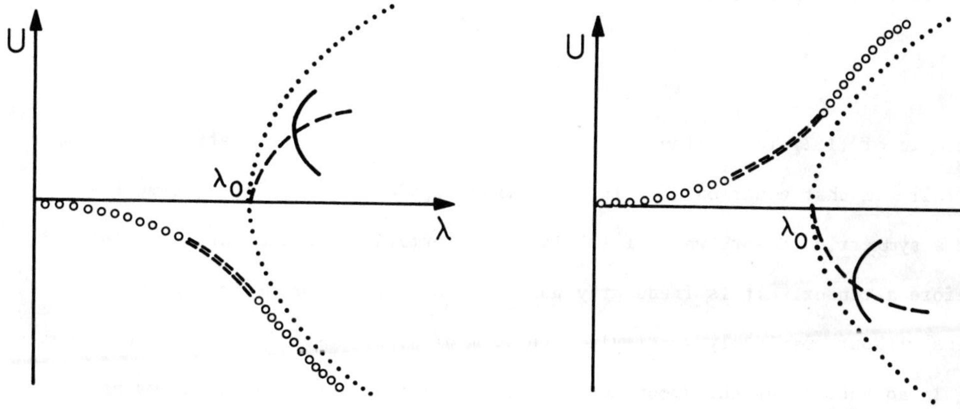

Case (i): sgn(τ) = sgn(T)   Case (ii): sgn(τ) = − sgn(T)

Fig. V.6. Solution branches of (II.1.1) when $f_{uu}(\lambda_0,0,0) = 0$ and $\langle f_{uuu}(\lambda_0,0,0)\phi_0^3, \phi_0^*\rangle \cdot \langle f_{\lambda u}(\lambda_0,0,0)\phi_0, \phi_0^*\rangle < 0$ for τ sufficiently small. $T = \langle f_{uuu}(\lambda_0,0,0)\phi_0^3, \phi_0^*\rangle \cdot \langle f_\tau(\lambda_0,0,0), \phi_0^*\rangle$.

FADING MEMORY AND FUNCTIONAL-DIFFERENTIAL EQUATIONS

Bernard D. Coleman
Department of Mathematics and Center for Special Studies
Carnegie-Mellon University, Pittsburgh, Pennsylvania

1. Introduction

Often in physics when one uses the word process, one means a collection $\mathcal{C}$ of functions f,g,..., over the real axis $\mathbb{R} = (-\infty,\infty)$. The argument t of these functions is the time; the values f(t),g(t),..., lie in normed vector spaces. Given any f in $\mathcal{C}$, the function f^t on $\mathbb{R}^+ = [0,\infty)$, defined by

$$f^t(s) = f(t-s), \qquad s\in\mathbb{R}^+, \tag{1.1}$$

is called the history of f up to t. Different materials and physical systems are distinguished by different constitutive assumptions which place restrictions on the class $\mathcal{C}$. In one type of constitutive assumption it is supposed that there is assigned a function $\mathfrak{g}$ which gives, for each t, the "present value" g(t) of g when the history of f up to t is specified:

$$g(t) = \mathfrak{g}(f^t). \tag{1.2}$$

An example of (1.2) in continuum mechanics is the constitutive equation of a simple material;# in that equation g(t) is the stress at a material point at time t and hence a symmetric tensor, while $f^t(s)$ is the deformation gradient at the point and therefore a tensor. It is frequently possible to prove theorems in a branch of physics without completely specifying the form of $\mathfrak{g}$, but usually one must know something in advance about the smoothness of $\mathfrak{g}$. For this reason several norms have been proposed for sets of histories f^t.##

The norm $\|\cdot\|$ on a space of histories f^t is subject to three elementary physical requirements:###

#Cf. Noll [1].

##E.g. Coleman & Noll [1]-[3]; Coleman [1]; Wang [1][2]; Coleman & Mizel [1]-[4].

###For a discussion of the physical significance of (1)-(3), see Coleman & Mizel [1].

(1) The history $f^{t+\sigma}$ of f up to time $t+\sigma$, $\sigma \geq 0$, in a process for which f has the history f^t up to time t and is held constant in the interval $[t,t+\sigma]$ is called the "static continuation of f^t by amount σ". It is required that if the norm of f^t is finite, then the norm of each static continuation of f^t must be finite also. Moreover, if the distance $\|f_1^t - f_2^t\|$ between two histories is zero, then there must be zero distance between their static continuations by any given amount.

(2) If f^t is the history of f up to time t, then the history of f up to the earlier time $t-\sigma$, $\sigma \geq 0$, is called the "σ-section of f^t". It is required that if f^t has finite norm, then the norm of each σ-section of f^t must be finite.

(3) In physical theories "equilibrium states" must have finite norm; that is, if $f^t(s) \equiv c$, a constant, then we require $\|f^t\| < \infty$.

V. J. Mizel and I have studied consequences of these physical requirements for a broad class of Banach function spaces,[#] and I should like to give here some of our results. The limitations of space, however, require that I omit proofs.

2. Influence Measures

Let μ be an influence measure; that is, a non-trivial (i.e. not identically zero), sigma-finite, positive, regular Borel measure on $\mathbb{R}^+$ and let $\mathcal{S}$ be the set of all μ-measurable functions y mapping $\mathbb{R}^+$ into $\mathbb{R}^+$. Let ν be a function on $\mathcal{S}$ such that for y (or y_i) in $\mathcal{S}$:

(i) $0 \leq \nu(y) \leq \infty$, and $\nu(y) = 0$ iff $y(s) = 0$ μ-a.e.;

(ii) $\nu(y_1+y_2) \leq \nu(y_1) + \nu(y_2)$, and $\nu(by) = b\nu(y)$ for all $b \in \mathbb{R}^+$;

(iii) if $y_1(s) \leq y_2(s)$ μ-a.e., then $\nu(y_1) \leq \nu(y_2)$;

(iv) there is at least one function y in $\mathcal{S}$ with $0 < \nu(y) < \infty$;

(v) if $y_\infty, y_1, y_2, \ldots, y_n, \ldots$ are in $\mathcal{S}$ and $y_n(s) \uparrow y_\infty(s)$ μ-a.e., then $\nu(y_n) \uparrow \nu(y_\infty)$.

[#] Coleman & Mizel [3][5].

Such a function ν is called a "non-trivial function norm, relative to μ, with the sequential Fatou property".[#] I write $\mathcal{A}$ for the set of functions y in $\mathcal{S}$ for which $\nu(y)$ is finite.

Let V be a non-trivial (i.e. of dimension > 0), separable, real, Banach space, with norm $|\cdot|$, and let $\mathcal{V}$ be the set of all (strongly) μ-measurable functions ϕ mapping $\mathbb{R}^+$ into V with $\nu(|\phi|)$ finite; i.e.

$$\mathcal{V} = \left\{\phi : \mathbb{R}^+ \to V \;\middle|\; |\phi| \in \mathcal{A}\right\}.$$

Clearly, the function $\|\cdot\| : \mathcal{V} \to \mathbb{R}^+$, defined by

$$\|\phi\| = \nu(|\phi|),$$

is a semi-norm on ϕ. In applications, the functions ϕ in $\mathcal{V}$ are called <u>histories</u>; their independent variable s is called the <u>elapsed time</u>. The value $\phi(0)$ of a history ϕ at $s = 0$ is the <u>present value of</u> ϕ and the <u>past values</u> are those for which $s > 0$. The function space $\mathfrak{B}$ obtained from V by calling the same those functions in V which are equal μ-a.e. is a Banach space with norm $\|\cdot\|$; one calls it a <u>history space</u> or, at length, a <u>Banach function space formed from histories with values in</u> V.

If ψ is a function on $\mathbb{R}^+$ and σ is in $\mathbb{R}^+$, then the <u>static continuation of</u> ϕ <u>by amount</u> σ is the function $\phi^{(\sigma)}$ on $\mathbb{R}^+$ defined by

$$\phi^{(\sigma)}(s) = \begin{cases} \phi(0), & s \in [0,\sigma], \\ \phi(s-\sigma), & s \in (\sigma,\infty); \end{cases}$$

the σ-<u>section of</u> ϕ is the function $\phi_{(\sigma)}$ on $\mathbb{R}^+$ given by

$$\phi_{(\sigma)}(s) = \phi(s+\sigma), \qquad s \in \mathbb{R}^+.$$

The physical requirements (1) and (2), stated in the introduction, here become the following two assumptions.

<u>Postulate</u> 1. If a given function ϕ is in $\mathcal{V}$, then so also are all the static continuations $\phi^{(\sigma)}$, $\sigma \geq 0$, of ϕ. Furthermore, if ϕ_1 and ϕ_2 in $\mathcal{V}$ are such that $\|\phi_1 - \phi_2\| = 0$, then $\|\phi_1^{(\sigma)} - \phi_2^{(\sigma)}\| = 0$ for all $\sigma \geq 0$.

<u>Postulate</u> 2. If ϕ is in $\mathcal{V}$, then so also are all its σ-sections, $\phi_{(\sigma)}$, $\sigma \geq 0$.

[#]Luxemburg & Zaanen [1].

Employing Postulate 1, one can prove

<u>Theorem</u> 1.# The influence measure μ has an atom at $s=0$ and is absolutely continuous on $\mathbb{R}^{++} = (0,\infty)$ with respect to Lebesgue measure.

Postulates 1 and 2, together, yield

<u>Theorem</u> 2.## Either $\mu(\mathbb{R}^{++}) = 0$ or Lebesgue measure is absolutely continuous on $\mathbb{R}^{++}$ with respect to μ.

Thus the μ-measure of the singleton $\{0\}$ is not zero, and if $\mu(\mathbb{R}^{++})$ is not zero, then an arbitrary subset of $\mathbb{R}^{++}$ has zero μ-measure if and only if it has zero Lebesgue measure. So as to have a non-trivial theory, <u>let us assume that</u> $\mu(\mathbb{R}^{++})$ <u>is not zero</u>.

The third physical requirement on the list given in the introduction corresponds to the following assumption.

<u>Postulate</u> 3. For each vector v in V, the constant function $v^{\dagger}$, defined by

$$v^{\dagger}(s) = v, \qquad s \in \mathbb{R}^{+},$$

is in $\mathcal{V}$.

It follows immediately from this assumption that given any functional g on $\mathfrak{B}$, we can define a function g° on V by the formula

$$g^{\circ}(v) = g(v^{\dagger}), \qquad \text{for all } v \in V.$$

The norm $\|\cdot\|$ on $\mathfrak{B}$ is said to have the <u>relaxation property</u>, if, for each function ϕ in $\mathcal{V}$,

$$\lim_{\sigma \to \infty} \|\phi^{(\sigma)} - \phi(0)^{\dagger}\| = 0,$$

where $\phi(0)^{\dagger}$ is the constant function on $\mathbb{R}^{+}$ with value $\phi(0)$. Hence assumption of

---

#Coleman & Mizel [3], Thm. 2.1.

##Coleman & Mizel [3], Thm. 2.2.

the relaxation property is equivalent to the assertion that every continuous functional $\mathfrak{g}$ on $\mathfrak{B}$ obeys the relation

$$\lim_{\sigma \to \infty} \mathfrak{g}(\phi^{(\sigma)}) = \mathfrak{g}^{\circ}(\phi(0)) \qquad \text{for all } \phi \text{ in } \mathfrak{B};$$

i.e. in the limit of large σ, the response $\mathfrak{g}(\phi^{(\sigma)})$ to the static continuation $\phi^{(\sigma)}$ of an arbitrary initial history ϕ depends on only the present value of ϕ and is given by the equilibrium response function $\mathfrak{g}^{\circ}$. As this is a property of a large class of constitutive functionals,[#] let us add the following postulate to the list.

Postulate 4. The norm $\|\cdot\|$ on $\mathfrak{B}$ has the relaxation property.

Postulates 1-4 yield the following two theorems.

Theorem 3.[##] Let f and g map $\mathbb{R}$ into V and be such that the histories f^t and g^t are in $\mathcal{V}$ for each t in $\mathbb{R}$. If $\lim_{t \to \infty} |f(t) - g(t)| = 0$, then $\lim_{t \to \infty} \|f^t - g^t\| = 0$.

Theorem 4.[###] Given any $\epsilon > 0$, there is a $\delta = \delta(\epsilon) > 0$, such that any Lebesgue measurable function $f: \mathbb{R} \to V$, with $|f(t)| < \delta$ for each $t > 0$ and $\|f^0\| < \delta$, must have $\|f^t\| < \epsilon$ for each $t \geq 0$.

For technical reasons one assumes, further,

Postulate 5. The space $\mathfrak{B}$ is separable.

The following two theorems can then be proven.

Theorem 5.[####] For each ϕ in $\mathfrak{B}$, the function $\sigma \mapsto \phi^{(\sigma)}$ is strongly continuous in the sense that

$$\lim_{\sigma \to \delta} \|\phi^{(\sigma)} - \phi^{(\delta)}\| \qquad \text{for each } \delta \in \mathbb{R}^+.$$

[#] Cf. Coleman & Noll [3]; Coleman [1]; Wang [1]; Coleman & Mizel [1].

[##] Coleman & Mizel [3], Thm. 5.1.

[###] Coleman & Mizel [5], Lemma 2.1.

[####] Cf. Coleman & Mizel [2], Appendix 1.

<u>Theorem</u> 6. Let $\mathfrak{g}$ be a continuous function mapping $\mathfrak{B}$ into a metric space, and suppose that f is a function on $\mathbb{R}$ with f^t in $\mathcal{V}$ for each t in $\mathbb{R}$. If f is a regulated function, i.e. if $\lim_{\tau \to t+} f(\tau)$ and $\lim_{\tau \to t^-} f(\tau)$ exist for each t in $\mathbb{R}$, then g, given by

$$g(t) = \mathfrak{g}(f^t), \qquad t \in \mathbb{R},$$

is also a regulated function. Moreover, g can suffer distontinuities only at those values of t at which f is discontinuous; at all other points in $\mathbb{R}$ the function g is continuous.

3. Functional-Differential Equations Involving History Spaces

Now, let $\mathfrak{B}$ be a history space of the type described in the previous section, i.e. obeying Postulates 1-5, and assume, further, that $\mathfrak{B}$ is formed from functions ϕ mapping $\mathbb{R}^+$ into a <u>finite-dimensional</u> vector space V. I write $0^\dagger$ for the zero vector in V, i.e. the constant function on $\mathbb{R}^+$ whose value is the zero element of V. For each $\delta > 0$, $\mathfrak{S}(\delta)$ denotes the open ball in $\mathfrak{B}$ about $0^\dagger$ with radius δ:

$$\mathfrak{S}(\delta) = \left\{ \phi \,\middle|\, \phi \in \mathfrak{B}, \ \|\phi\| < \delta \right\} .$$

Let $h > 0$ be given and let $\mathfrak{f}$ be a function mapping $\mathfrak{S}(h)$ into V and possessing the following properties:

(1) $\mathfrak{f}$ is continuous over $\mathfrak{S}(h)$;

(2) $\mathfrak{f}$ is bounded in the sense that $\mathfrak{f}(\mathfrak{S}(h))$ is a set in V with finite diameter;

(3) $\mathfrak{f}(0^\dagger) = 0$.

If ϕ is in $\mathfrak{B}$ and A is in $\mathbb{R}^{++}$, then one says that a function $x(\cdot): (-\infty, A) \to V$ is a <u>solution up to A of the equation</u>,

$$\dot{x}(t) = \mathfrak{f}(x^t), \tag{3.1}$$

<u>with initial history</u> ϕ, provided

(a) the history x^t is in $\mathfrak{B}$ for each t in $(-\infty, A)$, and x^t is in $\mathfrak{S}(h)$ for each t in $(0,A)$;

(b) $x(\cdot)$ is continuously differentiable (in the classical sense) on $(0,A)$ and has a right-hand derivative, $\dot{x}(0)$, at $t=0$;

(c) (3.1) holds for all t in $[0,A)$;

(d) x^0, the history of $x(\cdot)$ up to $t=0$, is equal to ϕ.

Remark. Properties (1)-(3) of $\mathfrak{f}$ insure that for each ϕ in $\mathfrak{S}(h)$ there exists an $A > 0$ and a function $x(\cdot)$ which is a solution up to A of (3.1) with initial history ϕ. This solution can be extended, i.e. A can be increased, until x^t reaches the boundary of $\mathfrak{S}(h)$.

It follows from Property (3) that $x \equiv 0$ [with 0 the zero vector in V] is a solution of (3.1) up to ∞ with initial history $0^\dagger$; one calls this the zero solution.

The zero solution is said to be stable if for every $\epsilon > 0$ there is a $\delta = \delta(\epsilon) > 0$ such that if ϕ is in $\mathfrak{S}(\delta)$, then every solution $x(\cdot)$ of (3.1) with initial history ϕ has $|x(t)| < \epsilon$ for each $t \geq 0$ in its domain of existence, and a solution with initial history ϕ exists for all t in $\mathbb{R}$. If, in addition, there is a $\zeta > 0$ such that for each solution $x(\cdot)$ with x^0 in $\mathfrak{S}(h)$ there holds $\lim_{t\to\infty} |x(t)| = 0$, then the zero solution is said to be asymptotically stable.

It follows from Theorems 1 and 4 and the Remark made above that the zero solution of (3.1) is stable if and only if for each ϵ in $(0,h)$ there is a $\delta = \delta(\epsilon) > 0$ such that ϕ in $\mathfrak{S}(\delta)$ implies that every solution $x(\cdot)$ of (3.1) with $x^0 = \phi$ has x^t in $\mathfrak{S}(\epsilon)$ for each $t > 0$.

A real-valued function Ψ on $\mathfrak{S}(h)$ is called a free energy functional for the equation (3.1) if

(i) Ψ is continuous over $\mathfrak{S}(h)$;

(ii) for each solution up to A of (3.1), $\Psi(x^t)$ is a non-increasing function of t for $t\in[0,A)$;

(iii) for each ϕ in $\mathfrak{S}(h)$ with $\phi(0)$ in the domain of Ψ°, the equilibrium response function corresponding to $\mathfrak{S}(h)$, there holds

$$\Psi(\phi) \geq \Psi^\circ\big(\phi(0)\big). \tag{3.2}$$

Functional-differential equations of the type (3.1), with free energy functionals, occur frequently in theories of materials with memory.[#] The following theorem, provable by elementary arguments,[##] shows that free energy functionals, when they exist, are useful.

Theorem 7. If there is a free energy functional Ψ for (3.1) whose equilibrium response function Ψ° has a strict local minimum at zero, then the zero solution of (3.1) is stable.

Of course, the presence of a strict local minimum for Ψ° at zero means that there exists an $\eta > 0$ such that

$$x \in V, \quad 0 < |x| < \eta \implies \Psi^\circ(x) > \Psi^\circ(0).$$

A function $\Psi : \mathfrak{S}(h) \to \mathbb{R}$ is called a strictly dissipative free energy functional for (3.1) if Ψ has the Properties (i)-(iii) of a free energy functional and, in addition,

(iv) for each solution $x(\cdot)$ of (3.1) that is not identically zero, there exists a $t_* > 0$ such that

$$\psi(x^0) > \psi(x^{t_*}).$$

In words: a free energy functional is strictly dissipative if it eventually decreases on each solution that differs at some time from the zero solution.

The following theorem goes much deeper than Theorem 7 and has a much longer proof.[###]

Theorem 8. If (3.1) has a strictly dissipative free energy functional Ψ whose equilibrium response function Ψ° has a strict local minimum at zero, then the zero solution of (3.1) is asymptotically stable.

[#] Cf. Coleman & Mizel [5], §6; Coleman & Dill [1], Coleman [2]; for the thermodynamical basis of the relation (3.2), see Coleman [1], Thm. 3, p. 26.

[##] Cf. Coleman & Mizel [4].

[###] See Coleman & Mizel [5], Thm. 5.1. Similar theorems for spaces of histories different from those described here were obtained earlier by Hale [1][2].

Acknowledgments

Although specific references are given in each case, I should like to emphasize that all of the results I have discussed here were obtained in collaboration with Professor Victor J. Mizel in the course of research supported by the Air Force Office of Scientific Research and the National Science Foundation.

I am grateful to Battelle Memorial Institute and the National Science Foundation for making possible my attendance at this summer program.

References

Coleman, B. D.,

[1] Arch. Rational Mech. Anal. **17**, 1-46, 230-254 (1964).

[2] Advances Chem. Phys. **24**, in press (1972).

Coleman, B. D., & E. H. Dill,

[1] Arch. Rational Mech. Anal. **30**, 197-224 (1968).

Coleman, B. D., & V. J. Mizel,

[1] Arch. Rational Mech. Anal. **23**, 87-123 (1966).

[2] ibid. **27**, 255-274 (1967).

[3] ibid. **29**, 18-31 (1968).

[4] ibid. **29**, 105-113 (1968).

[5] ibid. **30**, 173-196 (1968).

Coleman, B. D., & W. Noll,

[1] Arch. Rational Mech. Anal. **6**, 355-370 (1960).

[2] Reviews Mod. Phys. **33**, 239-249 (1961); erratum: ibid. **36**, 1103 (1964).

[3] Proc. Intl. Sympos. Second-Order Effects, Haifa, 1962, pp. 530-552.

Hale, J. K.,

[1] Proc. Natl. Acad. Sci. **50**, 942-946 (1963).

[2] J. Diff. Eqs. **1**, 452-482 (1965).

Luxemburg, W. A. J., & A. C. Zaanen,

[1] Math. Annalen **162**, 337-350 (1966).

Noll, W.,

[1] Arch. Rational Mech. Anal. **3**, 197-226 (1958).

Wang, C.-C.,

[1] Arch. Rational Mech. Anal. **18**, 117-126 (1965).

[2] ibid. **18**, 343-366 (1965).

SINGULAR PERTURBATION BY A QUASILINEAR OPERATOR

Paul C. Fife
Department of Mathematics
University of Arizona
Tucson, Arizona

1. INTRODUCTION

We examine the following boundary value problem for a function u defined in a bounded domain $\Omega \subset R^n$ with boundary $\partial\Omega$:

$$\varepsilon^2 \sum_{i=1}^{n} \partial_i[p(u)\partial_i u] - g(x,u) = 0 \,, \; u = f \quad \text{on} \quad \partial\Omega \,. \tag{1.1}$$

Here ε is a small parameter and $\partial_i = \partial/\partial x_i$. It will be assumed that the "degenerate problem"

$$g(x,u) = 0$$

has a solution $u = u^*(x)$; our main question is whether there exist solutions $u(x,\varepsilon)$ of the nondegenerate problem approaching $u^*(x)$ as $\varepsilon \to 0$ for every $x \in \Omega$. At the same time, we are concerned with the validity of expansion techniques for constructing approximate solutions and the stability of the exact solutions. Analogous problems in which p , g , and f depend also on ε constitute an easy generalization which we do not pursue.

A previous paper [2] treated the Dirichlet problem for the equation

$$\varepsilon^2 \Sigma\, \partial_i (a_{ik}(x,\varepsilon)\partial_k u) - g(x,u) = 0 \,.$$

Since quasilinear problems of the type (1.1) are relevant to the study of problems in chemical and biochemical reaction theory, it appears worthwhile to see whether the techniques used in [2] can be modified to apply in the present context. As we shall show, such a modification is possible, mainly because, after (1.1) is multiplied by the function $p(u)$, the new operator on the left has the convenient property that its linearization about any function is formally self-adjoint. At the same time, the device of multiplying by p allows the boundary layer constructions to proceed practically as easily as in the semilinear case. One of the principal tools in [2] was Lemma A.2. However, it was stated and proved in far greater generality than was needed there. The reason was to provide a result applicable in other situations, and, indeed, that lemma applies without change in the present context and so serves as a principal tool in the present paper as well.

The quasilinearity of the equation makes it appropriate to use Hölder norms in deriving the estimates, whereas the maximum norm sufficed in [2]. On the other hand, the isotropicity of the present equation effects a considerable saving in effort in deriving these estimates. Finally, the existence proof in the present case is based on the implicit function theorem rather than the Schauder fixed point theorem, which does not apply here.

It will be assumed throughout that p, g, u^*, and f are infinitely differentiable for $x \in \bar{\Omega}$ and $u \in R^1$. It will also be assumed that $\partial\Omega$ is infinitely smooth. In actuality, only a finite degree of smoothness is needed for our purposes; the infinite differentiability is assumed for convenience. Further assumptions are

(i) $p(u) > 0$ for $u \in R^1$,

(ii) $g_2(x, u^*(x)) > 0$ for all $x \in \bar{\Omega}$, where g_2 denotes the derivative with respect to the second argument,

(iii) for each $x \in \partial\Omega$, $\int_{u^*(x)}^{k} p(u)g(x,u)du > 0$ for all $k \neq u^*(x)$ in the closed interval between $u^*(x)$ and $f(x)$.

Under these assumptions it will be shown (1) that a natural expansion technique for constructing approximate solutions may be carried out to any order, with u^* the lowest order term in the interior, (2) that there exist exact solutions approximated by the functions constructed, and (3) that these exact solutions satisfy the linear stability criterion.

The procedure will be, first, to set up formal approximations by the time-honored method of "inner" and "outer" expansions, in a form near to that used by Vishik and Liusternik [4]. Next, it will be shown that the approximation constructed satisfies (1.1) with zero on the right replaced by a function bounded by $C\epsilon^{m+1}$, where m is the order of the approximation. Finally, this fact will be used to show the existence of an exact solution close to the approximate one. This existence proof relies on the invertibility of the operator obtained by linearizing the left side of (1.1) about the construced approximation.

2. THE INNER AND OUTER EXPANSIONS

As stated above, it will prove convenient to multiply (1.1) by $p(u)$, thus obtaining

$$N[u] \equiv \epsilon^2 p(u) \sum_{i=1}^{n} \partial_i (p(u)\partial_i u) - h(x,u) = 0 , \tag{2.1}$$

where $h = pg$. We seek a solution of (2.1) satisfying

$$u = f(x) \quad \text{on} \quad \partial\Omega . \tag{2.2}$$

The first step in constructing our approximate solution is to form a polynomial in ϵ, the coefficients depending on x, which will satisfy (2.1) approximately for all small ϵ. Thus for some $m > 0$, we set

$$U(x,\epsilon) \equiv \sum_{i=0}^{m} \epsilon^i u_i(x) . \tag{2.3}$$

Since, in general, it will be impossible to choose the u_i so that $N[U](x,\epsilon) \equiv 0$, we instead only require that

$$\partial_\epsilon^k N[U](x,0) = 0 , k = 0, \ldots, m . \tag{2.4}$$

In particular for $k = 0$, we have $h(x,u_0(x)) = 0$, which by assumption has a solution $u_0 = u^*$ (it may have other solutions too; under suitable conditions they could serve as first terms in other expansions). For $k \geq 1$, (2.4) takes the form $h_2(x,u_0(x))u_k(x)$ = a function expressible in terms of the u_j for $j < k$. The assumptions (i) and (ii) imply that $h_2(x,u_0(x)) > 0$, so the functions u_k, $k > 0$, are uniquely determinable in succession.

The function $U(x,\epsilon)$ is an approximate solution of (2.1). In fact, since $N[U]$ is infinitely differentiable in ϵ, we may form a Taylor series expansion, and (2.4) implies that $N[U](x,\epsilon) = \epsilon^{m+1}P(x,\theta\epsilon)$, where $P = \frac{1}{(m+1)!}\partial_\epsilon^{m+1}N[U]$ and $0<\theta<1$. Since P is bounded in $\bar{\Omega}\times[0,1]$, we have $|N[U]|_0 \equiv \sup|N[U](x,\epsilon)| \leq C_1\epsilon^{m+1}$, for some constant C_1 . For the same reason, we have, in fact,

$$|N[U]|_1 \equiv \sup_{\Omega}|N[U]| + \sup_{\Omega,i}|\partial_i N[U]| \leq C_2\epsilon^{m+1} . \tag{2.5}$$

However, the polynomial U will not, in general, satisfy the boundary conditions (2.2), even approximately. Therefore we supplement U with a boundary layer correction V so that $U+V$ satisfies (2.2) exactly, V decays rapidly for small ϵ at points not on the boundary, and $U+V$ is still an approximate solution of (2.1). For this purpose we need special coordinates. For points x near $\partial\Omega$, we let $t(x)$ = the distance from x to $\partial\Omega$, and $s(x)$ = the point on $\partial\Omega$ which is closest to x . We further define $\eta(x,\epsilon) = t(x)/\epsilon$. By convention, for any function $\phi(x,\epsilon)$, the corresponding function of η, s, ϵ is symbolized by means of a caret:

$$\phi(x,\epsilon) = \hat{\phi}(\eta,s,\epsilon) = \hat{\phi}(\eta(x,\epsilon),s(x),\epsilon) .$$

Furthermore the operator N , when written in terms of the variables η and s , is denoted by $\hat{N}$, so that

$$\hat{N}[\hat{u}](\eta,s,\epsilon) = N[u](x,\epsilon) .$$

In forming $\hat{N}$, for example, we use the fact that $\partial_i = \epsilon^{-1}\nu_i\partial_\eta + D_i$, where $\underset{\sim}{\nu} = (\nu_1,\ldots,\nu_n)$ is the unit normal in the direction of increasing t and D_i is a C^∞ first order linear differential operator in s .

In analogy to the determination of U , $V(\eta,s,\epsilon)$ is determined by setting

$$\hat{V}(\eta,s,\epsilon) = \sum_{i=0}^{m} \epsilon^i v_i(\eta,s) \tag{2.6}$$

and choosing the v_k so that

$$\partial_\epsilon^k \hat{N}[\hat{U}+\hat{V}](\eta,s,0)=0 \ , \ k=0,\ldots,m \ , \tag{2.7}$$

$$\partial_\epsilon^k[\hat{U}+\hat{V}-f](0,s,0)=0 \ , \ k=0,\ldots,m \ , \tag{2.8}$$

and

$$v_k(\infty,s)=0 \ . \tag{2.9}$$

We show that such functions v_k exist.

For $k=0$, (2.7) yields the following equation for v_0:

$$p(u^*(s)+v_0(\eta,s))\partial_\eta[p(u^*(s)+v_0(\eta,s))\partial_\eta v_0(\eta,s)]-\hat{h}(0,s,u^*(s)+v_0(\eta,s))=0 \ .$$

Suppressing the dependence on s and defining $p^*(v)=P(u^*+v)$, $h^*(v)=\hat{h}(0,s,u^*+v)$, we rewrite the equation in the form

$$p^*(v_0)\partial_\eta(p^*(v_0)\partial_\eta v_0)-h^*(v_0)=0 \ . \tag{2.10}$$

Also (2.8), (2.9) imply that

$$v_0(0,s)=f(s)-u^*(s) \tag{2.11}$$

$$v_0(\infty,s)=0 \ . \tag{2.12}$$

Since (2.10) remains invariant when η is replaced by $\eta_0-\eta$ for any η_0, we conclude that any solution of (2.10) satisfying $\partial_\eta v_0(\eta_0,s)=0$ will be symmetric about the line $\eta=\eta_0$. In particular, if there are two values of η at which $\partial_\eta v_0=0$, then v_0 will be periodic, which violates (2.12) unless $v_0\equiv 0$. Thus v_0 has at most one critical point (unless $f(s)-u^*(s)=0$, in which case $v_0\equiv 0$), and if η_0 is a critical point, then by the above symmetry property, $v_0(2\eta_0,s)=v_0(0,s)$. Therefore the function $v_0(\eta,s)=v_0(\eta+2\eta_0,s)$ will also satisfy (2.10-12) and, having no critical point in $(0,\infty)$, will furthermore be monotonic. Our conclusion is that the problem (2.10-12) has at most two solutions, and if it has one at all, then exactly one of them is monotone. Our function $v_0(\eta,s)$ will be taken to be this monotone solution.

Condition (iii) turns out to be not only sufficient, but also necessary for v_0 to exist. For assuming a monotone solution v_0 exists, we may use $V=v_0$ as an independent variable and $W(V) = p^*\partial_\eta v_0$ as a dependent variable. Then (2.10) becomes

$$W\frac{dW}{dV} = h^*(V) \quad \text{with} \quad W(0) = 0 \; . \tag{2.13}$$

Integrating, we find

$$\eta = \int_v^{f-u^*} \frac{p^*(\bar{v})}{\sqrt{G(\bar{v})}}\, d\bar{v} \;, \quad G(v) = 2\int_0^v h^*(\bar{v})\, d\bar{v} \; . \tag{2.14}$$

This necessitates $G > 0$ between 0 and $f(s) - u^*(s)$ which is (iii). Relation (2.13) is invertible and yields the required function $v_0(\eta, s)$. It can be checked that $v_0(\eta, s)$ and all its derivatives decay like $e^{-\sqrt{h'(0)}\,\eta/p(0)}$ as $\eta \to \infty$.

Differentiating (2.10) with respect to η and setting $\chi(\eta) = v_0'(\eta)$, we find that

$$L\chi \equiv \partial_\eta (p^{*2}(v_0(\eta,s))\partial_\eta \chi) + q(\eta,s)\chi = 0 \;, \tag{2.15}$$

where

$$q(\eta,s) = -h^{*\prime}(v_0(\eta,s)) + 2p^* p^{*\prime}(v_0(\eta,s))\partial_\eta^2 v_0(\eta,s) + (p^* p^{*\prime})'(\partial_\eta v_0(\eta,s))^2 \; .$$

At the same time, equation (2.7) with $k > 0$ can be written as

$$Lv_k = F_k(\eta,s) \;, \tag{2.16}$$

where the F_k are functions expressible in terms of the functions v_j for $j < k$. In fact, F_k will be a rather complex partial differential operator (in the variables η and s) applied to these previously found functions. The task is to solve (2.16) for v_k with the boundary condition (from (2.8), (2.9))

$$v_k(0,s) = \frac{1}{k!}\,\partial_\varepsilon^k \hat{V}(0,s,0) = -\frac{1}{k!}\,\partial_\varepsilon^k \hat{U}(0,s,0) \equiv b_k(s) \;, \tag{2.17}$$

$$v_k(\infty,s) = 0 \; . \tag{2.18}$$

Since we have one solution χ of $L\chi = 0$, the solutions of (2.16) may be found by a variation of constants formula. However, there is no guarantee that there will exist a solution satisfying (2.18), unless $F_k(\eta, s)$ decays exponentially as $\eta \to \infty$. This indeed turns out to be the case. In fact, the exponential decay of the F_k is an immediate consequence of Lemma A.2 in [2].

In this way, the function $\hat{V}(\eta, s, \varepsilon)$ is completely determined. However, the variable $t = \varepsilon\eta$ was only defined for small enough t, say, in $\Omega_d = \{x : t(x) < d\}$ for some $d > 0$. To overcome this technicality, we use a C^∞ cut-off function $\zeta(t) \equiv 1$ for $0 \leq t < d/2$, $\zeta(t) = 0$ for $t \geq d$. Then we define $V(x,\varepsilon) = \zeta(t(x))\hat{V}(\eta(x,\varepsilon), s(x), \varepsilon)$ in Ω_d and $V(x,\varepsilon) = 0$ in $\Omega \backslash \Omega_d$. The function V vanishes to the order $e^{-\gamma/\varepsilon}$, $\gamma > 0$, in subdomains of Ω bounded away from $\partial\Omega$.

This completes the determination of the approximation $U(x,\varepsilon) + V(x,\varepsilon)$.

3. EXISTENCE PROOF FOR EXACT SOLUTIONS

The symbol C will denote several constants depending only on p, h, f, u^*, and Ω.

The function $\hat{N}[\hat{U} + \hat{V}]$ is infinitely differentiable in ε and satisfies (2.7). It follows from Taylor's formula that

$$\hat{N}[\hat{U} + \hat{V}] = \frac{1}{(m+1)!}\, \varepsilon^{m+1} \partial_\varepsilon^{m+1} N[\hat{U} + \hat{V}](\eta, s, \theta\varepsilon) \;,\quad 0 < \theta < 1 \,.$$

Using Lemma A.2 of [2] again, it follows that the derivative is bounded uniformly in η, s, and small ε. Hence $|\hat{N}[\hat{U} + \hat{V}]| \leq C\varepsilon^{m+1}$.

The same argument yields

$$|D\hat{N}[\hat{U} + \hat{V}]| \leq C\varepsilon^{m+1} \tag{3.1}$$

for any C^∞ linear differential operator D in the variables η and s, whose coefficients are bounded together with their derivatives.

The operator ∂_i can be expressed as a first order operator D_i in η and s, the coefficients being bounded by $C\varepsilon^{-1}$. Hence if we define Q by

$$N[U + V] = \varepsilon^{m+1} Q \,, \tag{3.2}$$

we have from (3.1) $|\partial_i Q| \le C\epsilon^{-1}$ in $\Omega_{d/2}$. The function V is of order $e^{-\gamma/\epsilon}$ in $\Omega\backslash\Omega_{d/2}$; using this fact, the last estimate, and (2.5), we find

$$|Q|_1 \le C\epsilon^{-1} . \tag{3.3}$$

We shall show that $U+V$ differs by a quantity of order ϵ^{m+1} from an exact solution of (2.1), (2.2).

<u>Theorem</u>. <u>There exists a solution</u> $u(x,\epsilon)$ <u>of (2.1), (2.2) of the form</u>

$$u = U + V + R \tag{3.4}$$

<u>satisfying</u>

$$|R|_0 \le C\epsilon^{m+1} .$$

Substituting the expression (3.4) into (2.1) and subtracting (3.2), we obtain the following equation for R :

$$N[U+V+R] - N[U+V] = \epsilon^{m+1}Q \quad \text{in} \quad \Omega \tag{3.5}$$

$$R = 0 \quad \text{on} \quad \partial\Omega . \tag{3.6}$$

The difference on the left of (3.5) vanishes when $R=0$. It can be written as a linear part $N'[U+V;R]$ plus a nonlinear part which behaves quadratically in R for small R. We denote this nonlinear part by $\Psi[R]$. It is a function of x, η, s, ϵ, and the first and second derivatives of R. Thus (3.5) becomes

$$N'[U+V;R] = \epsilon^{m+1}Q - \Psi[R] . \tag{3.7}$$

The derivative N' is easily calculated; it is self-adjoint and is related to the operator L in (2.15):

$$N'[U+V;R] = \epsilon^2 \sum_1^n \partial_i[p^2(U(x,\epsilon)+V(x,\epsilon))\partial_i R] + \rho(x,\epsilon)R , \tag{3.8}$$

where

$$\rho(x,\epsilon) = -h_2(x,U(x,\epsilon)+V(x,\epsilon)) \qquad (3.9)$$
$$+\epsilon^2 \sum_{i=1}^{n} [(p^2)'(U+V)\partial_i^2(U+V) + \frac{1}{2}(p^2)''(U+V)(\partial_i(U+V))^2] .$$

Proving the theorem is therefore equivalent to finding a solution R of (3.7), (3.6) . A solution can be obtained by the implicit function theorem if it can be shown that N' is invertible in a suitable Banach space. The first step is to show that the eigenvalues of the problem

$$N'[U+V;\ \phi] + \lambda\phi = 0 \qquad (3.10a)$$

$$\phi = 0 \quad \text{on} \quad \partial\Omega \qquad (3.10b)$$

are positive and bounded away from zero, independently of ϵ .

Let λ_1 be the principal eigenvalue and ϕ_1 the associated non negative eigenfunction, normalized so that $\int_\Omega \phi_1^2 dx = 1$.

Lemma. For all ϵ , small enough

$$\lambda_1 \geq C > 0 . \qquad (3.11)$$

Proof. Now λ_1 can be characterized as the Rayleigh quotient

$$\lambda_1 = \epsilon^2 \int_\Omega \sum_i p^2(U(x,\epsilon)+V(x,\epsilon))(\partial_i\phi_1(x,\epsilon))^2 dx - \int_\Omega \rho(x,\epsilon)\phi_1^2(x,\epsilon)dx . \qquad (3.12)$$

It is evident from assumption (ii) that $h_2(x,u_0(x)) > 0$. From this fact, the exponential decay of V , and the fact that $|U - u_0| = O(\epsilon)$, we see that $\rho(x,\epsilon)$ in (3.9) is negative outside a boundary layer:

$$\rho(x,\epsilon) \leq -a < 0 \quad \text{in} \quad \Omega\setminus\Omega_d , \qquad (3.13)$$

for some a and for ϵ small enough depending on d . Therefore restricting the integration in (3.12) to Ω_d provides the inequality

$$\lambda_1 > \epsilon^2 \int_{\Omega_d} \sum p^2(U+V)(\partial_i\phi_1)^2 dx - \int_{\Omega_d} \rho\phi_1^2 dx \geq \int_{\Omega_d} [\epsilon^2 p^2(\partial_t\phi_1)^2 - \rho\phi_1^2]dx$$
$$= \int_{\partial\Omega} ds \int_0^d [\epsilon^2 p^2(U+V)(\partial_t\phi_1)^2 - \rho\phi_1^2]J(s,t)dt , \qquad (3.14)$$

where $J(s,t)$ is the ratio of volume elements $\frac{dx}{dsdt}$ and ds is the element of area on the hypersurface $\partial\Omega$. It also follows from (3.14) that $\epsilon^2 \int_{\Omega_d} (\partial_t \phi_1)^2 dx < C$ if $\lambda_1 < 1$ (which we may assume, since otherwise the lemma is true immediately). Therefore if, in the last integral of (3.14), we replace $p^2(U+V)$ by a function differing from it by a quantity of order ϵ and make a similar replacement for ρ, we find that the value of the right side changes by the same order of magnitude. In particular, we may replace $p^2(U+V)$ by $p^2(u_0+v_0) \equiv \tilde{p}^2(v_0)$ and $\rho(x,\epsilon)$ by $q(t/\epsilon,s)$, the function in (2.15), obtaining an error on the right of order ϵ. Similarly, if we replace $J(s,t)$ by $J(s,0)=1$, we obtain an error of the order d. Thus

$$\lambda_1 > \int_{\partial\Omega} ds \int_0^d [\epsilon^2 \tilde{p}^2(v_0(t/\epsilon,s))(\partial_t\phi_1)^2 - q(t/\epsilon,s)\phi_1^2]dt - k_1(\epsilon,d) , \tag{3.15}$$

where $k_1 \to 0$ as ϵ and $d \to 0$. The integral with respect to t is the Dirichlet form for the eigenvalue problem

$$\epsilon^2 \partial_t[\tilde{p}^2(v_0(t/\epsilon,s))\partial_t\psi] + q(t/\epsilon,s)\psi + \Lambda(s)\psi = 0 \tag{3.16}$$

$$\psi(0,s) = \psi(d,s) = 0 ,$$

and is thus bounded from below by $\Lambda_1(s) \int_0^d \phi_1^2 dt$, where Λ_1 is the lowest eigenvalue. If $\lambda_1 \leq a/2$ (which we may assume; otherwise the lemma is immediately true), then by Lemma 4.2 of [2] we have that $\int_{\Omega\setminus\Omega_d} \phi_1^2 = O(\epsilon^k)$ for all k. Using this fact and the above property of Λ_1, we find from (3.15)

$$\lambda_1 > \int_{\Omega_d} \Lambda_1(s(x))\phi_1^2(x)dx - k_2(\epsilon,d) > \Lambda_m \int_\Omega \phi_1^2 \, dx - k_3(\epsilon,d) = \Lambda_m - k_3(d,\epsilon) ,$$

where $\Lambda_m = \min_s \Lambda_1(s)$ and $k_3 \to 0$ as d, $\epsilon \to 0$.

Thus, to complete the proof we need to bound $\Lambda_1(s)$ from below. Let $\psi_1(t,s) \geq 0$ be the corresponding eigenfunction of (3.16). Multiplying (2.15) by ψ_1, integrating from 0 fo d by parts, and using (3.16) with $\psi = \psi_1$, we find

$$-\Lambda_1(s) \int_0^d \chi(t/\epsilon)\psi_1(t,s)dt + \epsilon^2[\epsilon^{-1}\psi_1\partial_\eta\chi - \chi\partial_t\psi_1]_{t=0}^{t=d} = 0 .$$

We further use the fact that $\psi_1 = 0$ at $t = d$ or 0, that χ is of one sign (assume it is positive; otherwise replace χ by $(-\chi)$ in the above), and that $\partial_t \psi_1(d,s) < 0$. We then divide by the integral on the left and use the Schwarz inequality on it to deduce

$$\Lambda_1(s) > \frac{\epsilon^2 \chi(0,s)\partial_t \psi_1(0,s)}{(\int_0^d \chi^2(t/\epsilon)dt)^{1/2}(\int_0^d \psi_1^2 dt)^{1/2}} .$$

Now $\chi(0,s)$ can be found explicitly from p, f, and h; it is positive and bounded away from 0. Also $\int_0^d \chi^2 dt < \epsilon \int_0^\infty \chi^2(\eta)d\eta$ and $\int_0^d \psi_1^2 dt = 1$. Hence

$$\Lambda_1(s) > C\epsilon^{3/2}\partial_t \psi_1(0,s) . \tag{3.17}$$

Again, we may assume that $\Lambda_1(s)$ is small enough so that the coefficient $\rho = q + \Lambda_1$ of (3.16) with $\psi = \psi_1$, $\Lambda = \Lambda_1$ is strictly negative in an interval $[c\epsilon, d]$. Otherwise, Λ_1 is bounded from below and nothing is left to prove. The proof of Lemma 2.4 in [2] may now be adapted to yield the fact that $\partial_t \psi_1(0,s) \geq C^{-1}\epsilon^{-3/2}$, which is the desired conclusion. This adaption is as follows. Whereas in [2] a comparison argument is used relating the solution of $\epsilon^2 \psi_1'' + \rho\psi_1 = 0 \geq \epsilon^2 \psi_1'' + a\psi_1$ to a solution ψ^* of $\epsilon^2 \psi^{*\prime\prime} + a\psi^* = 0$, for the present purpose we use a comparison argument relating the solution of $\epsilon^2(\tilde{p}^2 \psi_1')' + \rho\psi_1 = 0 \geq \epsilon^2(\tilde{p}^2 \psi_1') + a\psi_1$ to a function ψ^* satisfying $\epsilon^2(\tilde{p}^2 \psi^{*\prime}) + a\psi^* \geq 0$, constructed by setting $\tilde{p}^2 \psi^{*\prime} = A \sinh \frac{\sqrt{ax}}{\epsilon} + B \cosh \frac{\sqrt{ax}}{\epsilon}$, for appropriate constants A and B. With this result and (3.17), Λ_1, hence Λ_m, hence λ_1, is bounded from below for small ϵ and d. But since λ_1 is independent of d, this is true for small ϵ, and the proof of the lemma is complete.

Now consider the following linear problem for w, where f is a smooth function:

$$N'[U+V;\, w]\ f , \quad w = 0 \text{ on } \partial\Omega .$$

The above lemma implies that it has a unique solution with $|w|_{L_2} \leq C|f|_{L_2}$. Writing the equation in the form (from (3.8))

$$\Sigma\, \partial_i[p^2\partial_i w] = \epsilon^{-2}[f - \rho w] ,$$

we apply the L_p estimates for elliptic equations [1] and the Sobolev imbedding theorems in succession several times, if necessary, to obtain an estimate for $|w|_\alpha$ for some α, $0<\alpha<1$, in terms of $|f|_\alpha$, where $|w|_\alpha$ is the norm in the space C_α of functions Hölder continuous in $\bar{\Omega}$ with exponent α . Finally we apply the Schauder estimates [1] to obtain

$$|w|_{2+\alpha} \leq C\epsilon^{-\mu}|f|_\alpha$$

for some positive number μ .

Let $C^0_{2+\alpha}$ be the spcae of functions vanishing on $\partial\Omega$ with Hölder continuous second derivatives in $\bar{\Omega}$. Considered as a mapping from $C^0_{2+\alpha}$ to C_α , N' then has an inverse A with norm bounded as

$$|A| \leq C\epsilon^{-\mu} . \tag{3.18}$$

With this result about the inverse of N' , we are in a position to prove our main theorem. First we apply the operator $\epsilon^{1+\mu-m}A$ to (3.7), obtaining the following equivalent problem for $S = \epsilon^{1+\mu-m}R$:

$$S = \tilde{Q}(\epsilon) - \epsilon^{\pi}\tilde{\Psi}[\epsilon, S] , \tag{3.19}$$

where $\tilde{Q} = \epsilon^{\mu+2}AQ$, $\tilde{\Psi}[\epsilon,S] = \epsilon^{3\mu-2m+3}A\Psi[\epsilon^{m-\mu+1}S]$, and $\pi = m-2\mu-2$. We impose the condition $m > 2\mu+2$.

The operator Ψ can be written as a regular function of ϵ , s, η, x, R , and the first and second derivatives of R , which is at least quadratic in R and its derivatives as they approach zero. Thus $|\Psi[R]|_\alpha \leq C\epsilon^{-\alpha}|R|^2_{2+\alpha}$ for $|R|_{2+\alpha}<1$ (the factor $\epsilon^{-\alpha}$ comes from the dependence on $\eta = t(x)/\epsilon$) . From this and (3.19),

$$|\tilde{\Psi}[\epsilon,S]|_{2+\alpha} \leq C\epsilon^{2\mu-2m+3+\alpha}|\epsilon^{m-\mu+1}S|^2_{2+\alpha} \leq C\epsilon^{1-\alpha}|S|^2_{2+\alpha}$$

for $|S|_{2+\alpha}<1$. By similar reasoning, we have the Frechet derivatives,

$$|\partial_S \tilde{\Psi}[\epsilon,S]| < C\epsilon^{1-\alpha}|S|_{2+\alpha} , \qquad (3.20)$$

$$|\partial_S^2 \tilde{\Psi}[\epsilon,S]| < C\epsilon^{1-\alpha} ,$$

for $|S|_{2+\alpha} < 1$. Regarding Q , from (3.18) and (3.3) we have $|Q|_{2+\alpha} \leq C\epsilon^2 |Q|_\alpha \leq C\epsilon^2 |Q|_1 \leq C\epsilon$. We now write (3.19) in the form

$$f(\epsilon,S) = 0 ,$$

where f maps $R^1 \times C^0_{2+\alpha}$ into $C^0_{2+\alpha}$ with $f(0,0) = 0$, $\partial_S f = I + \epsilon^\pi \partial_S \tilde{\Psi}$, with the modulus of continuity of f with respect to ϵ at the origin and the modulus of continuity of $\partial_S f$ with respect to S bounded in the region $|S|_{2+\alpha} < 1$, $0 < \epsilon < 1$. Furthermore, $|\partial_S f^{-1}| < 2$ for $\epsilon^\pi |\partial_S \tilde{\Psi}| < \frac{1}{2}$, which according to (3.20) is satisfied for $|S|_{2+\alpha} < 1$, $0 < \epsilon < \epsilon_0$, ϵ_0 being some positive number. According to the implicit function theorem (see [3], for example), we have the existence of a solution $S(\epsilon)$ with $S(0) = 0$ in some interval $0 \leq \epsilon \leq \epsilon_1$, satifying $|S|_{2+\alpha} \leq 1$. This implies that $|R|_0 \leq |R|_{2+\alpha} \leq \epsilon^{m-1-\mu}$.

As in [2], the simple device of replacing m by $m+\mu+2$ in the above result and incorporating some of the terms in the expansion $U+V$ into the new remainder R actually yields $|R|_0 \leq C\epsilon^{m+1}$, and the theorem is proved.

Corollary. The solution $u(x,\epsilon) = U+V+R$ satisfies the following known linear stability criterion: all eigenvalues of the operator $N'[u]$, acting on functions vanishing on $\partial\Omega$, are negative and bounded away from 0 for $0 < \epsilon < \epsilon_1$.

Proof. Here the term eigenvalues must be understood as the negatives of the "eigenvalues" λ in (3.10) . The eigenvalues of a linear self-adjoint second order elliptic operator are continuous functions of the coefficients. The conclusion follows immediately from that fact, from the fact that we have proved (3.11) that the eigenvalues of $N'[U+V]$ satisfy the necessary inequality, and from the fact that $|R|_0 \leq C\epsilon^{m+1}$.

This research was supported by the National Science Foundation grant GP-30255X.

BIBLIOGRAPHY

1. Agmon, S., Douglis, A., and Nirenberg, L., Estimates near the boundary for solutions of elliptic partial differential equations satisfying general boundary conditions, I., Comm. Pure Appl. Math. 12, 623-727 (1959).

2. Fife, P., Semilinear elliptic boundary value problems with small parameters, Arch. Rat. Mech. and Anal. (to appear).

3. Liusternik, L., and Sobolev, V., Elements of Functional Analysis (translation). F. Ungar, New York, 1961.

4. Vishik, M., and Liusternik, L., Regular degeneration and boundary layer for linear differential equations with small parameters, Usp. Matem. Nauk S.S.S.R., 12, 3-122 (1957).

REMARKS ON BRANCHING FROM MULTIPLE EIGENVALUES

W. M. Greenlee
Department of Mathematics
University of Arizona
Tucson, Arizona 85721

INTRODUCTION

It is well known that the Lyapunov-Schmidt method for the solution of branching problems is rather easy to carry out for branching from a simple eigenvalue, while from multiple eigenvalues there may be no branching at all. We shall give some criteria for the existence of branching from multiple eigenvalues by fitting the Lyapunov-Schmidt technique into the general framework for the regular perturbation method as presented by J. B. Keller [3]. One novelty of this approach is that the theorems obtained apply in the case of eigenvalues of infinite, as well as finite, multiplicity.

For simplicity of exposition we treat a model problem in Hilbert space in which the linear part is selfadjoint and the nonlinear part is homogeneous. However, the methods apply in considerably greater generality. Theorem I is essentially a theorem to be found in Sather [5], p. 54, and V. M. Krasnoselskii [2]. It is obtained here in a very simple fashion and by a method which applies in the case of infinite multiplicity. Assumption I appears different from, but is equivalent to, a hypothesis in [5] and [2], which replaces the projection Q_2 by Q. In the present form, Assumption I is the same as a first order splitting hypotheis in analytic perturbation theory for linear operators.

(cf. Sz. Nagy [8]), whereas the analogue with Q is never satisfied in linear problems. Theorems II and III illustrate means of verifying the perturbation method when Assumption I does not hold. The development is written for a real Hilbert space, but the complex case is identical, if $\mathbb{R}^1$ is replaced by $\mathbb{C}^1$.

All results quoted in this development may be found in at least one of Pimbley [4], Stakgold [7], or Vainberg and Trenogin [9]. We wish to acknowledge helpful comments from Drs. P. C. Fife and J. R. Schulenberger. During the institute we learned that ideas similar to those presented here have been explored by Drs. N. Bazley and M. Reeken.

BRANCHING FROM MULTIPLE EIGENVALUES

Let H be a real Hilbert space with norm $|\cdot|$ and inner product $(\cdot,\cdot)$. Let B be a linear selfadjoint operator in H. Further, let m be a positive integer and $C: \prod_{i=1}^{m+1} H \to H$ be a continuous m + 1-linear operator, i.e., c is linear in each argument, and there exists $k > 0$ such that

$$|c(x_1, \ldots, x_{m+1})| \le k|x_1| \cdots |x_{m+1}|$$

$$\text{for all } (x_1, \ldots, x_{m+1}) \in \prod_{i=1}^{m+1} H.$$

Define $C: H \to H$ by

$$C(x) = c(x, \ldots, x) \quad \text{for all} \quad x \in H.$$

Then there exists $K > 0$ such that for any $r > 0$ and all $x, y \in H$ with $|x|, |y| \leq r$,

$$|C(x) - C(y)| \leq Kr^m|x - y|,$$

and obviously $C(\alpha x) = \alpha^{m+1}C(x)$ for all $\alpha \in \mathbb{R}^1$ and $x \in H$. Furthermore, C has continuous Fréchet derivatives of all orders; in fact, C is analytic.

Now let λ_0 be an isolated eigenvalue of B of multiplicity n, $1 \leq n \leq \infty$. We wish to discuss the perturbation theoretic technique known as the Lyapunov-Schmidt method for the following branching problem for the equation $(B - \lambda)x + C(x) = 0$.

I. <u>To find in</u> $\mathbb{R}^1 \times (H\setminus\{0\})$ <u>a sequence</u> $(\lambda_j, x_j) \to (\lambda_0, 0)$ <u>as</u> $j \to \infty$ <u>such that</u>

$$(B - \lambda_j)x_j + C(x_j) = 0; \quad j = 1, 2, \ldots .$$

The Lyapunov-Schmidt method begins as follows: let Q be the orthogonal projection on the λ_0-eigenspace and $P = I - Q$ where I is the identity operator. Write

$$x = \mu\hat{u} + w, \quad \lambda = \lambda_0 + \delta,$$

where $\hat{u} \in QH$, $|\hat{u}| \to 1$ as $\mu \to 0$, $w \in PH$, and $\mu, \delta \in \mathbb{R}^1$. Then Problem I is equivalent to the simultaneous solution of

i) $$(B - \lambda_0)w = \delta w - PC(\mu\hat{u} + w)$$

and

ii) $$\delta\mu\hat{u} = QC(\mu\hat{u} + w).$$

Since $B - \lambda_0$ is invertible on PH, equation i) is equivalent to

$$w = (B - \lambda_0)^{-1}\{\delta w - PC(\mu\hat{u} + w)\} \equiv T(w).$$

Then there exist μ_0, $\delta_0 > 0$ such that for $|\mu| \leq \mu_0$ and $|\delta| \leq \delta_0$, T is a contraction of the ball $|w| \leq \mu_0$ into itself. The unique fixed point $w = \overline{w}(\delta,\mu\hat{u})$ of T satisfies $|\overline{w}| = o(|\mu|^{m+1})$ as $\mu \to 0$ and $\overline{w}$ depends continuously on δ and $\mu\hat{u}$.

Thus we may, without loss of generality, set

$$\mu v = w, \quad \mu u = x$$

and convert Problem I to the following equivalent problem for the equation $(B - \lambda)u + \mu^m C(u) = 0$.

II. <u>To find in</u> $\mathbb{R}^1 \times \mathbb{R}^1 \times (H \setminus \{0\})$ <u>a sequence</u> (μ_j, λ_j, u_j) <u>such that</u> $(\mu_j, \lambda_j, |u_j|) \to (0, \lambda_0, 1)$ <u>as</u> $j \to \infty$ <u>and</u>

$$(B - \lambda_j)u_j + \mu_j^m C(u_j) = 0, \quad j = 1, 2, \ldots .$$

Formally Problem II is obtained from Problem I by setting $x = \mu u$.

We may now write

$$u = \hat{u} + v, \quad \lambda = \lambda_0 + \delta,$$

where $\hat{u} \in QH$, $|\hat{u}| \to 1$ as $\mu \to 0$, $v \in PH$ and $\delta \in \mathbb{R}^1$. Then Problem II is equivalent to the simultaneous solution of

iii) $$(B - \lambda_0)v = \delta v - \mu^m PC(\hat{u} + v)$$

and

iv) $$\delta\hat{u} = \mu^m QC(\hat{u} + v).$$

As previously, iii) is equivalent to

$$v = (B - \lambda_0)^{-1}\{\delta v - \mu^m PC(\hat{u} + v)\} \equiv T(v).$$

Then for any choice of $R_0 > 0$, there exist μ_0, $\delta_0 > 0$ such that if $|\mu| \leq \mu_0$ and $|\delta| \leq \delta_0$, T is a contraction of the ball $|v| \leq R_0$ into itself. Furthermore, the unique fixed point $v = \bar{v}(\delta,\mu,\hat{u})$ of T satisfies $|\bar{v}| = O(|\mu|^m)$ as $\mu \to 0$, and $\bar{v}$ depends continuously on δ, μ, and $\hat{u}$.

The next step in the Lyapunov-Schmidt method is to insert $\bar{v}(\delta,\mu,\hat{u})$ into iv) and attempt to show that there are sequences δ_j, $\mu_j \to 0$ and $\hat{u}_j$ with $|\hat{u}_j| \to 1$ such that iv) is satisfied for each $j = 1, 2, \ldots$. iv) becomes

v) $$\delta\hat{u} = \mu^m QC(\hat{u} + \bar{v}),$$

and it is obvious that this equation can be satisfied only if $\hat{u}$ and $\bar{v}$ are such that $QC(\hat{u} + \bar{v})$ is a multiple of $\hat{u}$. In case n, the dimension of the λ_0-eigenspace, is 1, this condition is trivially satisfied.

We shall look for branches parameterized by μ on some interval of the form $|\mu| \leq \mu_1$, $\mu_1 > 0$. This idea dates back to the work of Lyapunov (cf. [9], §1.) and always succeeds when $n = 1$ (cf. [1], [6]). Now if $C(\hat{u} + \bar{v})$ is a multiple of $\hat{u}$, v) becomes

$$\delta\hat{u} = \mu^m\,(C(\hat{u} + \bar{v}),\hat{u})\hat{u} = \mu^m((C(\hat{u}),\hat{u}) + \Gamma)\hat{u},$$

where $\Gamma \to 0$ as $\mu \to 0$. Thus we have the scalar equation

$$\delta = \mu^m\{(C(\hat{u}),\hat{u}) + \Gamma\},$$

and for all sufficiently small μ, the function

$$f(\delta) = \mu^m\{(C(\hat{u}),\hat{u}) + \Gamma\}$$

maps the interval $|\delta| \leq \delta_0$ into itself, since $(C(\hat{u}),\hat{u})$ is bounded. Since f is a continuous function of δ, it must have at least one fixed point satisfying

$$\delta = \mu^m(C(\hat{u}),\hat{u}) + o(\mu^m) \quad \text{as} \quad \mu \to 0,$$

which solves v).

The proportionality condition used to solve v) is of course highly impractical, though it is necessary and sufficient for branching, and such branching can be parameterized by μ. To progress further we shall look only for branching <u>directed by</u> $u_0 \in QH$, $|u_0| = 1$ (cf. [2]), i.e., we require that $\lim_{\mu\to 0} \hat{u} = u_0$. Under certain restrictions this is the only way in which branching can occur (cf. [5]). For directed branches the previous discussion implies that δ must have the form

vi) $$\delta = \mu^m (C(u_0), u_0) + o(\mu^m) \quad \text{as} \quad \mu \to 0.$$

Furthermore v) may be written as

$$[(C(u_0), u_0) + o(1)]\hat{u} = QC(\hat{u} + \bar{v}) \quad \text{as} \quad \mu \to 0,$$

and, letting $\mu \to 0$,

$$(C(u_0), u_0) u_0 = QC(u_0).$$

Thus we have the following condition (cf. also [5]),

Condition I. A branch directed by u_0, $|u_0| = 1$, exists only if $QC(u_0)$ is a multiple of u_0. Directed branches cannot exist unless there is a $u_0 \in QH$, $|u_0| = 1$, satisfying Condition I.

We must now find conditions with which we may supplement Condition I in order to obtain existence of a directed branch. Perhaps the simplest way to accomplish this is to look for formal Taylor approximate perturbation solutions of

vii) $$(B - \lambda)u + \mu^m C(u) = 0.$$

Set $\nu = \mu^m$ and

$$u = u_0 + \nu\dot{u}_0 + \frac{\nu^2}{2}\ddot{u}_0 + \ldots,$$

$$\lambda = \lambda_0 + \nu\dot{\lambda} + \frac{\nu^2}{2}\ddot{\lambda}_0 + \ldots .$$

To calculate the coefficients in these expansions, it is easiest to differentiate vii) with respect to ν and set $\nu = 0$. We do this along any solution (λ, u) of vii) which is smooth in ν by using the chain rule for Fréchet derivatives with the symbolism $\cdot = \frac{d}{d\nu}$. The left side of vii) is differentiated as an operator from the domain of B (with the graph norm) into H. Since derivatives are calculated only at elements in the domain of B, the operators $C'(u)$, $C''(u)$, ... are the same as those obtained by differentiating C as an operator on H.

Equation vii) and the first two differentiated equations are

viii) $$Bu - \lambda u + \nu C(u) = 0,$$

ix) $$B\dot{u} - \dot{\lambda}u - \lambda\dot{u} + C(u) + \nu C'(u)\dot{u} = 0,$$

and

x) $$B\ddot{u} - \ddot{\lambda}u - 2\dot{\lambda}\dot{u} - \lambda\ddot{u} + 2C'(u)\dot{u} + \nu C''(u)\dot{u}^2 + \nu C'(u)\ddot{u} = 0.$$

Setting $\nu = 0$ in viii) yields

$$Bu_0 - \lambda_0 u_0 = 0,$$

which (in order to avoid the trivial solution $(\lambda,0)$) requires that λ_0 be an eigenvalue of B (which we take to be isolated as previously) and u_0 a corresponding eigenvector. We take $|u_0| = 1$. Now, with $\nu = 0$, ix) becomes

$$B\dot{u}_0 - \lambda_0\dot{u}_0 = \dot{\lambda}_0 u_0 - C(u_0).$$

With P, Q as previously, this equation is solvable if and only if

$$Q(\dot{\lambda}_0 u_0 - C(u_0)) = \dot{\lambda}_0 u_0 - QC(u_0) = 0,$$

which is Condition I. So, assuming u_0 satisfies Condition I, $QC(u_0) = \dot{\lambda}_0 u_0$ which yields

$$\dot{\lambda}_0 = (C(u_0),u_0)$$

and

$$\dot{u}_0 = \dot{u}_{00} + \dot{u}_{01},$$

where $\dot{u}_{01} \in PH$ is uniquely determined and $\dot{u}_{00} \in QH$ is arbitrary. Observe that, without loss of generality, we may take $\dot{u}_{00}$ to be orthogonal to u_0. In fact we may require that $\dot{u}_0$, $\ddot{u}_0$, ... are all orthogonal to u_0, since this is only a choice of scale which fixes the parameter μ as the scalar projection of our perturbation solution for $x = \mu u$ in the direction of the "directing vector" u_0.

Now setting $\nu = 0$ in x) yields

$$B\ddot{u}_0 - \lambda_0 \ddot{u}_0 = \ddot{\lambda}_0 u_0 + 2\dot{\lambda}_0 \dot{u}_0 - 2C'(u_0)\dot{u}_0,$$

which is solvable if and only if

xi) $$Q(\ddot{\lambda}_0 u_0 + 2\dot{\lambda}_0 \dot{u}_0 - 2C'(u_0)\dot{u}_0)$$

$$= \ddot{\lambda}_0 u_0 + 2\dot{\lambda}_0 \dot{u}_{00} - 2QC'(u_0)\dot{u}_0 = 0.$$

To investigate this equation, first let Q_1 be the orthogonal projection onto the one-dimensional subspace spanned by u_0 and $Q_2 = Q - Q_1$. Since Q_1 and Q_2 are mutually orthogonal projections with $Q_1 + Q_2 = Q$, equation xi) is equivalent to the pair of equations

xii) $$\ddot{\lambda}_0 = 2(C'(u_0)\dot{u}_0, u_0)$$

and

$$\dot{\lambda}_0 \dot{u}_{00} - Q_2 C'(u_0)\dot{u}_0 = 0,$$

i.e.,

xiii) $$(C(u_0),u_0)\dot{u}_{00} - Q_2C'(u_0)\dot{u}_{00} = Q_2C'(u_0)\dot{u}_{01}.$$

Equation xii) merely determines the value of $\ddot{\lambda}_0$ once $\dot{u}_{00}$ is found from equation xiii). In the event that xiii) has no solution, there can be no solution of vii) which is directed by u_0 and smooth with respect to ν. The simplest hypothesis to make to solve both xii) and xiii) is the following:

Assumption I. The restriction to Q_2H of the operator $(C(u_0),u_0)I - Q_2C'(u_0)$, written as $[(C(u_0),u_0)I - Q_2C'(u_0)]|_{Q_2H}$, is invertible, since this guarantees a unique solution for $\dot{u}_{00}$, and therefore also for $\ddot{\lambda}_0$. Furthermore, an easy induction on the form of the equations for higher order derivatives of u and λ shows that Assumption I is sufficient to guarantee the existence of unique values of all derivatives of u and λ with respect to ν at $\nu = 0$.

We now wish to prove the following theorem (cf. also [2]; [5], p. 54).

Theorem I. If Condition I and Assumption I are satisfied, there exists a unique branch $(\lambda,u) \in \mathbb{R}^1 \times H$ directed by u_0. In addition (λ,u) is analytic in ν for small ν.

The existence of a branch which is smooth with respect to ν can be proved by an analysis similar to that previously sketched for Problem II with the modification that the appropriate pair of equations to study is obtained by applying the projections $P + Q_2$ and Q_1 instead of P and Q. The following approach (for the case $n = 1$, cf. also [1], [6]) is somewhat less arduous and also yields the uniqueness and analyticity statements.

We look for solutions of viii) in the form

$$u = u_0 + v, \quad \lambda = \lambda_0 + \delta = \lambda_0 + \nu\{(C(u_0),u_0) + \gamma\},$$

with u_0 satisfying Condition I, Assumption I (recall $|u_0| = 1$), v orthogonal to u_0, $v \to 0$ as $\mu \to 0$, and $\nu = \mu^m$. Now writing

$$C(u_0 + v) = C(u_0) + C'(u_0)v + E(v)$$

where $|E(v)| = O(|v|^2)$ as $v \to 0$, we find that equation viii) becomes

$$\text{xiv)} \qquad (B - \lambda_0)v - \nu\{(C(u_0),u_0) + \gamma\}(u_0 + v)$$

$$+ \nu\{C(u_0) + C'(u_0)v + E(v)\} = 0.$$

For $\nu \neq 0$, equation xiv) is equivalent to the following three equations:

$$(B - \lambda_0)Pv - \nu\{(C(u_0),u_0) + \gamma\}Pv$$

$$+ \nu P\{C(u_0) + C'(u_0)v + E(v)\} = 0,$$

i.e.,

xv)

$$Pv + \nu(B-\lambda_0)^{-1}\{-[(C(u_0),u_0)+\gamma]Pv + P[C(u_0)+C'(u_0)v+E(v)]\} = 0;$$

$$-[(C(u_0),u_0)+\gamma]Q_2v + Q_2[C(u_0)+C'(u_0)v+E(v)] = 0,$$

or, using Condition I in the form $Q_2C(u_0) = 0$,

xvi) $$[Q_2C'(u_0) - (C(u_0),u_0)]Q_2v$$

$$+ Q_2C'(u_0)Pv - \gamma Q_2v + Q_2E(v) = 0,$$

and

xvii) $$\gamma = (C'(u_0)v,u_0) + e(v),$$

where $|e(v)| = O(|v|^2)$ as $v \to 0$.

Now let $F : PH \times Q_2H \times \mathbb{R}^1 \times \mathbb{R}^1 \to PH \times Q_2H \times \mathbb{R}^1$ be defined by

$$\mathcal{F}(Pv, Q_2v, \gamma, \nu)$$

$$= (Pv + \nu(B-\lambda_0)^{-1}\{-[(C(u_0),u_0)+\gamma]Pv + P[C(u_0)+C'(u_0)v+E(v)]\},$$

$$[Q_2C'(u_0)-(C(u_0),u_0)]Q_2v + Q_2C'(u_0)Pv - \gamma Q_2v + Q_2E(v),$$

$$\gamma - (C'(u_0)v,u_0) - e(v)).$$

For $\nu = 0$ we may take $Pv = 0$, $Q_2v = 0$, and $\gamma = 0$ so that $\mathcal{F}(0,0,0,0) = 0$. Then the solutions of xiv) and the zeros of $\mathcal{F}$ coincide. The partial Fréchet derivative of $\mathcal{F}$ with respect to the first three arguments at $(0,0,0,0)$ is

$$\begin{pmatrix} P & 0 & 0 \\ Q_2C'(u_0)P & [Q_2C'(u_0) - (C(u_0),u_0)]Q_2 & 0 \\ -(C'(u_0)P\cdot,U_0) & -(C'(u_0)Q_2\cdot,u_0) & 1 \end{pmatrix},$$

which in view of Assuption I is invertible. Theorem 1 now follows from the implicit function theorem (cf. [9], §3).

We now wish to make a limited investigation of the "degenerate case", i.e., the case in which Assumption I is violated. To facilitate matters we assume that this is the case because of the following assumption.

<u>Assumption II</u>. $(C(u_0),u_0)$ <u>is an isolated eigenvalue of</u> $Q_2C'(u_0)|_{Q_2H}$, <u>and also</u> $C'(u_0)$ <u>is selfadjoint</u>.

Now let Q_{20} be the orthogonal projection on the null space of $[(C(u_0),u_0)I - Q_2C'(u_0)]|_{Q_2H}$ and $Q_{21} = Q_2 - Q_{20}$. Let us note in passing that the condition for solvability of equation xiii) is now that $Q_{20}C'(u_0)\dot{u}_{01} = 0$, and since $\dot{u}_{01} \varepsilon PH$, xiii) will certainly have a solution if $Q_{20}C'(u_0)P = 0$. Furthermore, $\ddot{\lambda}_0$ is uniquely determined even though $\dot{u}_0$ is not uniquely determined by xiii). This follows from the selfadjointness of $C'(u_0)$ for

$$\ddot{\lambda}_0 = 2(C'(u_0)\dot{u}_0,u_0) = 2(\dot{u}_{00},QC'(u_0)u_0) + 2(C'(u_0)\dot{u}_{01},u_0).$$

Now $QC'(u_0)u_0$ is the directional derivative of QC at u_0 in the direction u_0 which, by Condition 1, is a multiple of u_0. Since $\dot{u}_{00}$ is orthogonal to u_0,

$$\ddot{\lambda}_0 = 2(C'(u_0)\dot{u}_{01},u_0).$$

We shall now exhibit two techniques for obtaining conditions which together with Condition I and Assumption II will be sufficient to establish branching directed by u_0. We wish to use the implicit function theorem again, which means that we must somehow split up equation xvi) in order to remove degeneracy. Writing $Q_2 = Q_{21} + Q_{20}$ in xvi) yields

xviii) $$[Q_{21}C'(u_0) - (C(u_0),u_0)]Q_{21}v$$

$$+ Q_{21}C'(u_0)Pv - \gamma Q_{21}v + \nu Q_{21}E(v) = 0$$

and

$$Q_{20}C'(u_0)Q_{21}v + Q_{20}C'(u_0)Pv - \gamma Q_{20}v + \nu Q_{20}E(v).$$

It follows readily from the selfadjointness of $C'(u_0)$ that the first term in the latter equation is 0. Thus equation xvi) is equivalent to equations xviii) and

xix) $$Q_{20}C'(u_0)Pv - \gamma Q_{20}v + \nu Q_{20}E(v) = 0.$$

Equation xviii) presents no problem, and we shall exhibit two methods for handling equation xix). The first is a simple invariance hypothesis which, when satisfied, enables us to disregard equation xix) completely.

Assumption III. $Q_2C'(u_0)P = 0$ and $Q_{20}E|_{(P+Q_{21})H}$ $=0$.

Theorem II. If Condition I and Assumptions II and III are satisfied, there exists a unique branch $(\lambda,u) \in \mathbb{R}^1 \times (P + Q_{21} + Q_1)H$ directed by u_0. In addition (λ,u) is analytic in ν for small ν.

The proof is a repetition of the proof of Theorem I, requiring that $v \in (P + Q_{21})H$ and everywhere replacing Q_2 by Q_{21}. An example to which Theorem II applies is $C(u) = |u|^m u$ with m an even positive integer. In this case every normalized u_0 in the λ_0-eigenspace satisfies Condition I, $[(C(u_0),u_0)I - Q_2C'(u_0)]|_{Q_2H} = 0$, so that $Q_{20} = Q_2$, $Q_{21} = 0$, Assumptions II and III are satisfied, and $(\lambda,u) = (\lambda_0 + \nu,u_0) = (\lambda_0 + \mu^m,u_0)$.

A second way to deal with equation xix) is to divide by ν as was done to obtain equation xvi) in the proof of Theorem I. Observing that if a ν-smooth solution exists, we must have $\nu\gamma = \frac{\nu^2}{2}\ddot{\lambda}_0 + O(\nu^3)$ as $\nu \to 0$, a case in which this approach works is if the following assumption is satisfied.

<u>Assumption IV.</u> $Q_{20}C'(u_0)P = 0$ <u>and</u> $\lim_{\nu\to 0} \frac{\gamma}{\nu} = \frac{\ddot{\lambda}_0}{2} = (C'(u_0)\dot{u}_{01},u_0) \neq 0$.

<u>Theorem III. If Condition I and Assumptions II and IV are satisfied, there exists a unique branch</u> $(\lambda,u) \in \mathbb{R}^1 \times H$ <u>directed by</u> u_0 <u>with</u> $Q_{20}\dot{u}_0 = 0$. <u>In addition</u> (λ,u) <u>is analytic in</u> ν <u>for small</u> ν.

We seek simultaneous solutions of equations xv), xvii), xviii) and xix) with

xx) $$v = \nu\dot{u}_0 + \nu y, \quad \dot{u}_0 = \dot{u}_{00} + \dot{u}_{01},$$

where $\dot{u}_{01}$ and $Q_{21}\dot{u}_{00}$ are known, $\gamma = \nu(C'(u_0)\dot{u}_{01},\dot{u}_0) + \nu\zeta$, and $|y|$, $\zeta \to 0$ as $\nu \to 0$. We divide each of the resulting equations by ν. Thus xv) yields

xxi) $$Pw + (B-\lambda_0)^{-1}\{-[(C(u_0),u_0)+\nu(C'(u_0)\dot{u}_{01},u_0)+\nu\zeta]P(\nu\dot{u}_0+\nu y)$$

$$+ P[C'(u_0)(\nu\dot{u}_0+\nu y) + E(\nu\dot{u}_0+\nu y)]\} = 0,$$

since $P\dot{u}_0 + (B - \lambda_0)^{-1}PC(u_0) = 0$; xviii) becomes

xxii) $$[Q_{21}C'(u_0) - (C(u_0),u_0)]Q_{21}y$$

$$+ Q_{21}C'(u_0)Py - \nu((C'(u_0)\dot{u}_{01},u_0) + \zeta)Q_{21}(\dot{u}_0 + y)$$

$$+ Q_{21}E(\nu\dot{u}_0 + \nu y) = 0,$$

since $[Q_{21}C'(u_0) - (C(u_0),u_0)]Q_{21}\dot{u}_0 + Q_{21}C'(u_0)P\dot{u}_0 \doteq 0$; xix) yields

xxiii) $$[(C'(u_0)\dot{u}_{01},u_0) + \zeta]Q_{20}y - \frac{1}{\nu}Q_{20}E(\nu\dot{u}_0 + \nu y) = 0,$$

if we choose $Q_{20}\dot{u}_0 = 0$ which fixes $\dot{u}_0$; while xvii) becomes

xxiv) $$\zeta = (C'(u_0)y,u_0) + \frac{1}{\nu}e(\nu\dot{u}_0 + \nu y).$$

Taking $y = 0$ and $\zeta = 0$ when $\nu = 0$, it follows from the properties of E and e that xxi)-xxiv) are satisfied when $\nu = 0$. Now form a function $G: PH \times Q_{21}H \times Q_{20}H \times R^1 \times R^1 \to PH \times Q_{21}H \times Q_{20}H \times R^1$ from equations xxi)-xxiv) in the same fashion as F was constructed in the proof of Theorem I. Computation of the partial Fréchet derivative of G with respect to the first four arguments at $(0,0,0,0,0)$ completes the proof.

In the event that Condition I and Assumption II are satisfied, none of Assumptions I, III, or IV are satisfied, and non-existence of ν-smooth solutions has not been established, one can attempt to obtain solutions by the use of higher order Taylor expansions, proceeding analogously to the above.

This research was supported by NSF Grant GP-33652.

REFERENCES

1. Crandall, M. G., and Rabinowitz, P. H., Bifurcation from simple eigenvalues, *J. Functional Analysis*, 8, 321-340, (1971).

2. Krasnoselskii, V. M., Investigation of the bifurcation of small eigenfunctions in the case of multidimensional degeneration, *Soviet Math. Dokl.*, 11, 1609-1613, (1970).

3. Keller, J. B., *Perturbation Theory*, Notes on a series of lectures, Michigan State University, East Lansing, 1968.

4. Pimbley, G. H., Jr., *Eigenfunction Branches of Nonlinear Operators, and their Bifurcations*, Springer, New York, 1969.

5. Sather, D., Branching of solutions of an equation in Hilbert space, *Arch. Rational Mech. Anal.*, 36, 47-64, (1970).

6. Sattinger, D. H. Transition to instability, *these Proceedings*.

7. Stakgold, I. Branching of solutions of nonlinear equations, *SIAM Rev.*, 13, 289-332, (1971).

8. de Sz. Nagy, B., Perturbations des transformations autoadjointes dans l'espace de Hilbert, *Comment. Math. Helv.*, 19, 347-366, (1946-1947).

9. Vainberg, M., and Trenogin, V. A., The methods of Lyapunov and Schmidt in the theory of non-linear equations and their further development, *Russian Math. Surveys*, 17, 1-60, (1962).

ASYMPTOTIC ANALYSIS OF A CLASS OF NONLINEAR INTEGRAL EQUATIONS

Richard A. Handelsman*
University of Illinois at Chicago Circle
W. E. Olmstead*
Northwestern University

1. INTRODUCTION

The physical problem of determining the temperature in a conducting half-space which can radiate heat nonlinearly gives rise to the nonlinear Volterra equation,

$$(1.1)\qquad \epsilon\, u(t;\epsilon) = \pi^{-\frac{1}{2}}\int_0^t (t-s)^{-\frac{1}{2}}[f(s) - u^n(s;\epsilon)]ds, \quad t \geq 0, \quad n \geq 1 .$$

Here $\epsilon > 0$ is a constant and f is a given function. Some background and references related to this physical problem are given in [4].

For f bounded, locally integrable and positive, it is known that a unique positive solution u of (1.1) exists. This solution has been studied in various asymptotic limits [3], [4], [6]. Of particular interest here are the limits $t \to \infty$ for ϵ fixed and $\epsilon \to 0$ for all $t \geq 0$. We shall discuss both of these cases.

2. LARGE t EXPANSION**

We find that in the limit $t \to \infty$, it is just as convenient to consider the wider class of equations,

$$(2.1)\qquad \varphi(t) = h(t) - \int_0^t K(t-s)\varphi^n(s)ds, \quad t \geq 0, \quad n \geq 1 .$$

Since ϵ is fixed, we have set $\epsilon = 1$; its dependence can be recovered by a rescaling of (2.1). Also, for the moment we merely assume that h and K are such that a unique positive solution of (2.1) exists.

Let us now define the convolution

$$(2.2)\qquad \Psi(t) = \int_0^t K(t-s)\varphi^n(s)ds = K*\varphi^n ,$$

so that (2.1) becomes

*The authors' work was supported by the National Science Foundation under Grants GP-27953 and GP-29508, respectively.

**Some of the results of this section have been obtained by Shyam Johari, University of Illinois at Chicago Circle.

$$\varphi(t) = h(t) - \Psi(t) . \tag{2.3}$$

To determine the large t behavior of $\varphi(t)$, we need some knowledge of the large t behavior of each term in (2.3). Suppose that under certain conditions on φ and K an asymptotic expansion of Ψ as $t \to \infty$ could be obtained from (2.2). Suppose further that this could be accomplished from only having information about φ as $t \to \infty$. This suggests the following self-consistent procedure to analyze (2.3): One first assumes an asymptotic expansion for φ as $t \to \infty$. Then an asymptotic expansion of Ψ is derived from (2.2). Since f is known, then so is its asymptotic expansion. Therefore the unknown parameters in the assumed expansion of φ are determined by requiring both sides of (2.3) to be equal.

We see that the formal procedure just described hinges on our ability to find the large t behavior of Ψ. To consider this crucial aspect in more depth, we take the Laplace transform, $\mathcal{L}[\cdot;\omega] = \int_0^\infty e^{-\omega t}(\cdot)dt$, of (2.2) and find

$$\mathcal{L}[\Psi;\omega] = \mathcal{L}[K;\omega]\ \mathcal{L}[\varphi^n;\omega] . \tag{2.4}$$

We know that the large t behavior of Ψ is associated with the small ω behavior of $\mathcal{L}[\Psi;\omega]$. To establish the nature of this relationship, we need an appropriate Tauberian theorem.

Suppose that, as $t \to \infty$,

$$\Psi(t) \sim \sum_{m=0}^{\infty} \sum_{n=0}^{N(m)} c_{mn}\, t^{-r_m}(\log t)^n , \tag{2.5}$$

where $\mathrm{Re}(r_m) \uparrow \infty$ and $N(m)$ is finite for each m. Then we say $\Psi(t)$ has an asymptotic Mellin series as $t \to \infty$. Further suppose that, as $\omega \to 0+$,

$$\mathcal{L}[\Psi;\omega] \sim \sum_{m=0}^{\infty} \sum_{n=0}^{\overline{N}(m)} p_{mn}\, \omega^{a_m}(\log \omega)^n , \tag{2.6}$$

where $\mathrm{Re}(a_m) \uparrow \infty$ and $\overline{N}(m)$ is finite for each m. Then we say that $\mathcal{L}[\Psi;\omega]$ has an asymptotic Mellin series as $\omega \to 0+$. We can relate this through the following theorems which are offered without proof.

Theorem 1. (Handelsman and Lew [2]). *If $\Psi(t)$ is locally integrable on $[0, \infty)$ and has an asymptotic Mellin series as $t \to \infty$, then this series is uniquely recoverable from the asymptotic Mellin series of $\mathcal{L}[\Psi;\omega]$ as $\omega \to 0+$. Moreover, we have the* following correspondence:

$$\text{(2.7)}\quad \omega^{b}(\log \omega)^{n} \Leftrightarrow \begin{cases} \left(\frac{\partial}{\partial b}\right)^{n}\left[\frac{t^{-b-1}}{\Gamma(-b)}\right], & b \neq 0,1,2,\ldots \\ (-1)^{b+1}\sum_{j=0}^{[\frac{n-1}{2}]}(-1)^{j}\,\pi^{2j}\binom{n}{2j+1}\left(\frac{\partial}{\partial b}\right)^{n-2j-1}[\Gamma(b+1)t^{-b-1}], & b = 0,1,2,\ldots \end{cases}$$

Theorem 2. (Handelsman and Lew [2]). The space of functions locally integrable on $[0, \infty)$, having an asymptotic Mellin series as $t \to \infty$, is closed under Laplace convolution.

Returning to the convolution (2.2), if we assume that both $K(t)$ and $\varphi(t)$ have asymptotic Mellin series as $t \to \infty$, then by Theorem 2 we can conclude that $\Psi(t)$ does as well. By Theorem 1 we find that to recover this series we need only determine the asymptotic Mellin series of $\mathcal{L}[\varphi^{n};\omega]$ and $\mathcal{L}[K;\omega]$ as $\omega \to 0+$. The asymptotic expansion of Laplace transforms is considered in [1]. We shall quote here a special case of a more general theorem proved in [1].

Theorem 3. Let $g(t)$ be locally integrable on $[0, \infty)$ and have, as $t \to \infty$, the Mellin series

$$\text{(2.8)}\qquad g(t) \sim \sum_{m=0}^{\infty} c_{m}\, t^{-r_{m}} .$$

If r_m is not a positive integer for $m = 0,1,2, \ldots$, then as $\omega \to 0+$

$$\text{(2.9)}\qquad \mathcal{L}[g;\omega] \sim \sum_{m=1}^{\infty} \frac{(-\omega)^{m-1}}{(m-1)} M[g;m] + \sum_{m=0}^{\infty} \omega^{r_{m}-1} c_{m}\,\Gamma(1 - r_{m}) .$$

Here $M[g;m]$ denotes the analytic continuation of the Mellin transform $M[g;z] = \int_{0}^{\infty} t^{z-1} g(t)dt$ evaluated at $z = m$.

If any r_m is a positive integer, then a logarithm appears in the asymptotic expansion of $\mathcal{L}[g;\omega]$ to order ω^{r_m-1}. In particular if $r_0 = 1$ then

$$\text{(2.10)}\qquad \mathcal{L}[g;\omega] \sim - c_{0} \log \omega .$$

We now have enough information to determine the asymptotic expansion of the solution to (2.1) as $t \to \infty$. We assume that K and h have the asymptotic expansions,

$$\text{(2.11)}\qquad K(t) \sim \alpha_{0} t^{-\lambda_{0}},\quad \alpha_{0} > 0;\quad h(t) \sim \beta_{0} t^{-\mu_{0}},\quad \beta_{0} > 0 ,$$

while the unknown solution $\varphi(t)$ satisfies

(2.12) $$\varphi(t) \sim c_o t^{-r_o}, \quad c_o > 0 \text{ .}$$

Our objective is to determine the constants c_o and r_o. However, to completely solve this problem, there are numerous cases that must be treated separately. These various cases depend on the values of μ_o, λ_o and n. We shall not exhaust all possibilities here, but instead treat only one case to illustrate the procedure.

We shall only consider the case where $\mu_o = 0$ and $0 < \lambda_o < 1$. Further suppose that $0 < nr_o < 1$, an assumption that must be checked once r_o has been determined. From Theorem 3 we have

(2.13) $$\mathcal{L}[K;\omega] \sim \alpha_o \Gamma(1 - \lambda_o)\omega^{\lambda_o - 1} \; ; \quad \mathcal{L}[\varphi^n;\omega] \sim (c_o)^n \Gamma(1 - nr_o)\omega^{nr_o - 1} \text{ .}$$

Hence

(2.14) $$\mathcal{L}[\Psi;\omega] \sim \alpha_o \Gamma(1 - \lambda_o)\, \Gamma(1 - nr_o)\, (c_o)^n\, \omega^{-2+\lambda_o+nr_o} \text{ .}$$

The asymptotic behavior of $\Psi(t)$, as $t \to \infty$, then follows from Theorem 1. In this way we find that equation (2.3) demands the asymptotic equality

(2.15) $$c_o t^{-r_o} \sim \beta_o - \alpha_o (c_o)^n\, B(1 - \lambda_o,\, 1 - nr_o) t^{1-\lambda_o - nr_o} \text{ .}$$

Here $B(x, y)$ denotes the beta function.

By equating the exponents in (2.15) we find that there are only three possibilities, namely,

(i) $r_o = 0, \quad \lambda_o \geq 1$,

(ii) $r_o = nr_o + \lambda_o - 1, \quad r_o < 0$,

(iii) $1 - \lambda_o - nr_o = 0, \quad r_o > 0$.

Because we have taken $0 < \lambda_o < 1$ and $n \geq 1$, only (iii) can hold. Therefore

(2.16) $$r_o = \frac{1-\lambda_o}{n} \text{ .}$$

We note that $0 < nr_o = 1 - \lambda_o < 1$ as assumed.

To determine c_o, we equate the coefficients of the two terms of matched order in (2.15). This yields $c_o^n = [\beta_o/\alpha_o B(1 - \lambda_o,\, 1 - nr_o)]$, where the positive root is chosen since we have assumed that (2.1) has a unique positive solution. Now we have determined, to the leading order, that the asymptotic expansion of φ is given by

$$(2.17) \qquad \varphi(t) \sim \left[\frac{\beta_o t^{-(1-\lambda_o)}}{\alpha_o B(1-\lambda_o,\ \lambda_o)}\right]^{\frac{1}{n}}$$

as $t \to \infty$. While this result has been obtained formally, we wish to emphasize that it becomes rigorous when it is known that φ does have an asymptotic Mellin series as $t \to \infty$.

This result for $\mu_o = 0$ is an illustrative example of our procedure to determine the asymptotic behavior of φ. To get some idea of the various other cases to be treated,we refer the reader to [3]. There we consider $\mu_o \le \frac{1}{2}$ and $\lambda_o = \frac{1}{2}$, and we list results for some twenty cases.

3. SMALL ε EXPANSION

To study (1.1) in the limit $\varepsilon \to 0$ for all $t \ge 0$, we find it useful to consider an alternative form of (1.1). An Abel inversion of the integral operator in (1.1) yields

$$(3.1) \quad u^n(t;\varepsilon) = f(t) - \varepsilon D_t^{\frac{1}{2}} u(t;\varepsilon); \quad D_t^{\frac{1}{2}} u(t;\varepsilon) \equiv \pi^{-\frac{1}{2}} \frac{d}{dt}\int_o^t (t-s)^{-\frac{1}{2}} u(s;\varepsilon)ds \ .$$

With f bounded, continuous and positive on $[0, \infty)$, we know [4] that there exists a unique solution u of (1.1) or equivalently (3.1). Further restrictions on f are needed for a small ε analysis.

When ε is small, we find that our problem reveals a singular character. To see this we note that if $\varepsilon > 0$ then (1.1) implies $u(0;\varepsilon) = 0$, while if $\varepsilon = 0$ then (3.1) implies that $u(t;0) = [f(t)]^{\frac{1}{n}}$ for all $t \ge 0$. Thus for $f(0) > 0$ we expect $u(t;\varepsilon)$ to have a nonuniform dependence upon ε.

To treat (1.1) as a singular perturbation problem, we have in [6] utilized a basic idea from [5] and [7]. We introduce an "inner" variable $\tau = \varepsilon^{-2}t$ and look for an asymptotic representation of u in the form

$$(3.2) \qquad u(t;\varepsilon) = S_M(t,\ \tau;\varepsilon) + R_M(t,\ \tau;\varepsilon) \ ,$$

where

$$S_M(t,\ \tau;\varepsilon) = \sum_{j=o}^{M} u_j(t,\ \tau)\,\varepsilon^j \ .$$

We regard R_M as the error made in approximating u by S_M. Our procedure like that of

[5] and [7] is to determine the functions u_j by studying the errors R_j in sequence. We want $|R_M|$ to be $O(\epsilon^{M+1})$ uniformly in t for all $t \geq 0$. While this procedure could in principle be carried out to any order, we will only determine u_0 and u_1 here.

To study the error R_M, M = 0, 1,..., we recast (1.1) into an integral equation for R_M and then obtain an estimate on $|R_M|$ for all $t \geq 0$. In [6] we have shown that if f(t) satisfies

$$C_1(1 + C_0 t)^{-a_0} \leq f(t) \leq 1, \; t \geq 0, \; 0 \leq a_0 \leq \frac{n}{2(n-1)} \tag{3.7}$$

and has the asymptotic behavior, as $t \to \infty$

$$f(t) \sim C_1(C_0 t)^{-a_o}, \tag{3.8}$$

then

$$|R_M| \leq K \max_{t \geq 0} |(1 + C_0 t)^{a_0 - \frac{a_0}{n}} \{f(t) - \epsilon D_t^{\frac{1}{2}} S_M(t, \tau; \epsilon) - S_M^n(t, \tau; \epsilon)\}|, \tag{3.9}$$

where K is a constant independent of ϵ. We have also assumed that $S_M \geq 0$ in obtaining (3.9).

Our task becomes that of finding appropriate $u_j(t, \tau)$, j = 0, 1,...,M such that (3.9) yields the estimate $|R_M| \leq C\epsilon^{M+1}$. We shall assume that

$$u_j(t, \tau) = y_j(t) + z_j(\tau), \quad j = 0, 1, 2\ldots. \tag{3.10}$$

To determine the leading term u_0, we consider (3.9) for M = 0 and express it in the form

$$|R_0| \leq K \max_{t \geq 0} |(1 + C_0 t)^{a_0 - \frac{a_0}{n}} \{f(t) - y_0^n(t) - D_\tau^{\frac{1}{2}} [z_o(\tau) + y_o(0)] \tag{3.11}$$

$$- [z_o(\tau) + y_o(0)]^n + y_0^n(0) - \epsilon D_t^{\frac{1}{2}} [y_0(t) - y_0(0)] + Q_0(t, \tau)\}|$$

where

$$Q_0(t, \tau) = [z_0(\tau) + y_0(0)]^n - [z_0(\tau) + y_0(t)]^n + y_0^n(t) - y_0^n(0). \tag{3.12}$$

Our motivation for expressing (3.11) in this way stems from some experience with the linear problem (note that if n = 1 then $Q_0 \equiv 0$).

In anticipation that the terms $\epsilon D_t^{\frac{1}{2}} [y_0(t) - y_0(0)]$ and $Q_0(t, \tau)$ are going to be $O(\epsilon)$, we require that

$$(3.13,14) \qquad y_o^n(t) = f(t) \; ; \quad [z_o(\tau) + y_o(0)]^n + D_\tau^{\frac{1}{2}}[z_o(\tau) + y_o(0)] = y_o^n(0) \; .$$

We can solve immediately for $y_o(t)$ as

$$(3.15) \qquad y_o(t) = [f(t)]^{\frac{1}{n}} \; .$$

From this and an Abel inversion of (3.14) we find that $z_o(\tau)$ must satisfy the nonlinear integral equation

$$(3.16) \qquad z_o(\tau) + [f(0)]^{\frac{1}{n}} = \pi^{-\frac{1}{2}} \int_o^\tau (\tau - \sigma)^{-\frac{1}{2}}[f(0) - \{z_o(\sigma) + [f(0)]^{\frac{1}{n}}\}^n]d\sigma.$$

This integral equation, even though nonlinear, is considerably simpler than the original problem (1.1). While we cannot give an explicit solution of (3.16), we can say a great deal about $z_o(\tau)$ from the results of [3] and [4]. In fact we know that there is a unique solution of (3.16), and it has the properties

$$- [f(0)]^{\frac{1}{n}} \le z_o(\tau) \le 0 \; , \quad z_o'(\tau) \ge 0 \; , \quad \tau \ge 0 \; .$$

$$(3.17) \qquad z_o(\tau) \sim \begin{cases} - [f(0)]^{\frac{1}{n}} + 2\pi^{-\frac{1}{2}} f(0) \tau^{\frac{1}{2}} \; , & \tau \to 0 \; , \\ - \dfrac{[f(0)]^{\frac{2-n}{n}}}{n\pi^{\frac{1}{2}}} \tau^{-\frac{1}{2}} - \dfrac{(n-1)[f(0)]^{\frac{3-2n}{n}}}{2n^2\pi} \tau^{-1} \; , & \tau \to \infty \; . \end{cases}$$

With these results we then have $u_o(t, \tau) = [f(t)]^{\frac{1}{n}} + z_o(\tau)$ as our candidate for the leading order approximation to $u(t;\epsilon)$. To establish the asymptotic validity of this $u_o(t, \tau)$, we appeal to our estimate (3.11) for the remainder. First we must check that $S_o = u_o \ge 0$, a sufficient condition for (3.11) to hold. That this is indeed true is seen from the lower bound in (3.7) for $f(t)$ and the properties (3.17) for $z_o(\tau)$. Secondly, we have shown in [6] that $|D_t^{\frac{1}{2}}[y_o(t) - y_o(0)]|$ and $|\epsilon^{-1}Q_o(t, \tau)|$ are bounded independent of ϵ for all $t \ge 0$ and are at least $O(t^{-\frac{1}{2}})$ as $t \to \infty$. So (3.11) gives $|R_o| \le C\epsilon$, and therefore we have established our leading asymptotic result

$$(3.18) \qquad u(t;\epsilon) \sim [f(t)]^{\frac{1}{n}} + z_o(\tau) + O(\epsilon) \; .$$

Now the procedure is repeated for R_1 so as to determine $u_1(t, \tau)$ and thereby obtain the next order in the asymptotic expansion of $u(t;\epsilon)$. The results for

$u_1(t, \tau) = y_1(t) + z_1(\tau)$ are derived in [6]. We find there that

$$(3.19,20) \quad y_1(t) = n^{-1} \{(t\pi)^{-\frac{1}{2}} [f(0)]^{\frac{2-n}{n}} - [f(t)]^{\frac{1-n}{n}} D_t^{\frac{1}{2}} [f(t)]^{\frac{1}{n}}\} \ ; \ z_1(\tau) = 0 \ .$$

Again we check that $S_1 = u_0(t, \tau) + \epsilon\, u_1(t, \tau) \geq 0$, and then by some laborious estimates find that $|R_1| \leq C\epsilon^2$. Thus we establish

$$(3.21) \quad u(t;\epsilon) \sim [f(t)]^{\frac{1}{n}} + z_0(\tau) + \epsilon n^{-1}\{(t\pi)^{-\frac{1}{2}}[f(0)]^{\frac{2-n}{n}} - [f(t)]^{\frac{1-n}{n}} D_t^{\frac{1}{2}}[f(t)]^{\frac{1}{n}}\} + O(\epsilon^2).$$

In establishing these results, it was crucial that we have the bounds (3.7) on f and its asymptotic behavior (3.8). Indeed if f did go to zero at some finite t or approached zero too rapidly as $t \to \infty$, then we could expect other nonuniformities (boundary layers) to appear as part of the asymptotic analysis.

REFERENCES

1. R. A. Handelsman and J. S. Lew, "Asymptotic expansion of a class of Laplace transforms near the origin", SIAM J. Math. Anal. 1 (1970).

2. R. A. Handelsman and J. S. Lew, "Asymptotic expansion of Laplace convolution for large argument and tail densities for certain sums of random variables", IBM Research Report RC 3869, 1972.

3. R. A. Handelsman and W. E. Olmstead, "Asymptotic solution to a class of nonlinear Volterra integral equations", SIAM J. Appl. Math. 22 (1972).

4. J. B. Keller and W. E. Olmstead, "Temperature of a nonlinearly radiating semi-infinite solid", Quart. Appl. Math. 29 (1972).

5. B. J. Matkowsky and E. L. Reiss, "On the asymptotic theory of dissipative wave motion", Arch. Rational Mech. Anal., 42 (1971).

6. W. E. Olmstead and R. A. Handelsman, "Singular perturbation analysis of a certain Volterra integral equation", Z. Angew. Math. Phys. (to appear).

7. E. L. Reiss, "On multivariable asymptotic expansions", SIAM Review, 13 (1971).

REMARKS ABOUT BIFURCATION AND STABILITY OF QUASI-PERIODIC SOLUTIONS WHICH BIFURCATE FROM PERIODIC SOLUTIONS OF THE NAVIER STOKES EQUATIONS

Daniel D. Joseph
Department of Aerospace Engineering and Mechanics
University of Minnesota

SUMMARY

L. D. Landau (1944) and E. Hopf (1948) have conjectured that the transition to turbulence may be described as repeated branching of quasi-periodic solutions into quasi-periodic solutions with more frequencies. The simplest case is the bifurcation of periodic solutions from steady solutions. The next hardest problem is the bifurcation of quasi-periodic solutions from basic time periodic solutions of fixed frequency. This problem is treated in the lecture by a generalization of the Poincaré-Lindstedt perturbation which is successful in the simplest case. It is assumed that the Floquet exponents are simple eigenvalues of the spectral problem for the basic flow.

If the Floquet exponent is zero at criticality, the formal construction gives two bifurcating solutions of the same frequency as the basic flow. The small amplitude solutions which bifurcate supercritically are stable; subcritical solutions with small amplitudes are unstable.

When the Floquet exponents at criticality are complex and rationally independent of the fixed frequency, the solution of the spectral problem is quasi-periodic. The formal construction then gives the natural frequency of the bifurcating solution as a power series in the amplitude. The stability of the two-frequency power series is studied using a Floquet representation, generalizing a suggestion of Landau and using the Poincaré-Lindstedt method. Again the bifurcating solution is stable when supercritical and is unstable when subcritical.

When the Floquet exponents and the fixed frequency are rationally dependent at criticality, the perturbation problems whose solutions give the coefficients of the Poincare-Lindstedt series cannot be solved unless certain additional orthogonality conditions are satisfied. Though these "extra" conditions do not arise exactly when the frequencies are rationally independent, they may be viewed as limiting forms of conditions associated with small divisors. In general, a bounded inverse cannot be expected even in the quasi-periodic case.

§1. INTRODUCTION

This paper treats the problem of stability and bifurcation of periodic solutions of the Navier-Stokes equations. Such periodic motions may be assumed to be forced by periodic external conditions. For example, periodic fluid motion could arise from the rotation of gears in a gear box full of oil, or it could arise as an oscillating pipe flow driven by a time periodic pressure gradient. For sufficiently small Reynolds numbers these forced periodic solutions are both stable and unique.* Time periodic motions also arise from instability and bifurcation of steady motion. However, we shall concentrate here on the instability of the forced periodic motions.

Time dependent basic flows are currently a subject of active inquiry. The interested reader can find a bibliography on the linear aspects of this problem as well as derivation of formal perturbation theory involving amplitude equations of the Stuart-Watson type in the paper of Davis (1971).

Mathematically rigorous demonstrations of the existence of bifurcating solutions have been given by Sattinger (1971), Yudovich (1971), and Iooss (1972). The Poincaré-Lindstedt series for constructing the Floquet theory of stability of bifurcating periodic solutions given first by Joseph and Sattinger (1972) are important in the present work. In this paper we generalize (without justification) the series construction to study solutions which bifurcate from forced periodic motions. These series form the basis for generalizing to solutions bifurcating from periodic motions results known to hold for solutions bifurcating from steady motions (see Fig. 1) when (a) the bifurcating solution is also steady (Joseph (1971) hereafter

*Existence of forced periodic solutions at small R is proved by Serrin (1960) and at all R by Yudovich (1960), Prodi (1960) and Prouse (1963). Serrin shows that the forced periodic solution is stable and unique when R is small. Fojas (1962) has proved existence, stability and uniqueness of forced almost periodic solutions at small R.

called (J)) and (b) when the bifurcating solution is time periodic (Joseph and Sattinger (1972), hereafter called (JS)). In the context of present work, in which we consider bifurcation and instability of forced periodic motions, the steady motions appear as a special case in the limit of zero forcing frequency.

A precise summary of the results to be obtained from our formal analysis will now be given. Let $\tilde{u}(x,t,R)$ and $\epsilon u(x,t,R) + \tilde{u}(x,t,R)$ be solutions of (1.1) below:

$$\frac{\partial V}{\partial t} + V \cdot \nabla V - \frac{1}{R}\nabla^2 V + \nabla p + F(x,t) = 0\Big|_{\Omega(t)} \quad , \tag{1.1a}$$

$$\text{div } V = 0\Big|_{\Omega(t)} \quad , \tag{1.1b}$$

and

$$V = f(x,t)\Big|_{\partial\Omega(t)} \quad , \tag{1.1c}$$

where F, $\tilde{u}$, p, f and $\Omega(t)$ are periodic functions of t of period $2\pi/\tilde{\omega}$ and u is either a periodic function of the same period or an almost periodic function of t. Since the fluid is incompressible, the total volume of the (otherwise arbitrary but periodically varying) domain $\Omega(t)$ is constant in time. The symbol F stands for body force and $F(x,t)$, $f(x,t)$ and $\Omega(t)$ are preassigned periodic functions. The symbol p stands for the pressure.

We have already noted that the existence of at least one periodic solution of (1.1) for all R has been established by various authors. However, when R is sufficiently small there can be only one periodic solution of (1.1). We designate this basic solution as $\tilde{u}$. The second solution $\tilde{u} + \epsilon u$ will be assumed to arise as an instability of $\tilde{u}$ when R is increased past a critical value R_c obtained from a linearized theory of stability.

The disturbance ϵu satisfies equations (1.2) below:

$$\frac{\partial u}{\partial t} + \tilde{u} \cdot \nabla u + u \cdot \nabla \tilde{u} + \epsilon u \cdot \nabla u - \frac{1}{R}\nabla^2 u + \nabla p = 0 \tag{1.2a}$$

$$\text{div } u = 0 \ , \quad u = 0\Big|_{\partial\Omega(t)} \ . \tag{1.2b,c}$$

We now assume that the stability of the periodic solution $\tilde{u}$, that is, the solution $\underset{\sim}{u} = p = 0$ of (1.2), can be studied by the method of Floquet (see Yudovich (1970 A,B)). We set $\epsilon = 0$ in (1.2) and introduce the Floquet representation

$$\underset{\sim}{u} = e^{-\gamma t}\underset{\sim}{\zeta}(\underset{\sim}{x}\ t)\ , \quad p = e^{-\gamma t}\Pi(\underset{\sim}{x},t) \tag{1.3}$$

where $\underset{\sim}{\zeta}$ is a periodic function of t with period $2\Pi/\tilde{\omega}$ and

$$\gamma = \xi + i\eta$$

is the Floquet exponent. The exponent is determined as an eigenvalue of the spectral problem (1.4) below:

$$-\gamma\underset{\sim}{\zeta} + \frac{\partial\underset{\sim}{\zeta}}{\partial t} + L(\underset{\sim}{\tilde{u}},\lambda)\underset{\sim}{\zeta} + \nabla\Pi = 0\ , \quad \operatorname{div}\underset{\sim}{\zeta} = 0\ , \quad \underset{\sim}{\zeta} = 0|_{\partial\Omega(t)} \tag{1.4a,b c}$$

where $\lambda = 1/R$ and the i-th component of $L\underset{\sim}{\zeta}$ is defined as

$$\begin{aligned}(L(\underset{\sim}{\tilde{u}},\lambda)\underset{\sim}{\zeta})_i &= (L(\underset{\sim}{\tilde{u}}(\lambda,t),\lambda)\underset{\sim}{\zeta})_i = \underset{\sim}{\tilde{u}}\cdot\nabla\zeta_i + \underset{\sim}{\zeta}\cdot\nabla\tilde{u}_i - \lambda\Delta\zeta_i \\ &= \underset{\sim}{\tilde{u}}\cdot\nabla\zeta_i + \zeta_j\Omega_{ji}(\tilde{u}) + \zeta_j S_{ji}(\tilde{u}) - \lambda\Delta\zeta_i\end{aligned} \tag{1.5}$$

and $\partial_j\tilde{u}_i = \Omega_{ji} + S_{ji}$ is the resolution into the vorticity and the rate of strain tensor.

Stability in the linear approximation is determined by the sign of $\xi(R) = \operatorname{re}\gamma(R)$. When R is small we have stability and $\xi(R) > 0$; for $R = R_c$ we have the marginal case $\xi(R_c) = 0$; and when $R > R_c$ there is instability and $\xi(R) < 0$.

The assumption that

$$\gamma(R) = \xi(R) + i\eta(R)$$

is a simple discrete eigenvalue of (1.4) when $R - R_c$ is small is central to our analysis. This is not likely to be a restrictive assumption in bounded domains when $R - R_c$ is small.

We shall distinguish two cases: (a) the Floquet exponent is zero at criticality and (b) the Floquet exponent is not zero at criticality. In case (a) (Section 5), $\operatorname{im}\gamma(R_c) = 0$ and is both simple and discrete. Therefore, the Floquet exponents $\gamma(R) = \xi(R)$ form a single curve through the origin of the complex γ plane.

No other eigenvalues of (1.4) lie in some circle centered at the origin. This case is the analogue for bifurcation from a real simple eigenvalue in the steady case (cf. J). We find that two periodic solutions of period $2\pi/\tilde{\omega}$ bifurcate from the basic periodic solution $\tilde{u}(x,t\ R)$ at $R = R_c$. Assuming that the bifurcating solutions $u(x,t,R(\epsilon),\epsilon)$, $p(x,t,R(\epsilon),\epsilon)$ and $R(\epsilon)$ are smooth functions of ϵ , one finds that, in general, $dR(0)/d\epsilon \neq 0$ and a supercritical solution $(R(\epsilon) > R_c)$ and a subcritical solution $R(\epsilon) < R_c$ will branch out at $R(0) = R_c$. (Fig. 1 (c)).

A different set of bifurcation results may be associated with case (b) which is considered in Sections 3 and 4. In this case, $\eta(R_c) = \omega_o$. The formal analysis leads to a single quasi-periodic solution which bifurcates from the basic periodic solution when $R = R_c$. This quasi-periodic solution is a composition of functions with two frequencies*: the fixed and externally imposed frequency and a natural frequency $\omega(\epsilon)$ which arises from the instability of the basic motion and depends on the amplitude ϵ of the bifurcating motion. When the fixed forced frequency is zero, we return to the case of loss of stability of a steady solution to a bifurcating time periodic solution which was treated in (JS). Again we find that supercritical bifurcating solutions are stable and subcritical solutions are unstable when ϵ is small.

The techniques used here to study the bifurcation stability problem for the Navier-Stokes equations would appear to give a further mathematical structure (cf. JS) to previously unexplored aspects of Landau's conjecture about the transition to turbulence through repeated supercritical branching. However, the perturbation procedure which could be expected to give the quasi-periodic bifurcating solutions runs into solvability difficulties. The relation of these difficulties to the Landau-Hopf conjecture is discussed in the conclusion to Section 4.

---

*Quasi-periodic functions are defined as the special class of almost periodic functions possessing only a finite basis of frequencies. In other words, we are studying oscillations containing finitely many (rationally independent) frequencies $\omega_1, \omega_2, \ldots, \omega_n$. For example, the function $f(t) = \cos t \cos \pi t$ is a quasi-periodic function with frequencies $\omega_1 = 2\pi$ and $\omega_2 = 2$. The value $f(t) = 1$ occurs when $t = 0$ but not again; though $f(t) < 0$ when $t \neq 0$, there is always $\tau(\epsilon) > 0$ such that $|f(\tau) - f(o)| < \epsilon$ for preassigned $\epsilon > 0$.

§2. THE SPECTRAL PROBLEM AND THE SOLVABILITY CONDITIONS

Ultimately we shall want to consider quasi-periodic vector fields $\underset{\sim}{q}(\underset{\sim}{x},s,t) = \tilde{\underset{\sim}{q}}(\underset{\sim}{x},\omega(\epsilon)t,\tilde{\omega}t)$ where $\tilde{q}$ is 2π periodic in the variables $s = \omega(\epsilon)t$ and $\tilde{\omega}t$, and $\tilde{\omega}$ are rationally independent.

We first need to define the operator

$$J = \frac{\partial}{\partial t} + L(\tilde{\underset{\sim}{u}},\lambda) , \tag{2.1}$$

whose domain is the set of smooth vector fields which are $2\pi/\tilde{\omega}$ periodic in t. For these periodic functions we define the scalar product

$$[\underset{\sim}{a},\underset{\sim}{b}] = \frac{2\pi}{\tilde{\omega}} \int_0^{2\pi/\tilde{\omega}} (\underset{\sim}{a},\underset{\sim}{b})dt \qquad ,$$

where (2.2)

$$(a,b) = \int_{\Omega(t)} \underset{\sim}{a} \cdot \bar{\underset{\sim}{b}}\, d\underset{\sim}{x}$$

and the overbar designates "complex conjugate".

By an admissible $2\pi/\tilde{\omega}$ periodic vector field, we mean a periodic field which is solenoidal and vanishes on $\partial\Omega(t)$.

Let $\underset{\sim}{a}$ and $\underset{\sim}{b}$ be arbitrary admissible vectors. We define the formal adjoint of J in the usual way*

$$[\underset{\sim}{a},J\underset{\sim}{b}] = [J^*\underset{\sim}{a},b] \qquad ,$$

and find that

$$J^* = -\frac{\partial}{\partial t} + L^*(\tilde{u};\lambda) \qquad , \tag{2.3}$$

where

$$(L^*(\tilde{\underset{\sim}{u}};\lambda)\underset{\sim}{a})_i = -\tilde{\underset{\sim}{u}} \cdot \nabla a_i - a_j\Omega_{ji}(\tilde{\underset{\sim}{u}}) + a_jS_{ji}(\tilde{\underset{\sim}{u}}) - \lambda\Delta a_i \qquad .$$

---

*To obtain the adjoint we have made use of the Reynolds transport theorem in the form

$$\frac{d}{dt}\int_{\Omega(t)} \underset{\sim}{a}\cdot\bar{\underset{\sim}{b}}\, d\underset{\sim}{x} \int_{\Omega(t)} \partial(\underset{\sim}{a}\cdot\bar{\underset{\sim}{b}})/\partial t\, d\underset{\sim}{x} + \int_{\partial\Omega(t)} (\underset{\sim}{f}\cdot\underset{\sim}{n})\, \underset{\sim}{a}\cdot\underset{\sim}{b}\, d\underset{\sim}{x} \ .$$

The boundary integral vanishes since $\underset{\sim}{a} \cdot \bar{\underset{\sim}{b}} = 0$ when $\underset{\sim}{x} \in \partial\Omega$.

The adjoint eigenvalue problem is

$$-\bar{\gamma}\underset{\sim}{\zeta}^* + J^*\underset{\sim}{\zeta}^* + \nabla\Pi^* = 0 \ , \quad \text{div}\ \underset{\sim}{\zeta}^* = 0 \ , \quad \underset{\sim}{\zeta}^* = 0|_{\partial\Omega(t)} \quad . \tag{2.4}$$

Let $\underset{\sim}{\varphi}$ be any periodic admissible vector. Then

$$-\gamma[\underset{\sim}{\varphi},\underset{\sim}{\zeta}] + [\underset{\sim}{\varphi},J\underset{\sim}{\zeta}] = 0 \tag{2.5a}$$

and

$$-\bar{\gamma}[\underset{\sim}{\varphi},\underset{\sim}{\zeta}^*] + [\underset{\sim}{\varphi},J^*\underset{\sim}{\zeta}^*] = 0 \quad . \tag{2.5b}$$

Choose $\underset{\sim}{\varphi} = \underset{\sim}{\zeta}$ in (2.5b) and $\underset{\sim}{\varphi} = \bar{\underset{\sim}{\zeta}}^*$ in (2.5a) and use (2.2) to find that

$$[\underset{\sim}{\zeta},\bar{\underset{\sim}{\zeta}}^*] = 0 = [\underset{\sim}{\zeta}\ \underset{\sim}{\zeta}^*] \quad . \tag{2.6a}$$

By normalization

$$[\underset{\sim}{\zeta},\underset{\sim}{\zeta}^*] = [\bar{\underset{\sim}{\zeta}},\bar{\underset{\sim}{\zeta}}^*] = 1 \quad . \tag{2.6b}$$

When γ is real (2.6a) does not hold. Then it is sufficient to consider real-valued fields $\underset{\sim}{\zeta}$. Of course, we may always normalize as in (2.6b).

We next demonstrate that

$$[L_\lambda\underset{\sim}{\zeta},\underset{\sim}{\zeta}^*] = \frac{d\gamma}{d\lambda} \quad , \tag{2.7}$$

where

$$L_\lambda = \frac{d}{d\lambda} J = \frac{d}{d\lambda} L(\underset{\sim}{\tilde{u}}(\lambda),\lambda)$$

is a differential operator of order two. Let $\gamma' = \partial\gamma/\partial\lambda$. Then

$$-\gamma\underset{\sim}{\zeta}_\xi + J\underset{\sim}{\zeta}_\xi - \gamma'\underset{\sim}{\zeta} + L_\lambda\underset{\sim}{\zeta} + \nabla\Pi_\xi = 0$$

where $\underset{\sim}{\zeta}_\xi$ is solenoidal and vanishes on the boundary. Since

$$[\underset{\sim}{\zeta}^*,J\underset{\sim}{\zeta}_\xi] = [J^*\underset{\sim}{\zeta}^*,\underset{\sim}{\zeta}_\xi] = \gamma[\underset{\sim}{\zeta}^*,\underset{\sim}{\zeta}_\xi]$$

using (2.6), we find (2.7).

When $\omega_o \neq 0$ the solutions of the linearized problem (1.2 with $\varepsilon = 0$) are not necessarily periodic with period $2\Pi/\tilde{\omega}$ in t . If at criticality, $i\omega_o$ is a simple eigenvalue, there are two solutions of the linearized problem,

$$\underset{\sim}{Z}_1 = e^{-is}\underset{\sim}{\zeta}(\underset{\sim}{x},t,R_c) , \qquad \underset{\sim}{Z}_2 = \bar{Z}_1 \tag{2.8a}$$

with $s = \omega_o t$, and two solutions of the adjoint problem

$$\underset{\sim}{Z}_1^* = e^{-is}\underset{\sim}{\zeta}^*(\underset{\sim}{x},t,R_c) , \qquad \underset{\sim}{Z}_2^* = \bar{\underset{\sim}{Z}}_1^* \quad . \tag{2.8b}$$

These solutions are quasi-periodic with two frequencies.

For the quasi-periodic functions it is useful to introduce two times s and t corresponding to the two frequencies: the fixed forced frequency $\tilde{\omega}$ and the natural frequency $\omega(\epsilon)$. Associated with these quasi-periodic functions of two variables is the operator

$$\hat{J} = \frac{\partial}{\partial t}\Big|_{\underset{\sim}{x}} + L(\tilde{\underset{\sim}{u}},\lambda) = \omega\frac{\partial}{\partial s}\Big|_{t,x} + \frac{\partial}{\partial t}\Big|_{s,\underset{\sim}{x}} + L(\tilde{u},\lambda) \tag{2.9}$$

and the scalar product

$$[\![\underset{\sim}{a},\underset{\sim}{b}]\!] = \frac{1}{2\pi}\int_0^{2\pi} [\underset{\sim}{a},\underset{\sim}{b}]ds \quad . \tag{2.10}$$

When the domains of the operators $\hat{J}$, $[\![\cdot]\!]$ are restricted to the $2\pi/\tilde{\omega}$ periodic functions, these operators reduce to J and $[\cdot]$. The operator

$$\hat{J}^* = -\omega\frac{\partial}{\partial s} - \frac{\partial}{\partial t} + L^*(\tilde{\underset{\sim}{u}},\lambda)$$

is the adjoint to (2.9) relative to (2.10) over solenoidal vector fields vanishing on the boundary.

At criticality $\lambda = \lambda_o$, $\omega = \omega_o$, $J = J_o$ and $\hat{J} = \hat{J}_o$. It is readily verified that

$$\hat{J}_o\underset{\sim}{Z}_1 + e^{-is}\nabla\Pi = 0 \tag{2.11a}$$

and

$$\hat{J}_o^*\underset{\sim}{Z}_1^* + e^{-is}\nabla\Pi^* = 0 \quad , \tag{2.11b}$$

so that

$$[\![\underset{\sim}{Z}_i,\bar{\underset{\sim}{Z}}_j^*]\!] = \delta_{ij} \qquad (i,j = 1,2) \quad .$$

As a notational convenience we define

$$\|\underset{\sim}{a}\|_i \equiv \|\underset{\sim}{a}, \bar{Z}_i^*\| \tag{2.12}$$

for doubly periodic vector fields of s and t and

$$[\underset{\sim}{b}] \equiv [\underset{\sim}{b}, \bar{\underset{\sim}{\zeta}}^*] \tag{2.13}$$

for singly periodic vector fields $b(\underset{\sim}{x}, t)$ of t. If $\underset{\sim}{a}$ is a real-valued field then $\overline{\|\underset{\sim}{a}\|_1} = \|\underset{\sim}{a}\|_2$. For real-valued fields we shall take $\|\cdot\| \equiv \|\cdot\|_1$.

Lemma 1. Let $\underset{\sim}{b}(\underset{\sim}{x}, s, t)$ be a doubly periodic vector field with period 2π in s and $2\pi/\tilde{\omega}$ in t. The problem

$$\hat{J}_o \underset{\sim}{q} + \nabla p = \underset{\sim}{b}\,, \quad \text{div}\, \underset{\sim}{q} = 0\,, \quad \underset{\sim}{q} = 0|_{\partial\Omega(t)} \tag{2.14a}$$

can have doubly periodic solutions only if

$$\|\underset{\sim}{b}\|_i = 0 \quad . \tag{2.14b}$$

Proof.

$$\|\hat{J}_o \underset{\sim}{q}\|_i = \|\hat{J}_o \underset{\sim}{q}, \bar{Z}_i^*\| = \|\underset{\sim}{q}, \hat{J}_o^* \bar{Z}_i^*\| = 0 \quad .$$

The solvability Lemma 1 may be reduced to a solvability lemma involving only the $2\pi/\tilde{\omega}$ periodic functions by expanding $\underset{\sim}{b}$, $\underset{\sim}{q}$, and p into Fourier series

$$\begin{bmatrix} \underset{\sim}{b} \\ \underset{\sim}{q} \\ p \end{bmatrix} = \sum_{-\infty}^{\infty} e^{i\ell s} \begin{bmatrix} \underset{\sim}{b}^{(\ell)} \\ \underset{\sim}{q}^{(\ell)} \\ p^{(\ell)} \end{bmatrix} \quad . \tag{2.15}$$

The Fourier coefficients are $2\pi/\tilde{\omega}$ periodic functions of t. This expansion is always possible when s and t are independent; s and t are independent when $\omega(\epsilon)$ and $\tilde{\omega}$ are rationally independent.* Using the Fourier series and assuming

---

*Our basic decomposition into two times and the Fourier decomposition is not valid when $\omega(0)/\tilde{\omega}$ is rationally dependent.

rational independence, we find from (2.14a) that

$$i\ell\omega_o \underset{\sim}{q}^{(\ell)} + J_o \underset{\sim}{q}^{(\ell)} + \nabla p^{(\ell)} = \underset{\sim}{b}^{(\ell)} \ , \quad \text{div } \underset{\sim}{q}^{(\ell)} = 0 \quad , \qquad (2.15a,b)$$

$$q^{(\ell)} = 0_{|\partial\Omega(t)} \ , \quad q^{(\ell)}(\underset{\sim}{x},t) = q^{(\ell)}(\underset{\sim}{x},t + 2\Pi/\tilde{\omega}) \quad , \qquad (2.15c,d)$$

and from (2.14b) that

$$[\underset{\sim}{b}^{(1)}] = \underset{\sim}{b}^{(-1)}] = 0 \quad . \qquad (2.15e)$$

To find more general solvability conditions than (2.15e) which will also clearly bring out the problem of small divisors, we note that the eigenvalue problem for J_o implies that

$$-i(\omega_o - n\tilde{\omega})\underset{\sim}{f}_n + J_o \underset{\sim}{f}_n + e^{i\tilde{\omega}nt}\nabla\Pi = 0 \quad ,$$

where

$$\underset{\sim}{f}_n = e^{in\tilde{\omega}t}\underset{\sim}{\zeta}(\underset{\sim}{x},t) \ , \qquad n = 0, \pm 1, \pm 2, \ldots \quad ,$$

and $\underset{\sim}{f}_n$ satisfies (2.15b,c d). The adjoint eigenfunction is

$$\underset{\sim}{f}_n^* = e^{-in\tilde{\omega}t}\underset{\sim}{\zeta}^*(\underset{\sim}{x},t) \quad ,$$

and

$$i(\omega_o - n\tilde{\omega})\underset{\sim}{f}_n^* + J_o^* f_n^* + e^{-i\tilde{\omega}nt}\nabla\Pi^* = 0 \quad . \qquad (2.16)$$

Comparison of (2.16) and its complex conjugate with (2.15) shows that

$$[\underset{\sim}{b}^{(\ell)}, \underset{\sim}{f}_n^*] = i(\ell\omega_o + \omega_o - n\tilde{\omega})[\underset{\sim}{q}^{(\ell)}, \underset{\sim}{f}_n^*] \qquad (2.17a)$$

and

$$[\underset{\sim}{b}^{(\ell)}, \bar{\underset{\sim}{f}}_n^*] = i(\ell\omega_o - \omega_o + n\tilde{\omega}[\underset{\sim}{q}^{(\ell)}, \bar{\underset{\sim}{f}}_n^*] \quad . \qquad (2.17b)$$

Equations (2.17a) and (2.17b) may be regarded as conditions on the amplitude of $\underset{\sim}{q}^{(\ell)}$. Given non-zero values on the left side, the ℓth iterate $\underset{\sim}{q}^{(\ell)}$ will be unbounded when one of the numbers $\ell\omega_o \pm \omega_o \mp n\tilde{\omega} \to 0$.

Lemma 2. Suppose that $\omega_o \neq 0$ and

$$\ell\omega_o \pm \omega_o \mp n\tilde{\omega} = 0 \quad , \tag{2.18}$$

for some ℓ and n where

$$\ell \text{ or } n = 0, \pm 1, \pm 2, \ldots \quad .$$

Then (2.15a,b,c,d) cannot be solved unless

$$[\underset{\sim}{b}^{(\ell)}, e^{-in\tilde{\omega}t}\bar{\underset{\sim}{\zeta}}^*] = 0 \quad , \tag{2.19a}$$

and

$$[\underset{\sim}{b}^{(\ell)}, e^{-in\tilde{\omega}t}\underset{\sim}{\zeta}^*] = 0 \quad . \tag{2.19b}$$

Remark. In our construction we shall consider real-valued vector fields $\underset{\sim}{b}$. For these

$$\underset{\sim}{b}^{(\ell)} = \bar{\underset{\sim}{b}}^{(-\ell)}$$

and (2.18) and (2.19a) imply (2.19b).

When $\tilde{\omega} = 0$ (2.19a) reduces to

$$[\underset{\sim}{b}^{(1)}, \bar{\underset{\sim}{\zeta}}^*] = 0 \quad .$$

This was the case treated in (JS), and (2.19) is then a sufficient as well as a necessary condition for the solvability of (2.15).

Equation (2.18) cannot be satisfied when ω_o and $\tilde{\omega}$ are rationally independent. However, in this case we may always choose integers ℓ and n to make $\ell\omega_o \pm \omega_o \mp n\tilde{\omega}$ arbitrarily small. This is the small divisor problem, and a bounded inverse for all values of ℓ is not to be expected.

Only slight changes of interpretation of Lemmas 1 and 2 are needed when $\omega_o = 0$. In this case $\gamma(R_c) = 0$, and we may assume that the eigenfunctions of (1.4) and (2.4) are real-valued. Thus,

$$\underset{\sim}{Z}_1 = \bar{\underset{\sim}{Z}}_1 = \underset{\sim}{\zeta}, \quad \underset{\sim}{Z}_2 = \bar{\underset{\sim}{Z}}_2 = \underset{\sim}{\zeta}^* \quad .$$

When $\omega_o = 0$ we define

$$[\underset{\sim}{u}] = [\underset{\sim}{u}, \underset{\sim}{\zeta}^*]$$

and find that (2.6), (2.7) and (2.13) hold with the new understanding.

§3. QUASI-PERIODIC BIFURCATIONS OF PERIODIC SOLUTIONS

It will be convenient to treat the more general problem of almost periodic bifurcations (case (b) first. The simpler problem (a) in which the bifurcating solution is time periodic is considered in Section 5.

The solution of the bifurcation problem (1.2) which we are going to construct can be called a quasi-periodic solution with two frequencies: the fixed forced frequency $\tilde{\omega}$ and the natural frequency $\omega(\epsilon)$. The natural frequency is introduced at the point of instability and bifurcation of the forced periodic solution $\underset{\sim}{u}(\underset{\sim}{x},t)$. When $\epsilon = 0$ and $\lambda = \lambda_o = 1/R_c$, we get an infinitesimal almost periodic solution

$$\epsilon \underset{\sim}{u}_o = 2\epsilon \, \mathrm{re}(\underset{\sim}{Z}_1) = \epsilon(e^{-is}\underset{\sim}{\zeta}(\underset{\sim}{x},t) + e^{+is}\bar{\underset{\sim}{\zeta}}(\underset{\sim}{x},t)) \quad . \tag{3.1}$$

A crucial part of our construction of the quasi-periodic bifurcating solution for small $\epsilon \neq 0$ is the Poincare mapping

$$\omega(\epsilon)t = s \quad . \tag{3.2}$$

To understand the role of this mapping, we consider the case $\tilde{\omega} = 0$. This leads to time periodic bifurcations of steady solutions, a problem which is thoroughly treated in (JS). In the case $\tilde{\omega} = 0$, one constructs a power series solution of the following type:

$$\begin{bmatrix} \underset{\sim}{u}(\underset{\sim}{x},s\ \epsilon) \\ p(\underset{\sim}{x},s,\epsilon) \\ \lambda(\epsilon) \\ \omega(\epsilon) \end{bmatrix} = \sum_{n=0} \epsilon^n \begin{bmatrix} \underset{\sim}{u}_n(\underset{\sim}{x},s) \\ p_n(\underset{\sim}{x},s) \\ \lambda_n \\ \omega_n \end{bmatrix} \tag{3.3}$$

where, for example,

$$\underset{\sim}{u}_n(x,s) = \sum_{\ell=-N}^{N} \underset{\sim}{u}_{n\ell}(x)\, e^{i\ell s} \tag{3.4}$$

is a 2π periodic Fourier polynomial. The Poincare mapping is crucial here because in terms of the variable t,

$$\underset{\sim}{u}_n(\underset{\sim}{x},\epsilon,t) = \sum_{\ell=-N}^{N} \underset{\sim}{u}_{n\ell}(\underset{\sim}{x})\, e^{i\ell t \Sigma \epsilon^n \omega_n} \tag{3.5}$$

cannot be approximated by its power series in ϵ uniformly in time.

Now we shall formally construct a quasi-periodic bifurcating series solution with two frequencies. The solution to be constructed is in the form

$$\begin{bmatrix} \underset{\sim}{u}(\underset{\sim}{x},s,t,\epsilon) \\ p(x,s,t,\epsilon) \\ \lambda(\epsilon) \\ \omega(\epsilon) \end{bmatrix} = \sum_{n=0} \epsilon^n \begin{bmatrix} \underset{\sim}{u}_n(\underset{\sim}{x},s,t) \\ p_n(x,s,t) \\ \lambda_n \\ \omega_n \end{bmatrix} \tag{3.6}$$

where (3.2) holds and ϵ is defined by the projection

$$\epsilon = [\epsilon \underset{\sim}{u}]$$

or

$$[\underset{\sim}{u}] = [\underset{\sim}{u}, \underset{\sim}{Z}_1^*(\underset{\sim}{x},s,t)] = 1 \quad . \tag{3.7}$$

When $\epsilon = 0$ and $\lambda = \lambda_o = 1/R_c$, we get an infinitesimal almost periodic solution

$$\epsilon \underset{\sim}{u}_o = 2\epsilon\, \mathrm{re}(\underset{\sim}{Z}_1) = \epsilon(e^{-is}\underset{\sim}{\zeta}(\underset{\sim}{x},t) + e^{+is}\bar{\underset{\sim}{\zeta}}(\underset{\sim}{x},t)) \quad . \tag{3.8}$$

We shall find the $\underset{\sim}{u}_n(x,s,t)$ (and p_n) in the form

$$\underset{\sim}{u}_n = \sum_{\ell=-N_n}^{N_n} e^{i\ell s} \underset{\sim}{u}_{n\ell}(\underset{\sim}{x},t) \quad , \tag{3.9}$$

where $\underset{\sim}{u}_{n\ell} = \bar{\underset{\sim}{u}}_{n(-\ell)}$ and $\underset{\sim}{u}_{n\ell}$ is $2\pi/\tilde{\omega}$ periodic in t.

There are two times in this problem: t with period $2\pi/\tilde{\omega}$ and $s = \omega(\epsilon)t$ with period 2π. We use the notion of two times because it is essential that the period of the natural oscillation $2\pi/\omega(\epsilon)$ be mapped into the fixed 2π periodic (in s) domain, whereas the period $2\pi/\tilde{\omega}$ is given externally and is fixed.

With these preliminaries aside, we are ready to construct the series (3.5) solution of

$$\hat{J}\underset{\sim}{u} + \epsilon\underset{\sim}{u} \cdot \nabla\underset{\sim}{u} + \nabla p = 0 \ , \quad \text{div}\ \underset{\sim}{u} = 0 \ , \quad \underset{\sim}{u} = 0_{\partial\Omega} \ , \quad [\![\underset{\sim}{u}]\!] = 1 \quad . \tag{3.10}$$

When $\epsilon = 0$ we have $\hat{J}_o\underset{\sim}{u}_o + \nabla p = 0$, $[\![\underset{\sim}{u}_o]\!] = 1$ and the other conditions. The solution is given by (3.8). At order m we must solve the problem

$$\hat{J}_o\underset{\sim}{u}_m + \omega_m \frac{\partial \underset{\sim}{u}_o}{\partial s} + L_m\underset{\sim}{u}_o + \underset{\sim}{F}_m + \nabla p_m = 0 \quad , \tag{3.11a}$$

$$\text{div}\ \underset{\sim}{u}_m = 0 \ , \quad \underset{\sim}{u}_m = 0_{\partial\Omega(t)} \quad \text{and} \quad [\underset{\sim}{u}_m] = 0 \quad . \tag{3.11b,c,d}$$

Here,

$$\begin{aligned} L_m &= \frac{1}{m!}\frac{\partial^m}{\partial\lambda^m} L = (\lambda_1 L_\lambda)_{m-1} = \sum_{\ell+\nu=m-1} \lambda_{\ell+1}(L_\lambda)_\nu \quad , \\ \underset{\sim}{F}_m &= \sum_{\ell+j=m-1} (\underset{\sim}{u}_\ell\cdot\nabla)\underset{\sim}{u}_j + \tilde{\sum_{\ell+j=m}} L_\ell \underset{\sim}{u}_j + \tilde{\sum_{\ell+j=m}} \omega_\ell \frac{\partial \underset{\sim}{u}_j}{\partial s} \quad , \end{aligned} \tag{3.12}$$

and the summation with the tilda overbar denotes a sum over all non-negative integers $\ell + j = m$ minus terms of order m . The unknowns of problem (3.11) are $\underset{\sim}{u}_m$, p_m, λ_m and ω_m . The vector $\underset{\sim}{F}_m$ contains only terms known from lower order computation (order $\ell < m$).

The unknown coefficients λ_m and ω_m are determined by application of the solvability condition of Lemma 2. First, we note that (see 2.7)

$$[\![L_\lambda \underset{\sim}{u}_o]\!] = [\![L_\lambda \underset{\sim}{\zeta}, \underset{\sim}{\zeta}^*]\!] = \gamma' \quad .$$

The condition $\xi(\lambda_o) > 0$ means that the periodic solution loses stability as R is increased past R_c . Noting now that

$$[\![\frac{\partial \underset{\sim}{u}_o}{\partial s}]\!] = -i$$

and using (2.4b), we have

$$-i\omega_m + [\![L_m \underset{\sim}{u}_o]\!] + [\![\underset{\sim}{F}_m]\!] = 0 \tag{3.13}$$

where

$$\| L_m \underset{\sim}{u}_o \| = \lambda_m \gamma' + \overset{\sim}{\underset{\ell+\nu=m-1}{\Sigma}} \lambda_{\ell+1} \| (L_\lambda) \underset{\nu}{u}_o \| \qquad . \tag{3.14}$$

We see that the real and imaginary parts of (3.13) determine the values λ_m and ω_m :

$$-\lambda_m \xi' = \mathrm{re} \left\{ \| \underset{\sim}{F}_m \| + \overset{\sim}{\underset{\ell+\nu=m-1}{\Sigma}} \lambda_{\ell+1} \| (L_\lambda) \underset{\nu}{u} \|_o \right\}$$

and

$$\omega_m - \lambda_m \; \mathrm{im} \; \gamma' = \mathrm{im} \left\{ \| \underset{\sim}{F}_m \| + \overset{\sim}{\underset{\ell+\nu=m-1}{\Sigma}} \lambda_{\ell+1} \| (L_\lambda) \underset{\nu}{u}_o \| \right\} \qquad .$$

For $m = 1$ we find

$$\underset{\sim}{F}_1 = \underset{\sim}{u}_o \cdot \nabla \underset{\sim}{u}_o$$

and

$$\lambda_1 = \omega_1 = 0 \qquad . \tag{3.15}$$

When $m = 2$ we have

$$\lambda_2 = -\mathrm{re} \| \underset{\sim}{F}_2 \| / \xi' \tag{3.16a}$$

and

$$\omega_2 - \lambda_2 \; \mathrm{im} \; \gamma' = \mathrm{im} \| F_2 \| \qquad . \tag{3.16b}$$

We next want to show that quasi-periodic solutions of (3.12) are in the form (3.9) and also to give conditions under which the $\underset{\sim}{u}_{n\ell}(\underset{\sim}{x},t)$ may be uniquely determined. We may write (3.12) as

$$\hat{J}_o \underset{\sim}{a} + \nabla p = \underset{\sim}{b} \; , \quad \mathrm{div} \; \underset{\sim}{a} = 0 \; , \quad \underset{\sim}{a} = 0 \big|_{\partial\Omega} \; , \quad \| \underset{\sim}{a} \| = 0 \qquad . \tag{3.17a,b,c,d}$$

We argue by induction. Assuming that (3.8) holds for $\underset{\sim}{u}_m = \underset{\sim}{a}$ when $m < n$, we have $\underset{\sim}{b}$ in the form

$$\underset{\sim}{b} = \Sigma \; e^{i\ell s} \underset{\sim}{b}_\ell(\underset{\sim}{x},t) \qquad ,$$

where the summation is over a finite number of terms and $\underset{\sim}{b}_\ell(x,t)$ is $2\pi/\tilde{\omega}$ periodic in t . The induction may be started because $\underset{\sim}{b}_1 = \underset{\sim}{u}_o \cdot \nabla \underset{\sim}{u}_o$ is in the required form. Since (3.17) is a linear problem, the full solution is obtained as a sum of the solutions of problems with

$$\underset{\sim}{b} = e^{i\ell s} \underset{\sim}{b}_\ell(\underset{\sim}{x}, t) \quad . \tag{3.18}$$

Noting that through $L(\tilde{\underset{\sim}{u}};\lambda)$ the operator $\hat{\underset{\sim}{J}}_o$ is $2\pi/\omega$ periodic in t we seek quasi-periodic solutions of (3.17) with $\underset{\sim}{b}$ in the form (3.18) as Fourier polynomials

$$\underset{\sim}{a} = \Sigma\, e^{i\ell s} \underset{\sim}{a}_\ell(\underset{\sim}{x}, t) \ , \quad p = \Sigma\, e^{i\ell s} p_\ell(\underset{\sim}{x}, t)$$

where a_ℓ, p_ℓ are $2\pi/\tilde{\omega}$ periodic in t . Substituting (3.18) into (3.17), we arrive at the system of equations considered in Lemma 2.

$$[i\ell\omega + \frac{\partial}{\partial t} + L(\tilde{\underset{\sim}{u}};\lambda)]\ \underset{\sim}{a}_\ell + \nabla p = \underset{\sim}{b} \quad , \tag{3.19a}$$

$$\text{div}\ \underset{\sim}{a}_\ell = 0 \ , \quad \underset{\sim}{a}_\ell = 0\big|_{\partial\Omega} \ , \quad [\underset{\sim}{a}_\ell]_i = 0 \quad . \tag{3.19b,c,d}$$

We note that the reduction of the quasi-periodic problem (3.17) to the periodic problem (3.19) can be carried out even when $\omega(\epsilon)$ and ω are rationally dependent. Then, for a dense set of values ϵ near $\epsilon = 0$, our quasi-periodic solution is periodic with period $2\pi/\omega$. What is crucial here, however, is that $\omega(0) = \omega_o$, and ω should be rationally independent.

Assuming the validity of the series construction, we can prove the following lemma:

<u>Lemma 3</u>. $\lambda(\epsilon)$ <u>and</u> $\omega(\epsilon)$ <u>are even functions</u>.

It follows from this that in the case of a simple eigenvalue, the bifurcation of periodic solutions into quasi-periodic solutions with two frequencies is one-sided. To prove the lemma we shall consider polynomials of Floquet type

$$\underset{\sim}{G} = \sum_\ell e^{i\ell s} \underset{\sim}{g}_\ell(\underset{\sim}{x}, t) \quad ,$$

where $\underset{\sim}{g}_\ell(\underset{\sim}{x}, t)$ is periodic in t with period $2\pi/\tilde{\omega}$ and $\omega/\tilde{\omega}(\epsilon)$ is not rational. We call $\underset{\sim}{G}$ an even (odd) polynomial if the integers are even (odd). If $\underset{\sim}{G}$ is an even polynomial, then

$$[\underset{\sim}{G}] = 0 \quad .$$

By the nature of the perturbation construction, $\underset{\sim}{u}_m$ is a polynomial of the Floquet type. The hypothesis of the induction is that

$$\begin{aligned} &\underset{\sim}{u}_{2\ell} \text{ is an odd polynomial} \quad , \\ &\underset{\sim}{u}_{2\ell+1} \text{ is an even polynomial} \quad , \end{aligned} \tag{3.20a}$$

and

$$\lambda_{2\ell+1} = \omega_{2\ell+1} = 0 \tag{3.20b}$$

when $\ell < m$. When $\ell = 0$ we find that $\underset{\sim}{u}_0 = e^{-is}\underset{\sim}{\zeta} + e^{is}\bar{\underset{\sim}{\zeta}}$ is an odd polynomial. Since $\underset{\sim}{F}_1 = \underset{\sim}{u}_0 \cdot \nabla\underset{\sim}{u}_0$ is an even polynomial, $[\underset{\sim}{F}_1] = 0$ and by (3.15) $\lambda_1 = \omega_1 = 0$. Then $L_1 = 0$ and the only inhomogeneous term in (3.20a) when $m = 1$ is the even polynomial $\underset{\sim}{F}_1$. It follows that $\underset{\sim}{u}_1$ is an even polynomial and the hypotheses (3.20) of the induction are true when $m = 1$. Noting now that

$$L_{2\ell+1} = (\lambda_1 L_\lambda)_{2\ell} = \lambda_{2\ell+1}L_\lambda + \lambda_{2\ell}\lambda_1 L_{\lambda\lambda} + --- \quad ,$$

we verify that if $\lambda_{2\ell+1} = 0$ for $\ell \leq m$, $L_{2\ell+1} = 0$ for $\ell \leq m$. Given (3.20a,b) and $L_{2\ell+1}$ for $\ell \leq m$, we verify that $\underset{\sim}{F}_{2m}$ is odd. Returning now to (3.12a) we note that the inhomogeneous terms in the equation governing $\underset{\sim}{u}_{2m}$ are odd polynomials; hence $\underset{\sim}{u}_{2m}$ is odd. Now evaluating (3.13) with $m = 2m+1$, $[\underset{\sim}{F}_{2m+1}] = 0$, and

$$[L_{2m+1}\underset{\sim}{u}_0] = \lambda_{2m+1}[L_\lambda \underset{\sim}{u}_0] = \lambda_{2m+1}\left(\frac{d\lambda}{d\xi}\right)^{-1} \quad ,$$

we get

$$\omega_{2m+1} = \lambda_{2m+1} = 0 \quad .$$

Returning again to (3.12a) with m replaced by $2m+1$, we note that the inhomogeneous terms are even; hence $\underset{\sim}{u}_{2m+1}$ is even.

§4. STABILITY OF THE QUASI-PERIODIC BIFURCATING SOLUTION WITH TWO FREQUENCIES

In trying to determine the stability of the almost periodic motion with two frequencies, we are led to a generalization of Floquet theory which develops an idea suggested by Landau.* Let $\underset{\sim}{\nu}(\underset{\sim}{x},s,t;\epsilon)$ be a small disturbance of $\epsilon\underset{\sim}{u}$. Then

$$\omega\frac{\partial\underset{\sim}{\nu}}{\partial s} + \frac{\partial\underset{\sim}{\nu}}{\partial t} + L(\lambda;t)\underset{\sim}{\nu} + \epsilon(\underset{\sim}{u}\cdot\nabla\underset{\sim}{\nu} + \underset{\sim}{\nu}\nabla\underset{\sim}{u}) + \nabla p' = 0 \tag{4.1a}$$

$$\text{div}\ \underset{\sim}{\nu} = 0\ , \quad \underset{\sim}{\nu} = 0\big|_{\partial\Omega} \quad . \tag{4.1b,c}$$

The coefficients of (4.1a) are almost periodic functions with two frequencies. Landau (1944; 1959, p. 106) suggests that we look for solutions of (4.1) of the form

$$\underset{\sim}{u} = e^{-\sigma s}\underset{\sim}{\Gamma}(\underset{\sim}{x},s,t;\epsilon)\ , \quad p' = e^{-\sigma s}p(x,s,t;\epsilon) \tag{4.2}$$

where $\sigma = \sigma(\lambda)$ is a (possibly) complex number, called the Floquet exponent, and $\underset{\sim}{\Gamma}$ and p' are 2π periodic in $s = \omega(\epsilon)t$ and $2\pi/\tilde{\omega}$ periodic in t. With the "two" times so introduced, the rest of the argument is essentially that given in Section 5 of (JS).

Substitution of (4.2) into (4.1) leads to

$$-\sigma\omega(\epsilon)\underset{\sim}{\Gamma} + \omega\frac{\partial\underset{\sim}{\Gamma}}{\partial s} + \frac{\partial\underset{\sim}{\Gamma}}{\partial t} + L\,\underset{\sim}{\Gamma} + \epsilon(\underset{\sim}{u}\cdot\nabla\underset{\sim}{\Gamma} + \underset{\sim}{\Gamma}\cdot\nabla\underset{\sim}{u}) + \nabla p = 0 \quad . \tag{4.3}$$

When $\epsilon = 0$, $\omega(0) = \omega_o$, $\sigma = \sigma_o$, $\lambda = \lambda_o$, $L = L_o$ and two solutions of (4.3) are given by

$$\sigma_o = 0 \ \text{ and } \ \Gamma = e^{-is}\zeta(x,t)\ , \ \text{ or } \ \bar{\Gamma} = e^{is}\bar{\zeta}(x,t) \quad .$$

Thus for $\epsilon = 0$, there is a double Floquet exponent at the origin. We get all the Floquet exponents by looking for solutions of the form $\underset{\sim}{\Gamma}_o(\underset{\sim}{x},s,t) = e^{iks}\underset{\sim}{\Phi}(x,t)$, where k is an integer and $\underset{\sim}{\Phi}$ is $2\pi/\tilde{\omega}$ periodic in t. This leads us to the eigenvalue problem

$$(-\sigma\omega_o + ik)\underset{\sim}{\Phi} + J_o\underset{\sim}{\Phi} + \nabla p = 0 \quad . \tag{4.4}$$

---

*See the extended quote by Landau at the end of this section.

Since all eigenvalues $\gamma = \xi \pm i\omega_o$, except the two for which $\xi = 0$ are positive $(\xi > 0)$ at criticality and

$$-\gamma\underset{\sim}{\zeta} + J_o\underset{\sim}{\zeta} + \nabla p = 0 \quad ,$$

we have

$$\sigma = ik/\omega_o + \gamma \quad , \tag{4.5}$$

and all other Floquet exponents have positive real parts and lead to stability.

Now consider the case $\epsilon = 0$. Assuming that the perturbation theory holds for the Floquet exponents, we have only to worry about the critical ones, namely, those at the origin. One solution of (4.1)

$$\underset{\sim}{v} = \frac{\partial \underset{\sim}{u}}{\partial s} , \quad \sigma = 0$$

always exists even with $\epsilon \neq 0$. To check this, differentiate (3.10) with respect to s (holding t constant)

$$\omega \frac{\partial^2 \underset{\sim}{u}}{\partial s^2} + L(\tilde{\underset{\sim}{u}},\lambda) \frac{\partial \underset{\sim}{u}}{\partial s} + \epsilon \underset{\sim}{u} \cdot \nabla \frac{\partial \underset{\sim}{u}}{\partial s} + \epsilon \frac{\partial \underset{\sim}{u}}{\partial s} \cdot \nabla \underset{\sim}{u} + \nabla \frac{\partial p}{\partial s} = 0 \quad . \tag{4.6}$$

Therefore, of the two eigenvalues $\sigma = 0$ at the origin when $\epsilon = 0$, one remains at the origin even when $\epsilon \neq 0$.

The problem now is to calculate the second Floquet exponent. We look for an eigenfunction of (4.3) in the form

$$\underset{\sim}{\Gamma}(\underset{\sim}{x}, s, \epsilon) = a(\epsilon)\frac{\partial \underset{\sim}{u}}{\partial s} + \underset{\sim}{\gamma}(\underset{\sim}{x}, s, \epsilon) \quad , \tag{4.7}$$

where

$$\underset{\sim}{\gamma}(\underset{\sim}{x}, s, \epsilon) = \underset{\sim}{u}_o(\underset{\sim}{x}, s) + \epsilon \underset{\sim}{\gamma}_1(\underset{\sim}{x}, s, \epsilon)$$

and $a(\epsilon)$ is a coefficient to be determined. Substituting (4.7) into (4.3) and using (4.6), we get for a, σ, and $\underset{\sim}{\gamma}$ the equation

$$\omega \frac{\partial \underset{\sim}{\gamma}}{\partial s} - \sigma\omega\underset{\sim}{\gamma} + \frac{\partial \underset{\sim}{\gamma}}{\partial t} + L\underset{\sim}{\gamma} + \epsilon(\underset{\sim}{u} \cdot \nabla\underset{\sim}{\gamma} + \underset{\sim}{\gamma} \cdot \nabla\underset{\sim}{u}) - \sigma\omega a \frac{\partial \underset{\sim}{u}}{\partial s} + \nabla p = 0 \quad . \tag{4.8}$$

We again seek a solution in series

$$\begin{bmatrix} \underset{\sim}{\gamma}(\underset{\sim}{x},s,\epsilon) \\ p(\underset{\sim}{x},s,\epsilon) \\ \sigma(\epsilon) \\ a(\epsilon) \end{bmatrix} = \sum_{\ell=0} \epsilon^{\ell} \begin{bmatrix} \underset{\sim}{\gamma}_{\ell}(\underset{\sim}{x},s) \\ p_{\ell}(\underset{\sim}{x},s) \\ \sigma_{\ell} \\ a_{\ell} \end{bmatrix}$$

We shall require, as in the construction of the quasi-periodic solutions that

$$[\underset{\sim}{\gamma}_o] = 1 , \quad [\underset{\sim}{\gamma}_{\ell}] = 0 , \quad \ell > 0 \quad .$$

As in Section 3, we get

$$-\sigma_0\omega_0(a_0\frac{\partial \underset{\sim}{u}_0}{\partial s} + \underset{\sim}{\gamma}_o) + J_o\underset{\sim}{\gamma}_o + \nabla p_0 = 0 \quad ,$$

$$\text{div } \underset{\sim}{\gamma}_o = 0 , \quad \underset{\sim}{\gamma}_o = 0|_{\partial\Omega} \quad .$$

Using (3.11) and (4.7), we get

$$-\sigma_0\omega_0(-ia_0 + 1) = 0 \quad . \tag{4.9}$$

Hence, $\sigma_0 = 0$ and $\underset{\sim}{\gamma}_o = \underset{\sim}{u}_0$.

At first order we must solve the problem

$$-\sigma_0\omega_0(a_0\frac{\partial \underset{\sim}{u}_0}{\partial s} + \underset{\sim}{u}_0) + J_o\underset{\sim}{\gamma}_1 + 2\underset{\sim}{u}_0 \cdot \nabla\underset{\sim}{u}_0 + \nabla p_1 = 0 \tag{4.10}$$

and

$$\text{div } \underset{\sim}{\gamma}_1 = 0 , \quad \underset{\sim}{\gamma}_1 = 0|_{\partial\Omega} \text{ and } [\underset{\sim}{\gamma}_1] = 0 \quad .$$

Applying the orthogonality conditions to (4.10), we get $\sigma_1 = 0$ since $[\underset{\sim}{u}_0 \cdot \nabla\underset{\sim}{u}_0] = 0$. Now, comparing (4.10) for $\underset{\sim}{\gamma}_1$ with (3.12) and (3.15) for $\underset{\sim}{u}_1$, we see that $\underset{\sim}{\gamma}_1 = 2\underset{\sim}{u}_1$, since $[\underset{\sim}{u}_1] = [\underset{\sim}{\gamma}_1] = 0$. This relation enables us to establish the stability of supercritical bifurcation and the instability of subcritical bifurcation.

At second order

$$J_o\underset{\sim}{\gamma}_2 - \sigma_0\omega_0(a_0\frac{\partial \underset{\sim}{u}_0}{\partial s} + \underset{\sim}{u}_0) + \omega_2\frac{\partial \underset{\sim}{u}_0}{\partial s} - \lambda_2\Delta\underset{\sim}{u}_0 + \underset{\sim}{T}_2 + \nabla p_2 = 0 \quad , \tag{4.11}$$

$$\text{div } \underset{\sim}{\gamma}_2 = 0 \quad \underset{\sim}{\gamma}_2 = 0|_{\partial\Omega} \text{ and } [\underset{\sim}{\gamma}_2] = 0 \quad ,$$

where, since $\underset{\sim}{\gamma}_1 = \underset{\sim}{u}_1$

$$\underset{\sim}{T}_2 = 3(\underset{\sim}{u}_1 \cdot \nabla \underset{\sim}{u}_o + \underset{\sim}{u}_o \cdot \nabla \underset{\sim}{u}_1) = 3\underset{\sim}{F}_2 \quad ,$$

where $\underset{\sim}{F}_2$ is given by (3.13) using (3.15).

Applying the solvability conditions to (4.11) using (3.14)

$$-\sigma_o \omega_o (1 - i a_o) - i\omega_2 + \lambda_2 \gamma' + 3[\underset{\sim}{F}_2] = 0 \quad .$$

Taking real and imaginary parts, we get using (3.16a,b)

$$\sigma_2 \omega_o + 2\lambda_2 \xi' = 0 \quad ,$$

$$\sigma_2 \omega_o a_o + 2\omega_2 - \lambda_2 \text{ im } \gamma' = 0 \quad .$$

Hence,

$$\sigma_2 = -\frac{2\lambda_2 \xi'}{\omega_o} \tag{4.12}$$

and assuming $\lambda_2 \neq 0$,

$$a_o = -\frac{2\omega_2 - \lambda_2 \text{im}\gamma'}{\sigma_2} \quad .$$

Equation (4.12) contains an important result. Recall that solutions of the linearized equation decay when $\sigma > 0$ and that $\xi' > 0$.

Subcritical quasi-periodic motions with two frequencies $(\lambda_2 > 0)$ are unstable $(\sigma_2 > 0)$ and supercritical motions $(\lambda_2 > 0)$ are stable $(\sigma_2 > 0)$ in the linearized theory.

The same formal perturbation construction which was just used for quasi-periodic bifurcations of periodic solutions could be applied to the problem of bifurcation of solutions with n frequencies into solutions with $n + 1$ frequencies. Our construction is close to Landau's sketch of the way repeated bifurcation leads to turbulence. In this process new frequencies are introduced by instability when the Floquet exponents are complex. In Landau's view, a steady motion loses its stability to time periodic motion. (In the words of Landau (1944, p. 343),

"As Re is further increased, this periodic motion, too eventually becomes unsteady. The investigation of its unsteadiness should be conducted in a manner analogous to that described above. The role of the principal motion is now played by the periodic motion $\underset{\sim}{v}_o(\underset{\sim}{x},t)$ of frequency ω_1 . Substituting $\underset{\sim}{v} = \underset{\sim}{v}_o + \underset{\sim}{v}_2$ with small $\underset{\sim}{v}_2$ into the equation of motion, we shall again obtain for $\underset{\sim}{v}_2$ a linear equation, but this time the coefficients of this equation are not only functions of the coordinates, but of time also; with respect to time, they are are periodic functions with a period $2\pi/\omega_1$. The solution of such an equation should be sought in the form $\underset{\sim}{v}_2 = \underset{\sim}{\Pi}(\underset{\sim}{x},t)e^{-\Omega t}$ where $\underset{\sim}{\Pi}(\underset{\sim}{x},t)$ is a periodical function of time (with a period $2\pi/\omega_1$) . Unsteadiness sets in again when the frequency $\Omega_2 = \omega_2 + i\gamma_2$ turns up whose imaginary part γ_2 is positive and the corresponding real part ω_2 determines then the newly appearing frequency."

"The result is a quasi-periodic motion characterized by two different periods. It involves two arbitrary quantities (phases), i.e., has two degrees of freedom.

"In course of the further increase of the Reynolds number, new and new periods appear in succession, and the motion assumes an involved character typical of a developed turbulence. For every value of Re the motion has a definite number of degrees of freedom; in the limit, as Re tends to infinity, the number of degrees of freedom becomes likewise infinitely large."

Unfortunately, this very plausible description of transition to turbulence encounters difficulties when one tries to make it precise.* Though our construction is natural for Landau's conjectures (and works when $\tilde{\omega} = 0$), it cannot be generally carried out when $\tilde{\omega}/\omega_o$ is rational, and it leads to small divisors when $\tilde{\omega}/\omega_o$ is irrational (see Lemma 2). The first difficulty is perhaps superficial since in the rationally dependent case, the basic decomposition into two times does not follow from the solution at zeroth order. In the rationally independent case, the possibility remains that our series is asymptotic and that an asymptotic series

---

*It is argued in (JS) that the instability of subcritical bifurcating periodic is inconsistent with a bifurcation description of transition to turbulence. In the subcritical case the disturbances escape the domain of attraction of the basic flow and must snap through the small norm unstable, bifurcating, periodic solution to a solution with a larger norm. It follows that the Landau-Hopf conjecture can hold only in the case of supercritical bifurcations. We have seen that the conjecture may fail also in the supercritical case. This failure can stem from considerations associated with the solvability Lemma 2 and with small divisors. A similar argument against the Landau-Hopf conjecture has recently been given by D. Ruelle and F. Takens (1971).

for $\omega(\epsilon)$ has a physical relevance. In reservation, we note that any result which depends in a strong way on rational independence must be questioned on physical grounds, and in the rational case the series cannot be constructed unless exceptional special conditions are satisfied (Lemma 2).

§5. THE PERIODIC BIFURCATING SOLUTIONS AND THEIR STABILITY

In the previous section we assumed that the basic periodic solution with frequency ω loses stability by a complex pair of eigenvalues γ of (1.4) crossing the $\mathrm{im}\,\gamma = \eta(R_c) = \omega_0$ axis from the right. This was the case (b) (complex eigenvalue) of the introduction, and the solution which bifurcates off is quasi-periodic. Now we consider the easier case (a) in which the basic periodic solution loses its stability when a simple real eigenvalue $\gamma = \xi$ crosses through the origin of the complex γ plane as R is increased through R_c.

By our assumption, $\gamma = \xi$ $(\eta = 0)$, the spectrum of the operator J_o is real-valued and

$$\underset{\sim}{Z}_1 = \bar{\underset{\sim}{Z}}_1 = \underset{\sim}{\zeta}\,, \quad \underset{\sim}{Z}_2 = \bar{\underset{\sim}{Z}}_2 = \underset{\sim}{\zeta}^* \quad .$$

The perturbation problem to be considered is again (1.2) supplemented by the normalizing condition

$$1 = [\underset{\sim}{u}] = [\underset{\sim}{u}, \underset{\sim}{\zeta}^*] \quad . \tag{5.1}$$

The proposed solutions

$$\begin{bmatrix} \underset{\sim}{u}(x,t,\epsilon) \\ p(x,t,\epsilon) \\ \lambda(\epsilon) \end{bmatrix} = \sum_{n=0} \epsilon^n \begin{bmatrix} \underset{\sim}{u}_n(x,t) \\ p_n(x,t) \\ \lambda_n \end{bmatrix} \tag{5.2}$$

of (1.2) and (5.1) are to be periodic in t with period $2\pi/\tilde{\omega}$. Substitution of (5.2) into (1.2) and (5.1) leads to

$$J_o \underset{\sim}{u}_o + \nabla p_o = 0\,, \quad [\underset{\sim}{u}_o] = 1 \tag{5.3}$$

and

$$J_o \underset{\sim}{u}_m + L_m \underset{\sim}{u}_o + \underset{\sim}{\tilde{F}}_m + \nabla p_m = 0 \;, \quad [\underset{\sim}{u}_m] = 0 \qquad . \tag{5.4}$$

Here $\underset{\sim}{u}_o$ and $\underset{\sim}{u}_m$ are solenoidal and vanish on the boundary $\partial\Omega(t)$ of Ω and $\underset{\sim}{u}_o$, $\underset{\sim}{u}_m$, p_o, p_n are $2\pi/\tilde{\omega}$ periodic in t. The vector field $\underset{\sim}{\tilde{F}}_m$ is the same as $\underset{\sim}{F}_m$ defined by (3.13) when $\omega_\ell = 0$.

At zeroth order

$$\lambda_o = \frac{1}{R_c} \;, \quad \underset{\sim}{u}_o(\underset{\sim}{x}, t) = \underset{\sim}{\zeta} \qquad . \tag{5.5}$$

To solve (5.4) it is necessary to have

$$[L_m \underset{\sim}{u}_o] + [\underset{\sim}{F}_m] = 0 \qquad . \tag{5.6}$$

If the operator $L(\lambda_o, t)$ were independent of time, the condition (5.6) would also be sufficient for bounded invertibility of the operator J_o. Assuming this, we could apply here the implicit function method used in (JS) to prove convergence. The value of λ_m is fixed by (5.6) (see (3.14a)). Hence, equations (5.4) are uniquely solvable.

When $m = 1$ we find, using (3.14a) and (3.13), that when $\epsilon = 0$,

$$\frac{d\lambda}{d\epsilon} \Big/ \frac{d\lambda}{d\xi} + [\underset{\sim}{\zeta}^*, \underset{\sim}{\zeta} \cdot \nabla \underset{\sim}{\zeta}] = 0 \qquad . \tag{5.7}$$

Several interesting consequences follow from (5.7). First we note that if the bracket in (5.7) is non-zero,

$$\frac{d\lambda}{d\epsilon} = - \frac{1}{R^2} \frac{dR(0)}{d\epsilon} \neq 0 \qquad .$$

In this case

$$R(\epsilon) = R_c + \frac{dR(0)}{d\epsilon} \epsilon + \frac{1}{2} \frac{d^2R(0)}{d\epsilon^2} \epsilon^2 + \ldots \tag{5.8}$$

has different values for small positive and negative values of ϵ. Returning to (5.1) we see that

$$\epsilon = [\underset{\sim}{V} - \underset{\sim}{\tilde{u}}, \underset{\sim}{\zeta}^*]$$

gives the sign of the projection of $\underset{\sim}{V}(\epsilon) - \underset{\sim}{u}$ onto the eigensubspace of J_o and can be interpreted physically as the sign of the averaged motion (see (J)). Now (5.8) shows that if a motion of one sign branches off supercritically motion of the other sign branches subcritically. There are, therefore, two physically distinct solutions.

When $\lambda_1 = 0$ the sign of the bifurcation is determined by the sign of

$$\frac{d^2R(0)}{d\epsilon^2} = -R_c^2 \frac{d^2\lambda}{d\epsilon^2} ,$$

provided that this derivative does not vanish.

We turn now to the study of the stability of the periodic bifurcating solution. Setting

$$\underset{\sim}{V} = \underset{\sim}{\tilde{u}} + \epsilon \underset{\sim}{u} + \underset{\sim}{\nu} \tag{5.9}$$

into (1.1) and using the equations satisfied by the periodic basic solution u and the bifurcating periodic solution $\underset{\sim}{u}$, we find, after linearizing, that

$$J_0 \underset{\sim}{\nu} + \epsilon \underset{\sim}{u} \cdot \nabla \underset{\sim}{\nu} + \epsilon \underset{\sim}{\nu} \cdot \nabla \underset{\sim}{u} + \nabla p = 0 \tag{5.10a}$$

and

$$\text{div } \underset{\sim}{\nu} = 0 , \quad \underset{\sim}{\nu} = 0 \big|_{\partial\Omega(t)} , \quad [\underset{\sim}{\nu}] = 1 \quad . \tag{5.10b,c,d}$$

Again invoking Floquet theory

$$\underset{\sim}{\nu} = e^{-\sigma t} \underset{\sim}{\psi}(\underset{\sim}{x}, t, \epsilon) ,$$

where $\underset{\sim}{\psi}(\underset{\sim}{x}, t, \epsilon)$ is $2\pi/\tilde{\omega}$ periodic in t, we get

$$-\sigma \underset{\sim}{\psi} + J_0 \underset{\sim}{\psi} + \epsilon u \cdot \nabla \underset{\sim}{\psi} + \epsilon \underset{\sim}{\psi} \cdot \nabla \underset{\sim}{u} + \nabla p = 0 \quad . \tag{5.11}$$

We seek solutions of (5.10,11), $\underset{\sim}{\psi}(\underset{\sim}{x}, t, \epsilon)$, $p(\underset{\sim}{x}, t, \epsilon)$, $\sigma(\epsilon)$ as power series in ϵ. When $\epsilon = 0$, since zero is a simple eigenvalue of J_0 (cf. discussion associated with (1.6), we get

$$\sigma(0) = \sigma_0 = 0 , \quad \underset{\sim}{\psi}(\underset{\sim}{x}, t, 0) = \underset{\sim}{\psi}_0 = \underset{\sim}{\zeta} \quad . \tag{5.12}$$

At first order

$$J_o \underset{\sim}{\psi}_1 + \lambda_1 L_\lambda \underset{\sim}{\zeta} - \sigma_1 \underset{\sim}{\zeta} + 2\underset{\sim}{\zeta} \cdot \nabla \underset{\sim}{\zeta} + \nabla p_1 = 0 \quad .$$

Applying the solvability condition, we find that

$$\lambda_1 \;/\; \frac{d\lambda}{d\xi} - \sigma_1 + 2[\underset{\sim}{\zeta}^*, \underset{\sim}{\zeta} \cdot \nabla \underset{\sim}{\zeta}] = 0 \quad .$$

Using (5.7) we get

$$\sigma_1 = -\lambda_1 \;/\; \frac{d\lambda}{d\xi} = - \frac{dR(0)}{d\epsilon} \;/\; \frac{dR(0)}{d\xi} \quad . \qquad (5.13)$$

Since $dR(0)/d\xi < 0$, we have two cases:

$$\frac{dR}{d\epsilon} > 0 \, , \quad \sigma_1 > 0 \quad ,$$

$$\frac{dR}{d\epsilon} < 0 \, , \quad \sigma_1 < 0 \quad .$$

We may conclude that in the case of a simple eigenvalue, when $\lambda_1 \neq 0$ supercritical bifurcating solutions are stable and subcritical bifurcating solutions are unstable.

When $\lambda_1 = 0$ we find that $\sigma_1 = 0$. Then

$$J_o \underset{\sim}{\psi}_1 + 2\underset{\sim}{\zeta} \cdot \nabla \underset{\sim}{\zeta} + \nabla p_1 = 0 \quad ,$$

and comparison with (5.4) when $m = 1$ shows that

$$\underset{\sim}{\psi}_1 = 2\underset{\sim}{u}_1 \quad . \qquad (5.14)$$

At second order using (5.12,14), we get

$$-\sigma_2 \underset{\sim}{\zeta} + J \underset{\sim}{\psi}_2 + \lambda_2 L_\lambda \underset{\sim}{\zeta} + 3\underset{\sim}{u}_o \cdot \nabla \underset{\sim}{u}_1 + 3\underset{\sim}{u}_1 \cdot \nabla \underset{\sim}{u}_o + \nabla p_2 = 0 \quad .$$

Noting that

$$\underset{\sim}{u}_1 \cdot \nabla \underset{\sim}{u}_o + \underset{\sim}{u}_o \cdot \nabla \underset{\sim}{u}_1 = \underset{\sim}{F}_2 \qquad \text{and}$$

applying the solvability condition, we get

$$-\sigma_2 + \lambda_2 \;/\; \frac{d\lambda}{d\xi} + 3\, [\underset{\sim}{F}_2] = 0 \quad .$$

We note that when $\lambda_1 = 0$, equation (5.6) shows that

$$\lambda_2 \,/\, \frac{d\lambda}{d\xi} + [\underset{\sim}{F}_2] = 0 \qquad .$$

Hence

$$\sigma_2 = -2\lambda_2 \,/\, \frac{d\lambda}{d\xi} = -2\frac{d^2R(0)}{d\epsilon^2} \,/\, \frac{dR(0)}{d\xi} \qquad .$$

We may again conclude that subcritical bifurcating solutions are unstable and supercritical bifurcating solutions are stable.

It is important that our demonstration of the instability of small subcritical bifurcating solutions implies that the branch of the solution on which $R(\epsilon)$ is decreasing is unstable. The instability of this decreasing branch might have been anticipated on physical grounds. For example, in the problem of convection the heat transported decreases when the temperature difference increases on this decreasing subcritical branch. Such solutions are not observed.

It is well-known, from energy estimates, that any single valued branch $R(\epsilon)$ must have a positive minimum. Therefore, it is anticipated that the subcritical branch will decrease with ϵ to a positive minimum and then turn up. This is just what happens in generalized convection problems (Joseph, 1971). There one can demonstrate that the subcritical branch regains its stability as it passes through its minimum.

Fig. 1: Bifurcation of a time periodic basic flow with a fixed frequency $\tilde{\omega}$

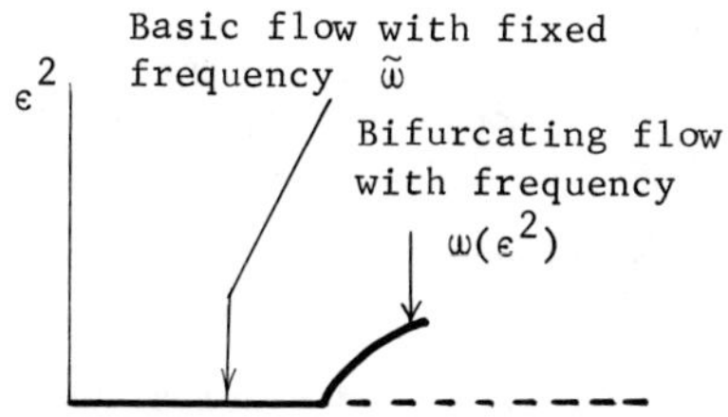

(a) Quasi-periodic super-critical bifurcation

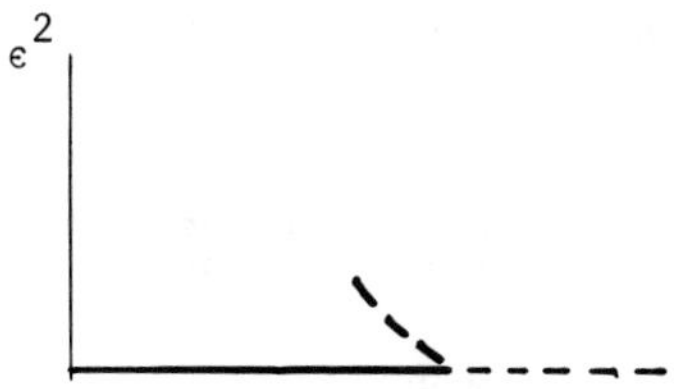

(b) Quasi-periodic subcritical bifurcation

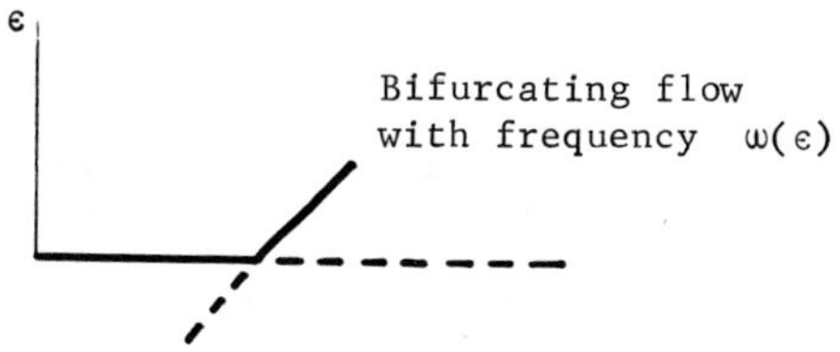

(c) Periodic bifurcation

CAPTION FOR FIG. 1

Fig. 1. Bifurcation diagrams for the solutions which bifurcate from a basic periodic solution of fixed frequency when the Floquet exponent is a simple eigenvalue of the linearized stability problem for the basic flow. The cases (a) and (b) represent the situation when the critical Floquet exponent is complex. In these cases the bifurcating solution is quasi-periodic; in addition to the fixed frequency ω a natural frequency $\omega(\epsilon)$ arises through instability. The bifurcation is symmetric about $\epsilon = 0$ for small ϵ, and only one-sided bifurcation can occur. Case (c) gives the typical situation which arises when the Floquet exponent is zero at criticality. In this case the bifurcating solution and basic solution have the same frequency, and the bifurcation can be two-sided.

Dashed lines show solutions which are unstable to small disturbances; heavy lines show solutions which are stable to small disturbances.

REFERENCES

1. Davis, S., Finite amplitude instability of time dependent flows. J. Fluid Mech. 45 (1971), 33-48.

2. Fojas, C., Essais dans l'étude des solutions des équations de Navier-Stokes dans l'espace. L'unicité et la presque-periodicité des solutions "petites". Rend. Sem. Mat. Univ. Padova, 32 (1962), 261-264.

3. Hopf, E., A mathematical example displaying features of turbulence. Comm. Pure Appl. Math. 1 (1948), 303. See also repeated branching through loss of stability, an example. Proc. of the Conf. on Diff. Eqs., University of Maryland, 49 (1956).

4. Joseph, D. D., Stability of convection in containers of arbitrary shape. J. Fluid Mech. 47 (1971), 257-282.

5. Joseph, D. D., and Sattinger, D., Bifurcating time periodic solutions and their stability. Arch. Rational Mech. Anal., 45 (1972), 79-109.

6. Iooss, G., Théorie non linéaire de la stabilité des écoulements laminaires dans le cas de "l'échange des stabilités", Arch. Rational Mech. Anal., 40 (1971), 166-208.

7. Landau, L., On the problem of turbulence. C. R. Acad. Sci., USSR, 44 (1944), 311. See also P. Landau and E. M. Lifschitz, "Fluid Mechanics", Pergamon Press, 1959.

8. Prodi, G., Qualche risultato riguardo alle equazoni di Navier-Stokes nel caso bidimensionale, Rend. Sem. Mat. Padova 30 (1960), 1-15.

9. Prouse, G., Soluzioni periodiche dell'equazione di Navier-Stokes. Rend. Acc. Naz. Lincei, 35 (1963).

10. Ruelle, D. and Takens, F., On the nature of turbulence. Comm. Math. Phys. 20 (1971), 167-192.

11. Sattinger, D., Bifurcation of periodic solutions of the Navier-Stokes equations. Arch. Rational Mech. Anal. 41 (1971), 66-80.

12. Serrin, J., A note on the existence of periodic solutions of the Navier-Stokes equations, 3 (1960), 120-122.

13. Yudovich, V., Periodic solutions of a viscous incompressible fluid. Doklady, Amer. Math. Soc. Trans. (1960), 168-171. The Russian original is in 29 (1960), 1214-1217.

14. Yudovich, V., On the stability of forced oscillations of a liquid. Doklady, Amer. Math. Soc. Trans., 11 (1971), 1473-1480. The Russian original is in 195 (1970), 292-295.

15. Yudovich, V., On the stability of self-oscillations of a liquid. Doklady, Amer. Math. Soc. Trans. 11 (1971), 1543-1546. The Russian original is in 195 (1970), 574-573.

BIFURCATION OF PERIODIC SOLUTIONS INTO INVARIANT TORI: THE WORK OF RUELLE AND TAKENS

Oscar E. Lanford, III
Department of Mathematics
University of California, Berkeley

The purpose of this lecture is to discuss a theorem of D. Ruelle and F. Takens [4] which extends Hopf's bifurcation theory. Hopf's bifurcation theory analyzes what happens when an equilibrium point of a differential equation depending on a parameter changes from stable to unstable at a critical value of the parameter. We shall be concerned with what happens when a stable periodic solution becomes unstable. What Ruelle and Takens show is that, under suitable technical conditions, a stable invariant two-dimensional torus appears in the vicinity of the unstable closed orbit. It may be helpful to visualize this torus as a tube containing the closed orbit in its interior,* although this picture is not quite accurate in more than three dimensions (since a torus in n-dimensional space, $n \geq 4$, does not divide the space into an inside and an outside). The result of Ruelle and Takens is incomplete in that it leaves the nature of the motion on the invariant torus completely obscure.

We shall begin by giving the stability analysis for periodic solutions in a form slightly different from that given in Sattinger's lectures. We shall describe the analysis for a smooth ordinary differential equation on Euclidean space; it will be apparent that it applies to at least some partial differential equations as well. Consider the equation

$$\frac{dx}{dt} = X(x) ,$$

where X is a smooth function of x, and suppose that this equation has a periodic solution which sweeps out a closed orbit Γ. Choose a point p on Γ,

*To avoid confusion, we should say explicitly that by the torus we mean the <u>surface</u> of the tube, not the solid it encloses.

and choose a small smooth surface Σ , of dimension one less than the number of components of x , and transversal to Γ at p . Define a mapping Φ of Σ into itself which sends a point q of Σ to the first point where the solution curve through q hits Σ again; Φ is called the Poincaré map. Of course, Φ may not be defined on all of Σ , but it is defined and smooth on a neighborhood of p ; note that $\Phi(p) = p$.

The Poincaré map provides a powerful tool for analyzing the qualitative behavior of the solution curves of the differential equation near Γ . For example, if Φ has a fixed point q in addition to p , then the solution curve through q is again periodic. More generally, if q is a periodic point of period n for Φ (i.e., $\Phi^n(q) = q$ but $\Phi^j(q) \neq q$ for $j = 1, 2, \ldots, n-1$), the solution curve through q is again periodic but follows Γ around n times before it closes. To find an invariant two-dimensional torus for the differential equation, we have only to find a circle which is mapped onto itself by Φ ; then the set of solution curves passing through that circle is the desired invariant torus.

Similarly, the stability of the periodic solution can be analyzed in terms of Φ . Suppose that some neighborhood V of p in Σ has the property that, for all q in V , $\Phi^n(q)$ is defined for all n and converges to p as $n \to \infty$. (We say in this case that the fixed point p of Φ is (asymptotically) stable.) Then every solution curve $q(t)$ starting near enough to Γ passes through V and can easily be shown to approach Γ as $t \to \infty$, i.e., Γ is asymptotically orbitally stable. A simple way to check whether p is stable is to use a linearized stability analysis. We let $D\Phi(p)$ denote the derivative of Φ at the fixed point p ; it is a linear mapping of the tangent space to Σ at p into itself. If the spectral radius of $D\Phi(p)$ is less than 1 (i.e., if all the eigenvalues of $D\Phi(p)$ are strictly less than 1 in absolute value), it is nearly immediate that p is a stable fixed point for Φ .

As another example of the usefulness of the Poincaré map, we shall indicate how it can be used to prove the persistence of closed orbits under small perturbations of the differential equation. To do this, we shall assume that 1 is not an eigenvalue of $D\Phi(p)$. Let us consider a smooth one-parameter family of differential equations

$$\frac{dx}{dt} = X_{\mu}(x)$$

with $X_0 = X$. Even without knowing that X_{μ} has a periodic solution, we can define the Poincaré map Φ_{μ} for X_{μ} on the same surface Σ as for X, provided that μ is small enough. Then X_{μ} will have a closed orbit near Γ if Φ_{μ} has a fixed point p_{μ} near p. But by our assumption on $D\Phi(p)$, the mapping

$$q \to q - \Phi(q)$$

has an invertible derivative at p, so, by the implicit function theorem, the equation

$$q - \Phi_{\mu}(q) = 0$$

has a smooth one-parameter family of solutions p_{μ} (defined for μ sufficiently small) with $p_0 = p$. Hence, the differential equation has, for each sufficiently small μ, a periodic solution passing through p_{μ}.

It is fairly clear that the Poincaré map can sometimes be constructed for periodic solutions of partial differential equations which can be treated as first order differential equations on Banach spaces. Such equations will usually not have the smoothness properties we assumed for X, because most interesting infinitesimal generators are unbounded. Nevertheless, if the partial differential equation has the property of smoothing initial data, which is characteristic of parabolic equations, the Poincaré map may well exist and be differentiable. For example, if the solution mapping

$$(q,t) \to q(t)$$

is a smooth function of the initial data q and the time t in some neighborhood of (p,T) (T the period of Γ), then the Poincaré map can be constructed and much of the above analysis goes through. It would be interesting to know whether the periodic solutions of the Navier-Stokes equations fit into this framework.

We now focus our attention on the Poincaré map Φ_μ rather than the differential equation. By subtracting out the fixed point p_μ, we can assume that $p_\mu = 0$. Thus, we consider a smooth one-parameter family of mappings Φ_μ of a neighborhood of zero in a Banach space Z into Z_1 with $\Phi_\mu(0) = 0$ for all μ. We want to see what happens as we change μ so that 0 changes from a stable to an unstable fixed point of Φ_μ, i.e., so that some of the spectrum of $D\Phi_\mu(0)$ crosses the unit circle. This can evidently happen in various ways, but we shall consider only the case of a single complex-conjugate pair of simple non-real eigenvalues crossing the unit circle; the remainder of the spectrum of $D\Phi_\mu$ will be assumed to stay strictly inside the unit circle. For simplicity of notation we shall assume that the crossing takes place at $\mu = 0$.

One of the novel and surprising features of the argument of Ruelle and Takens is that they are able at this point to reduce the general problem to a two-dimensional one. They do this by using a powerful technical result, called the center manifold theorem. There are several forms of this theorem, of varying generality; the form we need is the following:

Center Manifold Theorem. *Let Ψ be a mapping of a neighborhood of zero in a Banach space Z into Z. We assume that Ψ has a continuous $k+1$ derivative and that $\Psi(0) = 0$. We further assume that $D\Psi(0)$ has spectral radius 1 and that the spectrum of $D\Psi(0)$ splits into a part on the unit circle and the remainder which is at a non-zero distance from the unit circle.** *Let Y denote the generalized eigenspace of $D\Psi(0)$ belonging to the part of the spectrum on the unit circle; assume that Y has dimension $d < \infty$. Then there exist a neighborhood V of 0 in Z and a C^k submanifold M of V of dimension d, passing through 0 and tangent to Y at 0, such that*

a) (*Local Invariance*): *If $x \in M$ and $\Psi(x) \in V$, then $\Psi(x) \in M$*

b) (*Local Attractivity*): *If $\Psi^n(x) \in V$ for all $n = 0, 1, 2, \ldots,$ then, as $n \to \infty$, the distance from $\Psi^n(x)$ to M goes to zero.*

---

*This holds automatically if Z is finite dimensional or, more generally, if $D\Psi(0)$ is compact.

A manifold M of the sort described in the theorem is called a center manifold (for the fixed point 0 of Ψ.) It is not, in general, unique, but b) implies at least that it contains all ω-limit points of an orbit $\{\Psi^n(x): n = 0, 1, 2, \ldots\}$ which stays sufficiently near to zero. No accessible reference for this form of the center manifold theorem seems to exist; we have therefore given a proof of it in Appendix A.

Now comes a small but crucial trick. We apply the center manifold theorem, not to $D\Phi_\mu$, but to the mapping

$$\Psi: \ (x,\mu) \to (\Phi_\mu(x),\mu) \quad .$$

If $D\Phi_0$ has two simple complex-conjugate eigenvalues on the unit circle, then the Y for Ψ is three-dimensional, the extra dimension being in the μ direction. The theorem thus asserts the existence of a three-dimensional center manifold M. If we hold μ fixed and sufficiently small, we obtain a two-dimensional section M_μ of M which is locally invariant and attracting for Φ_μ. Since we are looking for recurrent behavior, we can restrict Φ_μ to M_μ. We are thus reduced to studying a smooth one-paremeter family of mappings, again denoted by Φ_μ, of a neighborhood of 0 in $\mathbb{R}^2$ into $\mathbb{R}^2$, with $\Phi_\mu(0) \equiv 0$ and with $D\Phi_0(0)$ having two distinct complex-conjugate eigenvalues on the unit circle. We want to find invariant circles for Φ_μ, $\mu > 0$.

We next assume that the eigenvalues of $D\Phi_\mu(0)$ pass through the unit circle with non-zero velocity as μ passes through 0. We can then reparametrize so the eigenvalues of $D\Phi_\mu(0)$ are $(1+\mu)e^{\pm i\theta(\mu)}$. By making a smooth μ-dependent change of coordinates, we can arrange that

$$D\Phi_\mu(0) = (1+\mu)\begin{pmatrix} \cos\theta(\mu) & -\sin\theta(\mu) \\ +\sin\theta(\mu) & \cos\theta(\mu) \end{pmatrix} .$$

The next step is to make a further change of coordinates to bring Φ_μ approximately into an appropriate canonical form. To be able to do this, we need a technical assumption:

A) $e^{im\,\theta(0)} \neq 1$, $m = 1, 2, 3, 4, 5$.

Proposition 1. Subject to Assumption A), we can make a smooth μ-dependent change of coordinates bringing Φ_μ into the form:

$$\Phi_\mu(x) = N\Phi_\mu(x) + O(|x|^5)$$

where, in polar coordinates,

$$N\Phi_\mu : (r,\varphi) \to ((1+\mu)r - f_1(\mu)r^3 \;,\; \varphi + \theta(\mu) + f_3(\mu)r^2) \;.$$

The proof of this proposition uses standard techniques and may be obtained, for example, from §23 of Siegel and Moser [5]. We give a straightforward and completely elementary proof in Appendix B. As indicated above, we think of $N\Phi_\mu$ as an approximate canonical form for Φ_μ. Note two special features of $N\Phi_\mu$:

i) The new r depends only on the old r, not on φ.

ii) The new φ is obtained from the old φ by an r-dependent rotation.

We now add a final assumption:

B) $f_1(0) > 0$.

This assumption implies that for small positive μ, $N\Phi_\mu$ has an invariant circle of radius r_0, where r_0 is obtained by solving

$$(1+\mu)r_0 - f_1(\mu)r_0^3 = r_0 \;,$$

i.e., $r_0 = \frac{\mu}{f_1(\mu)}$.

We shall verify shortly that this circle is attracting for $N\Phi_\mu$. Since Φ_μ differs only a little from $N\Phi_\mu$, it is not surprising that Φ_μ has a nearby invariance circle. This is, in fact, what Ruelle and Takens prove.

Theorem 2 (Ruelle-Takens). Assume A) and B). Then for all sufficiently small positive μ, Φ_μ has an attracting invariant circle.

Before giving the proof, let us look at what happens if B) is replaced by the assumption $f_1(0) < 0$. Then, for small positive μ, $N\Phi_\mu$ has no invariant sets except $\{0\}$ and $\mathbb{R}^2$. For $\mu < 0$, $N\Phi_\mu$ does have an invariant circle, but it is repelling rather than attracting. By applying the result of Ruelle and Takens to Φ_μ^{-1}, we prove that, in this case, Φ_μ has a nearby invariant circle. Thus, we again find the usual situation that supercritical branches are stable and subcritical branches unstable. What happens if $f_1(0) = 0$ remains to be investigated.

We turn now to the proof of Theorem 2. We have already indicated that we are going to look for an invariant circle for Φ_μ in the vicinity of the circle $r = \frac{\mu}{f_1(\mu)}$ which is invariant for $N\Phi_\mu$. It is therefore convenient to choose coordinates which exhibit as clearly as possible the behavior of Φ_μ near that circle. We first go to polar coordinates:

$$\Phi_\mu: (r,\varphi) \to ((1+\mu)r - f_1(\mu)r^3 + O(r^5) \;;\; \varphi + \theta(\mu) + f_3(\mu)r^2 + O(r^4)) \;.$$

(The terms $O(r^5)$, $O(r^4)$ depend on φ, μ as well as on r. Note also that the normal form is one power of r less accurate for the polar angle than for the rectangular coordinates.) Next we scale the radial coordinate so that the invariant circle for $N\Phi_\mu$ has radius 1, i.e., we introduce a new radial coordinate x by

$$r = \frac{\mu}{f_1(\mu)} x \;;$$

then

$$\Phi_\mu: (x,\varphi) \to ((1+\mu)x - \mu x^3 + \mu^2 O(x^3),\; \varphi + \theta(\mu) + \mu \frac{f_3(\mu)}{f_1(\mu)} x^2 + \mu^2 O(x^2)) \;.$$

Next, we let $x = 1 + y$, so that the circle $r = \frac{\mu}{f_1(\mu)}$ is translated to $y = 0$:

$$\Phi_\mu: (y,\varphi) \to ((1-2\mu)y - \mu(3y^2+y^3) + \mu^2 O(1) \;,$$

$$\varphi + \theta(\mu) + \mu \frac{f_3(\mu)}{f_1(\mu)} (1+y)^2 + \mu^2 O(1)) \;.$$

Finally, we scale y again by putting

$$y = \sqrt{\mu}\, z \;; \qquad \text{then}$$

$$\Phi_\mu: (z,\varphi) \to ((1-2\mu)z - \mu^{3/2}(3z^2 + \mu^{1/2}z^3) + \mu^{3/2}O(1) \;,$$

$$\varphi + \theta(\mu) + \mu \frac{f_3(\mu)}{f_1(\mu)} (1 + \mu^{1/2}z)^2 + \mu^2 O(1)) \;;$$

we rewrite this last formula as

$$(z,\varphi) \to ((1-2\mu)z + \mu^{3/2}H_\mu(z,\varphi),\ \varphi + \theta_1(\mu) + \mu^{3/2}K_\mu(z,\varphi)) \quad .$$

The functions $H_\mu(z,\varphi)$, $K_\mu(z,\varphi)$ are smooth in z, φ, μ on

$$-1 \leq z \leq 1,\ 0 \leq \varphi \leq 2\pi,\ 0 \leq \mu \leq \mu_0$$

for some sufficiently small μ_0 ; the region $-1 \leq z \leq 1$, $0 \leq \varphi \leq 2\pi$ corresponds to an annulus of width $O(\mu)$ about the invariant circle for $N\Phi_\mu$ (which has radius $O(\sqrt{\mu})$). We are going to produce an invariant circle inside this annulus.

The qualitative behavior of Φ_μ is now easy to read off; Φ_μ can be written as

$$(z,\varphi) \to ((1+2\mu)z,\ \varphi + \theta_1(\mu))$$

plus a small perturbation. The approximate Φ is simply a rotation in the φ direction and a contraction in the z direction. Note, however, that the strength of the contraction goes to zero with μ . If this were not the case, we could simply invoke known results about the persistence of attracting invariant circles under small perturbations. As it is, we need to make a slightly more detailed argument, exploiting the fact that the size of the perturbation goes to zero faster than the strength of the contraction.

We are going to look for an invariant manifold of the form

$$\{z = u(\varphi)\} \quad ,$$

where

i) $u(\varphi)$ is periodic in φ with period 2π

ii) $|u(\varphi)| \leq 1$ for all φ

iii) $u(\varphi)$ is Lipschitz continuous with Lipschitz constant 1 (i.e., $|u(\varphi_1) - u(\varphi_2)| \leq |\varphi_1 - \varphi_2|$).

The space of all functions u satisfying i), ii) and iii) will be denoted by U .

We shall give a proof based on the contraction mapping principle. In outline, the argument goes as follows: We start with a manifold

$$M = \{z = u(\varphi)\} \quad ,$$

with $u \in U$, and consider the new manifold $\Phi_\mu M$ obtained by acting on M with Φ_μ. We show that, for μ sufficiently small, $\Phi_\mu M$ again has the form $\{z = \hat{u}(\varphi)\}$ for some $\hat{u} \in U$. Thus, we construct a non-linear mapping $\mathcal{F}$ of U into itself by

$$\mathcal{F}u = \hat{u} \quad .$$

We then prove, again for small positive μ, that $\mathcal{F}$ is a contraction on U (with respect to the supremum norm) and hence has a unique fixed point u^*. The manifold $\{z = u^*(\varphi)\}$ is the desired invariant circle. As a by-product of the proof of contractivity, we prove this manifold is attracting in the following sense: Pick a starting point (z,φ) with $|z| \leq 1$, and let (z_n,φ_n) denote $\Phi_\mu^n(z,\varphi)$. Then

$$\lim_{n \to \infty} z_n - u^*(\varphi_n) = 0 \quad .$$

It is not hard to see that the domain of attraction is much larger than the annulus $|z| \leq 1$ - In particular, it contains everything inside the annulus except the fixed point at the center - but we shall not pursue this point.

To carry out the argument outlined above, we must first construct the non-linear mapping $\mathcal{F}$. To find $\mathcal{F}u(\varphi)$, we should proceed as follows:

i) Show that there is a unique $\tilde{\varphi}$ such that the φ-component of $\Phi_\mu(u(\tilde{\varphi}),\tilde{\varphi})$ is φ, i.e., such that

$$\varphi \equiv \tilde{\varphi} + \theta_1(\mu) + \mu^{3/2} K_\mu(u(\tilde{\varphi}),\tilde{\varphi})(2\pi) \; .^* \qquad (1)$$

ii) Put $\mathcal{F}u(\varphi)$ equal to the z-component of $\Phi_\mu(u(\tilde{\varphi}),\tilde{\varphi})$, i.e.,

$$\mathcal{F}u(\varphi) = (1 - 2\mu)u(\tilde{\varphi}) + \mu^{3/2} H_\mu(u(\tilde{\varphi}),\tilde{\varphi}) \quad . \qquad (2)$$

In the estimates we are going to make, it will be convenient to introduce

$$\lambda = \sup_{\substack{0 \leq \varphi \leq 2\pi \\ -1 \leq z \leq 1}} \{|H_\mu| \vee |K_\mu| \vee \frac{|\partial H_\mu|}{\partial z} \vee \frac{|\partial K_\mu|}{\partial z} \vee \frac{|\partial H_\mu|}{\partial \varphi} \vee \frac{|\partial K_\mu|}{\partial \varphi}\} \; ;$$

---

*i.e., φ differs from $\tilde{\varphi} + \theta_1(\mu) + \mu^{3/2} K_\mu(u(\tilde{\varphi}),\tilde{\varphi})$ by an integral multiple of 2π.

so defined, λ depends on μ but remains bounded as $\mu \to 0$

We now prove that (1) has a unique solution. To do this, it is convenient to denote the right-hand side of (1) temporarily by $x(\tilde{\varphi})$:

$$x(\tilde{\varphi}) = \tilde{\varphi} + \theta_1(\mu) + \mu^{3/2}K_\mu(u(\tilde{\varphi}),\tilde{\varphi}) \quad .$$

We want to show that, as $\tilde{\varphi}$ runs from 0 to 2π , $x(\tilde{\varphi})$ runs exactly once over an interval of length 2π . From the periodicity of $u(\tilde{\varphi})$, $K_\mu(z,\tilde{\varphi})$ in $\tilde{\varphi}$, it follows that

$$x(2\pi) = x(0) + 2\pi \ .$$

We therefore only have to show that x is strictly increasing. Let $\tilde{\varphi}_1 < \tilde{\varphi}_2$. Then

$$x(\tilde{\varphi}_2) - x(\tilde{\varphi}_1) = \tilde{\varphi}_2 - \tilde{\varphi}_1 + \mu^{3/2}[K_\mu(u(\tilde{\varphi}_2),\tilde{\varphi}_2) - K_\mu(u(\tilde{\omega}_2),\tilde{\varphi}_2)] \quad .$$

Now

$$|K_\mu(u(\tilde{\varphi}),\tilde{\omega}) - K_\mu(\tilde{\varphi}_1),\tilde{\varphi}_1)| \leq \lambda[|u(\tilde{\varphi}_2) - u(\tilde{\varphi}_1)| + |\tilde{\varphi}_2 - \tilde{\varphi}_1|]$$

$$\leq 2\lambda|\tilde{\varphi} - \tilde{\varphi}_1| = 2\lambda(\tilde{\varphi}_2 - \tilde{\varphi}_1) \quad .$$

(The second inequality follows from the Lipschitz continuity of u .) Thus

$$x(\tilde{\varphi}_2) - x(\tilde{\varphi}_1) \geq (1 - 2\lambda\mu^{3/2})(\tilde{\varphi}_2 - \tilde{\varphi}_1) \ , \quad \text{so, provided}$$

$$1 - 2\lambda\mu^{3/2} > 0 \ , \tag{3}$$

x is strictly increasing and (1) has a unique solution. We thus get $\tilde{\varphi}$ as a function of φ , and it follows from our above estimates that $\tilde{\varphi}$ is Lipshitz continuous:

$$|\tilde{\varphi}(\varphi_1) - \tilde{\varphi}(\varphi_2)| \leq (1 - 2\lambda\mu^{3/2})^{-1} |\varphi_1 - \varphi_2| \quad . \tag{4}$$

The definition (2) of $\mathcal{F}u$ therefore makes sense, and we next have to check that $\mathcal{F}u \in U$. Condition i) is immediate.

ii) $$|\mathcal{F}u(\varphi)| \leq (1 - 2\mu)|u(\tilde{\varphi})| + \mu^{3/2}|H_\mu(u(\tilde{\varphi}),\tilde{\varphi})|$$

$$\leq 1 - 2\mu + \mu^{3/2}\lambda \quad .$$

Thus, $|\mathcal{F}u(\varphi)| \leq 1$ for all φ provided

$$2\mu - \mu^{3/2}\lambda \geq 0 \quad . \tag{5}$$

iii)

$$|\mathcal{F}u(\varphi_1) - \mathcal{F}u(\varphi_2)| \leq (1 - 2\mu)|u(\tilde{\varphi}_1) - u(\tilde{\varphi}_2)|$$

$$+ \mu^{3/2}\lambda[|u(\tilde{\varphi}_1) - u(\tilde{\varphi}_2)| + |\tilde{\varphi}_1 - \tilde{\varphi}_2|]$$

$$\leq (1 - 2\mu + 2\mu^{3/2}\lambda)|\tilde{\varphi}_1 - \tilde{\varphi}_2|$$

by the Lipschitz continuity of u . Inserting estimate (4) for $|\tilde{\varphi}_1 - \tilde{\varphi}_2|$, we get

$$|\mathcal{F}u(\varphi_1) - \mathcal{F}u(\varphi_2)| \leq (1 - 2\mu + 2\mu^{3/2}\lambda)(1- 2\mu^{3/2}\lambda)^{-1}|\varphi_1 - \varphi_2| \quad ,$$

so $\mathcal{F}u$ is Lipschitz continuous with Lipschitz constant 1 provided

$$(1 - 2\mu + 2\mu^{3/2}\lambda)(1 - 2\mu^{3/2}\lambda)^{-1} \leq 1 \quad . \tag{6}$$

Evidently (6) holds for all sufficiently small positive μ .

The next step is to prove that $\mathcal{F}$ is a contraction. Thus, let $u_1, u_2 \in U$, choose φ , and let $\tilde{\varphi}_1, \tilde{\varphi}_2$ denote the solutions of

$$\varphi = \tilde{\varphi}_1 + \theta_1(\mu) + \mu^{3/2}K_\mu(u_1(\tilde{\varphi}_1),\tilde{\varphi}_1)$$

$$\varphi = \tilde{\varphi}_2 + \theta_1(\mu) + \mu^{3/2}K_\mu(u_2(\tilde{\varphi}_2),\tilde{\varphi}_2) \quad ,$$

respectively. Subtracting these equations, transposing, and taking absolute values yields

$$\begin{aligned} |\tilde{\varphi}_1 - \tilde{\varphi}_2| &\leq \mu^{3/2}|K_\mu(u_1(\tilde{\varphi}_1),\tilde{\varphi}_1) - K_\mu|u_2(\tilde{\varphi}_2),\tilde{\varphi}_2)| \\ &\leq \mu^{3/2}\lambda[|u_1(\tilde{\varphi}_1) - u_2(\tilde{\varphi}_2)| + |\tilde{\varphi}_1 - \tilde{\varphi}_2|] \quad . \end{aligned} \tag{7}$$

Now

$$|u_1(\tilde\varphi_1) - u_2(\tilde\varphi_2)| \le |u_1(\tilde\varphi_1) - u_2(\tilde\varphi_1)| + |u_2(\tilde\varphi_1) - u_2(\tilde\varphi_2)|$$

$$\le \|u_1 - u_2\| + |\tilde\varphi_1 - \tilde\varphi_2| \quad .$$

Inserting this inequality into (7), collecting all the terms involving $|\tilde\varphi_1 - \tilde\varphi_2|$ on the left, and dividing yields

$$|\tilde\varphi_1 - \tilde\varphi_2| \le (1 - 2\mu^{3/2}\lambda)^{-1}\ \mu^{3/2}\lambda \cdot \|u_1 - u_2\| \quad . \tag{8}$$

Now we use the definition (2) of $\mathcal{F}u$:

$$|\mathcal{F}u_1(\varphi) - \mathcal{F}u_2(\varphi)| \le (1 - 2\mu)|\tilde u_1(\tilde\varphi_1) - \tilde u_2(\tilde\varphi_2)|$$

$$+ \mu^{3/2}|H_\mu(u_1(\tilde\varphi_1),\tilde\varphi_1) - H_\mu(u_2(\tilde\varphi_2),\tilde\varphi_2)|$$

$$\le (1 - 2\mu)[\|u_1 - u_2\| + |\tilde\varphi_1 - \tilde\varphi_2|]$$

$$+ \mu^{3/2}\lambda[\|u_1 - u_2\| + 2|\tilde\varphi_1 - \tilde\varphi_2|]$$

$$\le \|u_1 - u_2\|\{(1-2\mu)(1 + \mu^{3/2}\lambda(1-2\mu^{3/2}\lambda)^{-1}]$$

$$+ \mu^{3/2}\lambda[1 + 2\mu^{3/2}\lambda(1 - 2\mu^{3/2}\lambda)^{-1}]\} \quad .$$

Let α denote the expression in braces. Then

$$\alpha = 1 - 2\mu + O(\mu^{3/2}) \ ,$$

so we can make $\alpha < 1$ by making μ small enough. If this is done, we have

$$\|\mathcal{F}u_1 - \mathcal{F}u_2\| \le \alpha \cdot \|u_1 - u_2\| \quad \text{with} \quad \alpha < 1 \quad , \tag{9}$$

i.e., $\mathcal{F}$ is a contraction on U and hence has a unique fixed point u^* .

To prove that the invariant manifold $\{z = u^*(\varphi)\}$ is attracting, we pick a point (z,φ) in the annulus $|z| \le 1$, and we let (z_1,φ_1) denote $\Phi_\mu(z,\varphi)$. Note that

$$|z_1| \le (1-2\mu)|z| + \mu^{3/2}\lambda \le 1 - 2\mu + \mu^{3/2}\lambda \le 1 \quad \text{(by (5)), so}$$

(z_1,φ_1) is again in the annulus. Now let $\tilde\varphi_1$ denote the solution of

$$\varphi_1 = \tilde{\varphi}_1 + \theta_1(\mu) + \mu^{3/2}K_\mu(u^*(\tilde{\varphi}_1),\tilde{\varphi}_1) \quad .$$

The definition of φ_1, on the other hand, needs

$$\varphi_1 = \varphi + \theta_1(\mu) + \mu^{3/2}K_\mu(z,\varphi) \quad .$$

Subtracting these equations and then estimating and re-arranging as in the proof of (6), we get

$$|\tilde{\varphi}_1 - \varphi| \leq \mu^{3/2}\lambda(1 - 2\mu^{3/2}\lambda)^{-1}|z - u^*(\varphi)| \quad .$$

Now subtract the equations

$$u^*(\varphi_1) = \mathcal{F}u^*(\varphi_1) = (1-2\mu)u^*(\tilde{\varphi}_1) + \mu^{3/2}H_\mu(u^*(\tilde{\varphi}_1),\tilde{\varphi}_1)$$

$$Z_1 \qquad = (1-2\mu)z + \mu^{3/2}H_\mu(z,\varphi)$$

and again imitate the proof that $\mathcal{F}$ is a contraction to get

$$|z_1 - u^*(\varphi_1)| \leq \alpha \cdot |z - u(\varphi)| \quad ,$$

with the same α as in (9). By induction,

$$|z_n - \varphi^*(\varphi_n)| \leq \alpha^n|z - u(\varphi)| \to 0 \quad \text{as} \quad n \to \infty \quad .$$

In our proof, we used only the continuity of H_μ, K_μ and their first derivatives, and we obtained a Lipschitz continuous u^*. Closer examination of the argument shows that we needed only Lipschitz continuity of H_μ, K_μ. If we have more differentiability of H_μ, K_μ, we would expect to obtain more differentiability for u^*. This is indeed the case. Specifically, let U_k denote the set of periodic functions $u(\varphi)$ of class C^k satisfying

i) $|u^{(j)}(\varphi)| \leq 1$, $\quad j = 0, 1, \ldots, k$; all φ .

ii) $|u^{(k)}(\varphi)|$ is Lipschitz continuous with Lipschitz constant one.

If H_μ, K_μ have Lipschitz continuous k^{th} derivatives, a straightforward generalization of the estimates we have given shows that for μ sufficiently small, $\mathcal{F}$

maps U_k into itself. It may be shown that U_k is complete in the supremum norm (see the proof of Proposition A1 in Appendix A), so the fixed point of $\mathcal{F}$ must be in U_k, i.e., u^* has Lipschitz continuous k^{th} derivative. If we make the weaker assumption that H_μ, K_μ have continuous k^{th} derivatives, slightly more complicated arguments show that u^* also has a continuous k^{th} derivative; we proceed with the proof by showing that the set of u's, whose k^{th} derivatives have an appropriately chosen modulus of continuity, is mapped into itself by $\mathcal{F}$.

APPENDIX A. THE CENTER MANIFOLD THEOREM

The version of the center manifold theorem which we used above is a very special case of a theorem due to H. Hirsch, Pugh, and M. Shub. It is also a mild generalization of the results of A. Kelley [3] for ordinary differential equations. Because the proof of Hirsch, Pugh, and Shub has yet to appear, and because a relatively simple proof can be constructed for the form of the theorem we used, it seems worthwhile to give that proof here. The technique of proof we use is modelled in the argument of Ruelle and Takens reproduced above, which in turn was inspired by the ideas of Hirsch Pugh, and Shub; most of the important ideas involved can be found in the paper of Kelley.

We begin by reformulating (in a slightly more general way) the theorem we want to prove. We have a mapping Ψ of a neighborhood of zero in a Banach space Z into Z, with $\Psi(0) = 0$. We assume that the spectrum of $\Psi(0)$ splits into a part on the unit circle and the remainder, which is contained in a circle of radius strictly less than one, about the origin. Elementary special theory (see, for example, Dunford and Schwartz [2], Chapter VII) guarantees the existence of a spectral projection P of Z belonging to the part of the spectral on the unit circle with the following properties:

i) P commutes with $D\Psi(0)$, so the subspaces PZ and $(I-P)Z$ are mapped into themselves by $D\Psi(0)$.

ii) The spectrum of the restriction of $D\Psi(0)$ to PZ lies on the unit circle.

iii) The spectral radius of the restriction of $D\Psi(0)$ to $(-P)Z$ is strictly less than one.

We let X denote $(-P)Z$, Y denote PZ , A denote the restriction of $D\Psi(0)$ to X and B denote the restriction of $D\Psi(0)$ to Y . Then $Z = X \oplus Y$ and

$$\Psi(x,y) = (Ax + X(x,y) \ , \ By + Y(x,y) \ ,$$

where

A is bounded linear operator on X with spectral radius strictly less than one.

B is a bounded operator on Y with spectrum on the unit circle. (All we actually need is that the spectral radius of B^{-1} is no larger than one.)

X is a C^{k+1} mapping of a neighborhood of the origin in $X \oplus Y$ into X with a second order zero at the origin.

Y is a C^{k+1} mapping of a neighborhood of the origin in $X \oplus Y$ into Y with a second-order zero at the origin.

(Here, k is an integer ≥ 1 .)

We want to find an invariant manifold for Ψ which is tangent to Y at the origin. Such a manifold will have the form $\{x = u(y)\}$, where u is a mapping of a neighborhood of the origin in Y into X , with $u(0) = 0$, $Du(0) = 0$.

In the version of the theorem we stated in the text, we assumed that Y was finite-dimensional. We can weaken this assumption, but not eliminate it entirely.

<u>Assumption</u>. There exists a C^{k+1} real-valued function X on Y which is 1 on a neighborhood of the origin and zero for $\|y\| > 1$. Perhaps, surprisingly, this assumption is actually rather restrictive. It holds trivially if Y is finite-dimensional or if Y is a Hilbert space; for a more detailed discussion of when it holds, see Bonic and Frampton [1].

We can now state the precise theorem we are going to prove.

Center Manifold Theorem. Let the notation and assumptions be as above. Then there exist $\epsilon > 0$ and a C^k-mapping u^* from $\{y \in Y: \|y\| < \epsilon\}$ into X, with a second-order zero at zero, such that

a) The manifold $\{(x = u^*(y); \|y\| < \epsilon\} \subset X \oplus Y$ is invariant for Ψ in the sense that, if $\|y\| < \epsilon$ and if $\Psi(u^*(y),y) = (x_1,y_1)$ with $\|y_1\| < \epsilon$ then $x_1 = u^*(y_1)$.

b) The manifold $\{x = u^*(y)\}$ is locally attracting for Ψ in the sense that, if $\|x\| < \epsilon$, $\|y\| < \epsilon$, and if $(x_n\ y_n) = \bar{\Psi}^n(x,y)$ are such that $\|x_n\| < \epsilon$, $\|y_n\| < \epsilon$ for all $n > 0$, then

$$\lim_{n \to \infty} \|x_n - u^*(y_n)\| = 0 .$$

Proceeding with the proof, it will be convenient to assume that $\|A\| < 1$ and that $\|B^{-1}\|$ is not much greater than 1. This is not necessarily true but we can always make it true by replacing the norms on X,Y by equivalent norms.* We shall assume that we have made this change of norm. It is unfortunately a

---

*This is a general fact about operators on Banach spaces. Let C be a bounded operator on a Banach space Z, and let σ be any number strictly greater than the spectral radius of C. Then the original norm on Z may be replaced by an equivalent norm with respect to which C has norm $\leq \sigma$. The proof goes as follows: By general spectral theory, the spectral radius of C is equal to $\lim_{n \to \infty} \|C^n\|^{1/n}$. Since σ is larger than the spectral radius of C, $\lim_{n \to \infty} \frac{\|C^n\|}{\sigma^n} = 0$, and in particular

$$\sup_{n \geq 0} \frac{\|C^n\|}{\sigma^n} < \infty$$

Define a new norm on Z by

$$|||z||| = \sup_{n \geq 0} \frac{\|C^n z\|}{\sigma^n} .$$

Then

$$\|z\| \leq |||z||| \leq \sup_n \{\frac{\|C^n\|}{\sigma^n}\} \cdot \|z\| ,$$

so the new norm is equivalent to the original one. Also,

$$|||Cz||| = \sup_{n \geq 0} \frac{\|C^{n+1} z\|}{\sigma^n} \leq \sigma \cdot \sup_{n \geq 0} \frac{\|C^n z\|}{\sigma^n} = \sigma \cdot |||z||| ,$$

so $|||C||| \leq \sigma$.

little awkward to explicitly set down exactly how close to one $\|B^{-1}\|$ should be taken. We therefore carry out the proof as if $\|B^{-1}\|$ were an adjustable parameter; in the course of the argument we shall find a finite number of conditions on $\|B^{-1}\|$. In principle, one should collect all these conditions and impose them at the outset.

The theorem guarantees the existence of a function u defined on what is perhaps a very small neighborhood of zero. Rather than work with very small values of x, y, we shall scale the system by introducing new variables x/ε, y/ε (and calling the new variables again x and y). This scaling does not change A, B , but, by taking ε very small, we can make X, Y , together with their derivatives of order $\leq k+1$, as small as we like on the unit ball. Then by multiplying $X(x,y)$, $Y(x,y)$ by the function $X(y)$ whose existence is asserted in the assumption preceding the statement of the theorem, we can also assume that $X(x,y)$, $y(x,y)$ are zero when $\|y\| > 1$. Thus, if we introduce

$$\lambda = \sup_{\substack{\|x\| \leq 1 \\ y \text{ unrestricted}}} \ \sup_{\substack{j_1, j_2 \\ j_1+j_2 \leq k+1}} \{\|D_x^{j_1} D_y^{j_2} X(x,y)\| y \|D_x^{j_1} D_y^{j_2} Y(x,y)\|\} ,$$

we can make λ as small as we like by choosing ε very small. The only use we make of our technical assumption on Y is to arrange things so that the supremum in the definition of λ may be taken over all y and not just over a bounded set.

Once we have done the scaling and cutting off by X , we can prove a global center manifold theorem. That is, we shall prove the following theorem.

Theorem 1. Keep the notation and assumptions of the center manifold theorem. If λ is sufficiently small (and if $\|B^{-1}\|$ is close enough to one), there exists a function u^* , defined and k times continuously differentiable on all of Y , with a second-order zero at the origin, such that

a) The manifold $\{x = u^*(y)\}$ is invariant for Ψ in the strict sense.

b) If $\|x\| \leq 1$, and y is arbitrary then $\lim_{n \to \infty} \|x_n - u^*(y_n)\| = 0$ (where $(x_n, y_n) = \Psi^n(x,y)$).

As with $\|B^{-1}\|$, we shall treat λ as an adjustable parameter and impose the necessary restrictions on its size as they appear. It may be worth noting that λ depends on the choice of norm; hence, one must first choose the norm to make

$\|B^{-1}\|$ close to one, then do the scaling and cutting off to make λ small. To simplify the task of the reader who wants to check that all the required conditions on $\|B^{-1}\|$, λ can be satisfied simultaneously, we shall number these conditions as (C1), (C2), etc.

The strategy of proof is very simple. We start with a manifold M of the form $\{x = u(y)\}$; we let ΨM denote the image of M under Ψ . With some mild restrictions on u , we first show that the manifold ΨM again has the form

$$\{x = \hat{u}(y)\}$$

for a new function $\hat{u}$. If we write $\mathcal{F}u$ for $\hat{u}$ we get a (nonlinear) mapping

$$u \to \mathcal{F}u$$

from functions to functions. The manifold M is invariant if and only if $u = \mathcal{F}u$, so we must find a fixed point of $\mathcal{F}$. We do this by proving that $\mathcal{F}$ is a contraction on a suitable function space (assuming that λ is small enough). The proof may be divided into a number of steps:

I) Derive heuristically a "formula" for $\mathcal{F}$.

II) Show that the formula obtained in I) yields a well-defined mapping of an appropriate function space U into itself.

III) Prove that $\mathcal{F}$ is a contraction on U and hence has a unique fixed point u^* .

IV) Prove that b) of Theorem A1 holds for u^* .

I) To construct $\mathcal{F}u(y)$, we should proceed as follows:

i) Solve the equation

$$y = B\tilde{y} + Y(u(\tilde{y}),\tilde{y}) \qquad (1)$$

for $\tilde{y}$. This means that y is the Y-component of $\Psi(u(\tilde{y}),\tilde{y})$.

ii) Let $\mathcal{F}u(y)$ be the X-component of $\Psi(u(\tilde{y}),\tilde{y})$, i.e.,

$$\mathcal{F}u(y) = Au(\tilde{y}) + X(u(\tilde{y}),\tilde{y}) \quad . \qquad (2)$$

II) We shall somewhat arbitrarily choose the space of functions u we want to consider to be

$$U = \{u : Y \to X \; ; \; D^{k+1}u \text{ continuous} \; ; \; \|D^j u(y)\| < 1 \text{ for}$$

$$j = 0, 1, \ldots, k+1 \; , \text{ all } y \; ; \; u(0) = Du(0) = 0\} \; .$$

i) Prove that, for any given $u \in U$, equation (1) has a unique solution $\tilde{y}$ for each $y \in Y$.

ii) Prove that $\mathcal{F}u$, defined by (2), is in U.

i) We rewrite (1) as a fixed-point problem:

$$\tilde{y} = B^{-1}y - B^{-1}Y(u(\tilde{y}),\tilde{y}) \; .$$

It suffices, therefore, to prove that the mapping

$$\tilde{y} \to B^{-1}y - B^{-1}Y(u(\tilde{y}),\tilde{y})$$

is a contraction on Y. We do this by estimating its derivative:

$$\|D_{\tilde{y}}[B^{-1}y - B^{-1}Y(u(\tilde{y}),\tilde{y})]\|$$

$$\leq \|B^{-1}\| \; \|D_1Y(u(\tilde{y}),\tilde{y})Du(\tilde{y}) + D_2Y(u(\tilde{y}),\tilde{y})\|$$

$$\leq 2\lambda\|B^{-1}\|$$

by the definitions of λ and U. If we require

$$2\lambda\|B^{-1}\| < 1 \; , \tag{C.1}$$

equation (1) has a unique solution $\tilde{y}$ for each y. Note that $\tilde{y}$ is a function of y, depending also on the function u. By the inverse function theorem, $\tilde{y}$ is a C^{k+1} function of y.

ii) By what we have just proved, $\mathcal{F}u \in C^{k+1}$. Thus to show $\mathcal{F}u \in U$, what we must check is

α) $\|D^j\mathcal{F}u(y)\| \leq 1$ for all y, $j = 0, 1, 2, \ldots, k+1$

β) $\mathcal{F}u(0) = 0$, $D\mathcal{F}u(0) = 0$.

α) First take $j = 0$:

$$\|\mathcal{F}u(y)\| \leq \|A\|\cdot\|u(\tilde{y})\| + \|X(u(\tilde{y}),\tilde{y})\| \leq \|A\| + \lambda \quad ,$$

so if we require

$$\|A\| + \lambda \leq 1 \quad , \tag{C.2}$$

then $\|\mathcal{F}u(y)\| \leq 1$ for all y .

To estimate $D\mathcal{F}u$ we must first estimate $D\tilde{y}(y)$. By differentiating (1), we get

$$= [B + DY^u(\tilde{y})]D\tilde{y} \quad , \tag{3}$$

where $Y^u: Y \to Y$ is defined by

$$Y^u(y) = Y(u(y),y) \quad .$$

By a computation we have already done

$$\|DY^u(\tilde{y})\| \leq 2\lambda \quad \text{for all} \quad \tilde{y} \quad .$$

Now

$$B + DY^u = B[I + B^{-1}DY^u] \quad ,$$

and since

$$2\lambda\|B^{-1}\| < 1 \qquad \text{(by (C.2))} \quad ,$$

$B + DY^u$ is invertible and

$$\|(B+DY^u)^{-1}\| \leq \|B^{-1}\|(1 - 2\lambda\|B^{-1}\|)^{-1} \quad .$$

The quantity on the right-hand side of this inequality will play an important role in our estimates, so we give it a name:

$$\gamma \equiv \|B^{-1}\|(1 - \lambda\|B^{-1}\|)^{-1} \quad . \tag{4}$$

Note that, by first making $\|B^{-1}\|$ very close to one and then by making λ small, we can make γ as close to one as we like.

We have just shown that

$$\|Dy(\tilde{y})\| \leq \gamma \quad \text{for all} \quad y \ . \tag{5}$$

Differentiating the expression (2) for $\mathcal{F}u(y)$ yields

$$D\mathcal{F}u(y) = [A\ Du(\tilde{y}) + DX^u(\tilde{y})]D\tilde{y}(y) \quad ; \tag{6}$$

$$(X^u(\tilde{y}) = X(u(\tilde{y}),\tilde{y}) \ ) \quad .$$

Thus

$$\|D\mathcal{F}u(y)\| \leq (\|A\| + 2\lambda) \cdot \gamma \quad , \tag{7}$$

so if we require

$$(\|A\| + 2\lambda)\gamma \leq 1 \quad , \tag{C.3$_1$}$$

we get

$$\|D\mathcal{F}u(y)\| \leq 1 \quad \text{for all} \quad y \ .$$

We shall carry the estimates just one step further. Differentiating (3) yields

$$0 = (B + DY^u(\tilde{y}))D^2\tilde{y} + D^2Y^u(D\tilde{y})^2 \quad .$$

By a straightforward computation

$$\|D^2Y^u(\tilde{y})\| \leq 5\lambda \quad \text{for all} \ \tilde{y} \quad ,$$

so

$$\|D^2\tilde{y}(y)\| = \|(B + DY^u(\tilde{y}))^{-1}D^2Y^u(D\tilde{y})^2\|$$

$$\leq \gamma \cdot 5\lambda \cdot \gamma^2 = 5\lambda\gamma^3 \quad .$$

Now, by differentiating the formula (6) for $D\mathcal{F}u$, we get

$$D^2\mathcal{F}u(y) = [A\ D^2u(\tilde{y}) + D^2X^u(\tilde{y})](D\tilde{y})^2$$

$$+ [A\ Du(\tilde{y}) + DX^u(\tilde{y})]D^2\tilde{y} \quad ,$$

so

$$\|D^2 \mathcal{F}u(y)\| \leq (\|A\| + 5\lambda)\gamma^2 + (\|A\| + 2\lambda) \cdot 5\lambda\gamma^3 \quad .$$

If we require

$$(\|A\| + 5\lambda)\gamma^2 + (\|A\| + 2\lambda) \cdot 5\lambda\gamma^3 \leq 1 \quad , \tag{C.3$_2$}$$

we have

$$\|D^2 \mathcal{F}u(y)\| \leq 1 \qquad \text{for all} \quad y \quad .$$

At this point it should be plausible by imposing a sequence of stronger and stronger conditions on γ, λ, that we can arrange

$$\|D^j \mathcal{F}(y)\| \leq 1 \qquad \text{for all} \quad y \ , \quad j = 3,\ 4,\ \ldots,\ k{+}1 \quad .$$

The verification that this is in fact possible is left to the reader.

β) To check that $\mathcal{F}u = 0$, $D\mathcal{F}u = 0$ (assuming $u = 0$, $DU = 0$) we note that

$$\tilde{y}(0) = 0 \text{ since } 0 \text{ is a solution of } 0 = B\tilde{y} + Y(u(\tilde{y}),\tilde{y}) \quad .$$

$$\mathcal{F}u(0) = Au(0) + X(u(0),0) = 0 \quad .$$

$$D\mathcal{F}u(0) = [A\ Du(0) + D_1X(0,0)Du(0) + D_2X(0,0)] \cdot D\tilde{y}(0)$$

$$= [A \cdot 0 + 0 + 0] \cdot D\tilde{y}(0) = 0 \quad .$$

III) We next show that $\mathcal{F}$ is a contraction and apply the contraction mapping principle. What we actually do is slightly more complicated.

i) We show that $\mathcal{F}$ is a contraction in the supremum norm. Since U is not complete in the supremum norm, the contraction mapping principle does not imply that $\mathcal{F}$ has a fixed point in U, but does imply that $\mathcal{F}$ has a fixed point in the completion of U with respect to the supremum norm.

ii) We show that the completion of U with respect to the supremum norm is contained in the set of cofunctions u from Y to X with Lipschitz-continuous k^{th} derivatives and with a second-order zero at the origin. Thus, the fixed point u^* of $\mathcal{F}$ has the differentiability asserted in the theorem.

i) Consider $u_1, u_2 \in U$, and let $\|u_1 - u_2\|_o = \sup_y \|u_1(y) - u_2(y)\|$. Let

$\tilde{y}_1(y)$, $\tilde{y}_2(y)$ denote the solution of

$$y = B\tilde{y}_i + Y(u_i(\tilde{y}_i),\tilde{y}_i) \qquad i = 1, 2 \ .$$

We shall estimate successively

$$\|\tilde{y}_1-\tilde{y}_2\|_o \ , \quad \|\mathcal{F}u_1-\mathcal{F}u_2\|_o \ .$$

Subtracting the defining equations for $\tilde{y}_1,\tilde{y}_2$, we get

$$B(\tilde{y}_1-\tilde{y}_2) = Y(u_2(\tilde{y}_2),\tilde{y}_2) - Y(u_1(\tilde{y}_1),\tilde{y}_1) \ ,$$

so that

$$\|\tilde{y}_1-\tilde{y}_2\| \leq \|B^{-1}\| \cdot \lambda \cdot [\|u_2(\tilde{y}_2)-u_1(\tilde{y}_1)\| + \|\tilde{y}_2-\tilde{y}_1\|] \ . \tag{8}$$

Since $\|Du_1\|_o \leq 1$, we can write

$$\begin{aligned}\|u_2(\tilde{y}_2)-u_1(\tilde{y}_1)\| &\leq \|u_2(\tilde{y}_2)-u_1(\tilde{y}_2)\| + \|u_1(\tilde{y}_2)-u_1(\tilde{y}_1)\| \\ &\leq \|u_2 - u_1\|_o + \|\tilde{y}_2 - \tilde{y}_1\| \ .\end{aligned} \tag{9}$$

Inserting (9) in (8) and rearranging yields

$$\begin{aligned}(1 - 2\lambda \cdot \|B^{-1}\|)\|\tilde{y}_1 - \tilde{y}_2\| &\leq \lambda \cdot \|B^{-1}\| \cdot \|u_2 - u_1\|_o \ , \qquad \text{or} \\ \|\tilde{y}_1 - \tilde{y}_2\|_o &\leq \lambda \cdot \gamma \cdot \|u_2 - u_1\| \ .\end{aligned} \tag{10}$$

Now insert estimates (9) and (10) in

$$\begin{aligned}\mathcal{F}u_1(y) - \mathcal{F}u_2(y) &= A[u_1(\tilde{y}_1) - u_2(\tilde{y}_2)] \\ &\quad + [X(u_1(\tilde{y}_1),\tilde{y}_1) - X(u_2(\tilde{y}_2),\tilde{y}_2)]\end{aligned}$$

to get

$$\begin{aligned}\|\mathcal{F}u_1 - \mathcal{F}u_2\|_o &\leq \|A\|[\|u_2-u_1\|_o + \|\tilde{y}_2-\tilde{y}_1\|_o] \\ &\quad + \lambda[\|u_2-u_1\|_o + 2\cdot\|\tilde{y}_2-\tilde{y}_1\|_o] \\ &\leq \|u_2-u_1\|_o\{\|A\|(1 + \gamma\lambda) + \lambda(1 + 2\gamma\lambda)\} \ .\end{aligned}$$

If we now require

$$\alpha = \|A\|(1 + \gamma\lambda) + \lambda(1 + 2\gamma\lambda) < 1 \quad , \tag{C.4}$$

$\mathcal{F}$ is a contraction in the supremum norm.

ii) The assertions we want all follow directly from the following general result.

Proposition A2. Let (u_n) be a sequence of functions on a Banach space Y with values on a Banach space X. Assume that, for all n, y

$$\|D^j u_n(y)\| \leq 1 \qquad j = 0, 1, 2, \ldots, k \quad ,$$

and that each $D^k u_n$ is Lipschitz continuous with Lipschitz constant one. Assume also that for each y, the sequence $(u_n(y))$ converges weakly (i.e., in the weak topology on X) to a unit $u(y)$. Then

a) u has a Lipschitz continuous k^{th} derivative with Lipschitz constant one.

b) $D^j u_n(y)$ converges weakly to $D^j u(y)$* for all y and $j = 1,2,\ldots,k$.

If X, Y are finite dimensional, all the Banach space technicalities in the statement of the proposition disappear, and the proposition becomes a straightforward consequence of the Arzela-Ascoli Theorem. We postpone the proof to the end of this appendix.

IV) We shall prove the following: Let $x \in X$ with $\|x\| \leq 1$ and let $y \in Y$ be arbitrary. Let $(x_1, y_1) = \Psi(x,y)$. Then

$$\|x_1\| \leq 1 \qquad \text{and}$$

$$\|x_1 - u^*(y_1)\| \leq \alpha \cdot \|x - u^*(y)\| \quad , \tag{11}$$

where α is as defined in (C.4). By induction,

---

*This statement may require some interpretation. For each n, y, $D^j u_n(y)$ is a bounded symmetric j-linear map from Y^j to X. What we are asserting is that, for each y, $y_1,\ldots,y_j$, the sequence $(D^j u_n(y)(y_1,\ldots,y_j))$ of elements of X converges in the weak topology on X to $D^j u(y)(y_1,\ldots,y_j)$.

$$\|x_n - u^*(y_n)\| \leq \alpha^n \|x - u^*(y)\| \to 0 \quad \text{as} \quad n \to \infty \quad ,$$

as asserted.

To prove $\|x_1\| \leq 1$, we first write

$$x_1 = Ax + X(x,y) \quad , \qquad \text{so that}$$

$$\|x_1\| \leq \|A\| \cdot \|x\| + \lambda \leq \|A\| + \lambda \leq 1 \quad \text{by} \quad (C.2) \quad .$$

To prove (11), we essentially have to repeat the estimates made in proving that $\mathcal{F}$ is a contraction. Let $\tilde{y}_1$ be the solution of

$$y_1 = B\tilde{y}_1 + Y(u^*(\tilde{y}_1), \tilde{y}_1) \quad .$$

On the other hand, by the definition of y_1 we have

$$y_1 = By + Y(x,y) \quad .$$

Subtracting these equations and proceeding exactly as in the derivation of (10), we get

$$\|\tilde{y}_1 - y\| \leq \lambda \cdot \gamma \cdot \|u^*(y) - x\| \quad .$$

Next, we write

$$u^*(y_1) = \mathcal{F}u^*(y_1) = Au^*(y_1) + X(u^*(\tilde{y}_1), \tilde{y}_1)$$

$$x_1 \qquad\qquad = Ax \qquad + X(x,y) \quad .$$

Subtracting and making the same estimates as before, we get

$$\|x_1 - u^*(y_1)\| \leq \alpha \cdot \|x - u^*(y)\|$$

as desired. Except for the proof of Proposition A2, this completes the proof of Theorem A1 and hence of the version of the Center Manifold Theorem used by Ruelle and Takens.

It may be noted that we seem to have lost some differentiability in passing from Ψ to u^* , since we assumed that Ψ is C^{k+1} and only concluded that u^*

is C^k . In fact, however, the u^* we obtain has a Lipschitz continuous k^{th} derivative, and our argument works just as well if we only assume that Ψ has a Lipschitz continuous k^{th} derivative. Moreover, if we make the weaker assumption that the kth derivative of Ψ is uniformly continuous on some neighborhood of zero, we can show that the same is true of u^* . (Of course, if X and Y are finite dimensional, continuity on a neighborhood of zero implies uniform continuity on a neighborhood of zero, but this is no longer true if X or Y is infinite dimensional.)

It does not seem to be known, if Ψ is infinitely differentiable, whether the center manifold can be taken to be infinitely differentiable. It is definitely not true, however, that, if Ψ is analytic there is an analytic center manifold. We shall give a counterexample in the context of equilibrium points of differential equations rather than fixed points of maps. Consider the system of equations:

$$\frac{dy_1}{dt} = -y_2 , \quad \frac{dy_2}{dt} = 0 , \quad \frac{dx}{dt} = -x + h(y_1) , \qquad (12)$$

where h is analytic near zero and has a second-order zero at zero. We claim that, if h is not analytic in the whole complex plane, there is no function $u(y_1,y_2)$, analytic in a neighborhood of $(0,0)$ and vanishing to second order at $(0,0)$, such that the manifold

$$\{x = u(y_1,y_2)\}$$

is locally invariant under the flow induced by the differential equation near $(0,0)$. To see this, we assume that we have an invariant manifold with

$$u(y_1,y_2) = \sum_{\substack{j_1,j_2 \\ j_1+j_2\geq 1}} c_{j_1j_2} y_1^{j_1} y_2^{j_2} .$$

Straightforward computation shows that the expansion coefficients c_{j_1,j_2} are uniquely determined by the requirement of invariance and that

$$c_{j_1,j_2} = \frac{(j_1+j_2)!}{(j_1)!} h_{j_1+j_2} ,$$

where

$$h(y_1) = \sum_{j \geq 2} h_j y_1^j .$$

If the series for h has a finite radius of convergence, the series for $u(0,y_2)$ diverges for all non-zero y_2.

The system of differential equations has nevertheless many infinitely differentiable center manifolds. To construct one, let $\tilde{h}(y_1)$ be a bounded infinitely differentiable function agreeing with h on a neighborhood of zero. Then the manifold defined by

$$u(y_1,y_2) = \int_{-\infty}^{0} d\sigma\, e^{\sigma}\, \tilde{h}(y_1-\sigma y_2) \tag{13}$$

is easily verified to be globally invariant for the system

$$\frac{dy_1}{dt} = -y_2 , \quad \frac{dy_2}{dt} = 0 , \quad \frac{dx}{dt} = -x + \tilde{h}(y_1) \tag{14}$$

and hence locally invariant at zero for the original system.

(To make our expression for u less mysterious, we sketch its derivation. The equations for y_1,y_2 do not involve x and are trivial to solve explicitly. A function u defining an invariant manifold for the modified system (14) must satisfy

$$\frac{d}{dt} u(y_1-ty_2,y_2) = -u(y_1-ty_2) + \tilde{h}(y_1-ty_2)$$

for all t,y_1,y_2. The formula (13) for u is obtained by solving this ordinary differential equation with a suitable boundary condition at $t = -\infty$.)

To complete this appendix we must prove Proposition A2. We shall give the argument only for $k = 1$; the generalization to arbitrary k is a straightforward induction argument.

We start by choosing $y_1,y_2 \in Y$ and $\varphi \in X^*$ and consider the sequence of real-valued functions of a real variable

$$t \to \varphi(u_n(y_1 + ty_2)) \equiv \psi_n(t) .$$

From the assumptions we have made about the sequence (u_n), it follows that

$$\lim_{n \to \infty} \psi_n(t) = \varphi(u(y_1 + ty_2)) \equiv \psi(t)$$

for all t, that $\psi_n(t)$ is differentiable, that

$$|\psi_n'(t)| \leq \|\varphi\| \cdot \|y_2\| \quad \text{for all } n, t$$

and that

$$|\psi_n'(t_1) - \psi_n'(t_2)| \leq \|\varphi\| \cdot \|y_2\|^2 \; |t_1 - t_2|$$

for all n, t_1, t_2. By this last inequality and the Arzela-Ascoli Theorem, there exists a subsequence $\psi_{n_j}'(t)$ which converges uniformly on every bounded interval. We shall temporarily denote the limit of this subsequence by $X(t)$. We have

$$\psi_{n_j}(t) = \psi_{n_j}(0) + \int_0^t \psi_{n_j}'(\tau)d\tau \quad ;$$

hence, passing to the limit $j \to \infty$, we get

$$\psi(t) = \psi(0) + \int_0^t X(\tau)d\tau \quad ,$$

which implies that $\psi(t)$ is continuously differentiable and that

$$\psi'(t) = X(t) \quad .$$

To see that

$$\lim_{n \to \infty} \psi_n'(t) = \psi'(t)$$

(i.e., that it is not necessary to pass to a subsequence), we note that the argument we have just given shows that any subsequence of $(\psi_n'(t))$ has a subsequence converging to $\psi'(t)$; this implies that the original sequence must converge to this limit.

Since

$$\psi_n'(0) = \varphi(Du_n(y_1)(y_2)) \quad ,$$

we conclude that the sequence

$$Du_n(y_1)(y_2)$$

converges in the weak topology on X^{**} to a limit, which we shall denote by $Du(y_1)(y_2)$; this notation is at this point only suggestive. By passage to a limit from the corresponding property of $Du_n(y_1)(y_2)$, we see that

$$y_2 \to Du_n(y_1)(y_2)$$

is a bounded linear mapping of norm ≤ 1 from Y to X^{**} for each y_1 . We denote this linear operator by $Du(y_1)$. Since

$$\|Du_n(y_1)(y_2) - Du_n(y_1')(y_2)\| \leq \|y_1 - y_1'\| \cdot \|y_2\| \quad ,$$

we have

$$\|Du(y_1) - Du(y_1')\| \leq \|y_1 - y_1'\| \quad ,$$

i.e., the mapping $y \to Du(y)$ is Lipschitz continuous from Y to $L(Y,X^{**})$.

The next step is to prove that

$$u(y_1+y_2) - u(y_1) = \int_0^1 d\tau \; Du(y_1+\tau y_2)(y_2) \quad ; \tag{15}$$

this equation together with the norm-continuity of $y \to Du(y)$ will imply that u is (Frechet)-differentiable. The integral in (15) may be understood as a vector-valued Riemann integral. By the first part of our argument,

$$\varphi(u(y_1+y_2) - \varphi(u(y_1)) = \int_0^1 d\tau \; \varphi(Du(y_1+\tau y_2)(y_2)) \quad ,$$

for all $\varphi \in X^{**}$, and taking Riemann integrals commutes with continuous linear mappings, so that

$$\varphi([u(y_1+y_2) - u(y_1) - \int_0^1 d\tau \; Du(y_1+\tau y_2)(y_2)]) = 0$$

for all $\varphi \in X^*$. Therefore (15) is proved.

The situation is now as follows: We have shown that, if we regard u as a mapping into X^{**} , which contains X , then it is Frechet differentiable with derivative Du . On the other hand, we know that u actually takes values in X and want to conclude that it is differentiable as a mapping into X . This is equivalent to proving that $Du(y_1)(y_2)$ belongs to X for all y_1, y_2 . But

$$Du(y_1)(y_2) = \text{norm} \lim_{t \to 0} \frac{u(y_1+ty_2)-u(y_1)}{t} \quad ;$$

the difference quotients on the right all belong to X, and X is norm closed in X^{**}. Thus, $Du(y_1)(y_2)$ is in X and the proof is complete.

APPENDIX B. THE CANONICAL FORM

We shall give here an elementary and straightforward derivation of the canonical form for the mapping $\Phi_\mu: \mathbb{R}^2 \to \mathbb{R}^2$. Recall that we have already arranged things so that

$$\Phi_\mu\binom{x}{y} = (1+\mu)\begin{pmatrix} \cos\theta(\mu) & -\sin\theta(\mu) \\ +\sin\theta(\mu) & \cos\theta(\mu) \end{pmatrix}\binom{x}{y} + O(r^2) \quad .$$

We want to organize the second, third, and fourth degree terms by making further coordinate changes. It will be convenient to identify $\mathbb{R}^2$ with the complex plane by writing

$$z = x + iy \quad .$$

Then

$$\Phi_\mu(z) = \lambda(\mu)z + O(|z|^2) \;, \quad \lambda(\mu) = (1+\mu)e^{i\theta(\mu)} \quad .$$

From now on we shall leave μ out of our notation as much as possible.

The higher-order terms in the Taylor series for Φ may be written as polynomials in z and $\bar{z}$, i.e.,

$$\Phi(z) = \lambda z + A_2(z) + A_3(z) + \ldots \;,$$

where, for example,

$$A_2(z) = \sum_{j=0}^{2} \alpha_j^2 z^j \bar{z}^{2-j} \quad .$$

Let us begin in a pedestrian way with A_2. We choose a new coordinate $z' = z + \gamma(z)$, where γ is homogeneous of degree 2, i.e., has the same form as A_2. We can invert the relation between z and z' as

$$z = z' - \gamma(z') + \text{higher order terms.}$$

Since for the moment we are only concerned with terms of degree 2 or lower, we calculate module terms of degree 3 and higher and replace equality signs by congruance signs (≡) . Thus we have

$$z \equiv z' - \gamma(z') = (I - \gamma)(z') \quad .$$

In terms of the new coordinate we have

$$\begin{aligned}
\Phi'(z') &\equiv (I + \gamma)\Phi(z' - \gamma(z')) \\
&\equiv (I + \gamma)[\lambda z' - \lambda\gamma(z') + A_2(z' - \gamma(z'))] \\
&\equiv (I + \gamma)[\lambda z' - \lambda\gamma(z') + A_2(z')] \\
&\equiv \lambda z' - \lambda\gamma(z') + A_2(z') + \gamma(\lambda z' - \lambda\gamma(z') + A_2(z')) \\
&\equiv \lambda z' + A_2(z') + \gamma(\lambda z') - \lambda\gamma(z') \quad .
\end{aligned}$$

Now

$$\gamma(z') = \gamma_2 z'^2 + \gamma_1 z'\overline{z'} + \gamma_0 \overline{z'}^2$$

$$\gamma(\lambda z') - \lambda\gamma(z') = \gamma_2(\lambda^2 - \lambda)z'^2 + \gamma_1(|\lambda|^2 - \lambda)z'\overline{z'} + \gamma_0(\bar{\lambda}^2 - \lambda)\overline{z'}^2 \ .$$

On the other hand,

$$A_2(z') = \alpha_2^2 z'^2 + \alpha_1^2 z'\overline{z'} + \alpha_0^2 \overline{z'}^2 \quad ,$$

so, if we put

$$\gamma_2 = \frac{-\alpha_2^2}{\lambda^2 - \lambda} \ , \quad \gamma_1 = \frac{-\alpha_1^2}{|\lambda|^2 - \lambda} \ , \quad \gamma_0 = \frac{-\alpha_0^2}{\bar{\lambda}^2 - \lambda} \ ,$$

we get

$$\Phi'(z') = \lambda z' + O(|z'|^3) \quad .$$

We must, of course, make sure that the denominators in our expressions for the γ_i do not vanish. Since $|\lambda| = 1+\mu$, there is no problem for $\mu \neq 0$, but we want our μ-dependent coordinate change to be well-behaved as $\mu \to 0$. This will be the case provided

$$e^{2i\theta(0)} \neq e^{i\theta(0)} \ , \quad 1 \neq e^{i\theta(0)} \ , \quad e^{-2i\theta(0)} \neq e^{i\theta(0)} \ ,$$

i.e., provided

$$e^{i\theta(0)} \neq 1 , \quad e^{3i\theta(0)} \neq 1 .$$

Thus, if these conditions hold, we can make a smooth μ-dependent coordinate change, bringing A_2 to zero. We assume that we have made this change and drop the primes:

$$\Phi(z) = \lambda z + A_3(z) + \dots .$$

(A_3 is <u>not</u> the original A_3) and see what we can do about A_3.

This time, we take a new coordinate $z' = z + \gamma(z)$, γ homogeneous of degree 3, and we calculate modulo terms of degree 4 and higher. Just as before, we have

$$\begin{aligned} \Phi'(z') &\equiv (\ + \gamma)\Phi(z' - \gamma(z')) \\ &\equiv \lambda z' + A_3(z') + \gamma(\lambda z') - \lambda\gamma(z') \ . \end{aligned}$$

Again, we write out

$$\begin{aligned} \gamma(z') &= \gamma_3 z'^3 + \gamma_2 z'^2\bar{z}' + \gamma_1 z\bar{z}'^2 + \gamma_0 \bar{z}'^2 \\ \gamma(\lambda z')-\lambda\gamma(z') &= \gamma_3(\lambda^3-\lambda)z'^3 + \gamma_2(|\lambda|^{-}-1)\lambda z'^2\bar{z}' \\ &\quad + \gamma_1(|\lambda|^2\bar{\lambda}-\lambda)z'\bar{z}'^2 + \gamma_0(\bar{\lambda}^3-\lambda)\bar{z}'^3 \\ A_3(z') &= \alpha_3^3 z'^3 + \alpha_2^3 z'^2\bar{z}' + \alpha_1^3 z'\bar{z}'^2 \div \alpha_0^3\bar{z}'^3 \ . \end{aligned}$$

By an appropriate choice of $\gamma_3, \gamma_1, \gamma_0$, we can cancel the $\alpha_3^3, \alpha_1^3, \alpha_0^3$ terms provided

$$e^{2i\theta(0)} \neq 1 , \quad e^{4i\theta(0)} \neq 1 .$$

The α_2^3 term presents a new problem. For $\mu \neq 0$, we can, of course, cancel it by putting

$$\gamma_3 = \frac{-\alpha_2^3}{\lambda(|\lambda|^2-1} .$$

This expression, however, diverges as $\mu \to 0$, independent of the value of $\theta(0)$.

For this reason, we shall not try to adjust this term and simply put $\gamma_3 = 0$. Then, in the new coordinates (dropping the primes)

$$\Phi(z) = (\lambda + \alpha_2^3 |z|^2)z + O(|z|^4) .$$

We next set out to cancel the 4th degree terms by a coordinate change $z' = z + \gamma(z)$, γ homogeneous of degree 4. A straightforward calculation of a by now familiar sort shows that such a coordinate change does not affect the terms of degree ≤ 3 and that all the terms of degree 4 can be cancelled provided

$$e^{5i\theta(0)} \neq 1 .$$

Thus we get

$$\Psi(z) = (\lambda + \alpha_2^3|z|^2)z + O(|z|^3) .$$

This is still not quite the desired form. To complete the argument, we write

$$\lambda + \alpha_2^3|z|^2 = (1+\mu)e^{i\theta(\mu)}[1 - \frac{f_1(\mu)}{1 + \mu} |z|^2 + i f_3(\mu)|z|^2]$$

$$(f_1, f_3 \text{ real})$$

$$= (1 + \mu - f_1(\mu)|z|^2)e^{i[\theta(\mu)+f_3(\mu)|z|^2]} + O(|z|^4) .$$

Thus

$$\Phi(z) = (1+\mu-f_1(\mu)|z|^2)e^{i[\theta(\mu)+f_3(\mu)|z|^2]} z + O(|z|^5) ;$$

when we translate back into polar coordinates, we get exactly the desired canonical form.

This work was performed while the author was an Alfred P. Sloan Foundation Fellow and was supported in part by N.S.F. grant GP 15735.

REFERENCES

1. Bonic, R. and Frampton, J., Smooth Functions on Banach Manifolds, J. Math. Mech. 15 (1966), 877-898.

2. Dunford, H., and Schwartz, J. T., Linear Operators I. New York: Interscience, 1958.

3. Kelley, A., The Stable, Center-Stable, Center, Center-Unstable, and Unstable Manifolds; Appendix C of R. Abraham and J. Robbins, Transversal Mappings and Flows. New York: W. A. Benjamin, 1967.

4. Ruelle, D. and Takens, F., On the Nature of Turbulence, Commun. Math. Phys. 20 (1971), 167-192.
   Note Concerning our Paper "On the Nature of Turbulence", Comm. Math. Phys. 23 (1971), 343-344.

5. Siegel, C. L. and Moser, J. K., Lectures on Celestial Mechanics. Berlin, Heidelberg, New York: Springer-Verlag, 1971.

ERGODIC THEORY AND STATISTICAL MECHANICS OF NON-EQUILIBRIUM PROCESSES

Joel L. Lebowitz
Belfer Graduate School of Science
Yeshiva University
New York, New York 10033

SUMMARY

I. INTRODUCTION

Statistical Mechanics of Non-Equilibrium Processes like other physical theories has two aspects: (1) the mathematical investigation of certain well-posed, that is, mathematically formulated problems and (2) the proper mathematical formulation of physical phenomena. It is wise, I believe, to separate these two parts of the problem, and this talk is primarily about some of the progress that has been made in recent years in the mathematical theory of measure preserving transformations. Since the dynamical flow in the phase space, which describes the time evolution of a Hamiltonian system, is an example of such a transformation, this work has, in my opinion, much relevance to statistical mechanics and to the question of irreversibility. The progress in this field is the result of the work of many people: Hopf, Kolmogoroff, Sinai and others, most of whose names and work I shall not have a chance to mention. (A small bibliography is given at the end.)

II. EQUILIBRIUM ENSEMBLES

Let me begin by recalling briefly the situation in equilibrium statistical mechanics: There is a purely macroscopic theory, thermodynamics, which states that a great variety of the properties of a large system, of a given quantity $N (N = \text{particle number} \sim 10^{26})$ contained in a volume V, which is in equilibrium, are determined once its energy E is known. This information about the system is contained in a function $S(E,N,V)$, the entropy, which is extensive, that is, $S(E,N,V) = V\ s(E/V,N/V) \equiv V\ s(e,\rho)$ and which has certain convexity properties that insure thermodynamic stability. From $s(e,\rho)$ the pressure, temperature, etc., can be found. (If, instead of specifying e, the energy per unit volume, we specify the temperature $T=\beta^{-1}$ (in units in which Boltzmann's constant is unity) the 'thermodynamics' is determined by the Helmholtz free energy $A(\beta,N,V)=Va(\beta,\rho)$, which is related to the entropy by a Legendre transformation.)

Now, according to statistical mechanics as developed by Boltzmann, Gibbs, Einstein and others, these thermodynamic functions for macroscopic systems are obtainable from a knowledge of the microscopic structure of the system by well known formulae. The microscopic nature of the system is specified by its Hamiltonian which we assume to be the sum of a kinetic energy term and a potential energy term

$$H_N(x) = \sum_{i=1}^{N} \frac{1}{2m} p_i^2 + U(\underline{r}_1,\ldots,\underline{r}_N) \quad , \tag{1}$$

where m is the mass of a particle, $\underline{r}_i$, $\underline{p}_i$ are the position and momentum vectors of the ith particle, $\underline{r}_i \in V$, $\underline{p}_i \in R^3$ and $x=(\underline{r}_1,\ldots\underline{r}_N, \underline{p}_1,\ldots,p_N)$ is a point in the phase space Γ of M (=6N) dimensions. We shall say that x specifies the 'dynamical state' of the system. To obtain the entropy $S(E,N,V)$ of this (classical) system, we define the energy surface S_E by the relation $H_N(x)=E$ for $x \in S_E$ and equate the entropy to the logarithm of its 'surface area,'

$$S(E,N,V) = \ell n|S_E| \; , \; |S_E| = \int_{S_E} \frac{d\sigma_E(x)}{|\nabla H|} \quad , \tag{2}$$

where $d\sigma_E$ is the $M-1$ dimensional surface area element on S_E induced by the Euclidian metric on Γ and $|\nabla H|$ is the length of the gradient of H. (Similarly, $A(\beta,N,V) \sim \ell n\{\int \exp[-\beta H_N(x)]dx\}$.)

The statistical mechanics of Gibbs does not stop at giving formulae for the thermodynamic potentials $S(E,N,V)$ (or $A(\beta,N,V)$); it interprets these formulas in terms of 'ensembles' or probabilities; i.e., if $f(x)$ is a real valued function (dynamical observable) of the dynamical state of the system which is accessible to macroscopic measurements, the results of measuring its values in an equilibrium system with energy E will have a probability distribution obtained from a normalized probability measure, given by

$$d\mu_o(x) = \begin{cases} |S_E|^{-1}\, d\sigma_E(x)/|\nabla H(x)| & , \quad x \in S_E \\ 0 & , \quad x \notin S_E \end{cases} \quad . \tag{3}$$

The probability density μ_o given in (3) (which is left invariant by the dynamical flow in Γ) is usually called the Gibbs micro-canonical ensemble density. Thus, according to statistical mechanics, if we make many observations on one system or on a collection (ensemble) of equilibrium systems with energy E (having, of course, the same Hamiltonian), then the average and mean square deviation of f are given by

$$\langle f(x)\rangle = \int f(x)d\mu_o(x) \quad , \quad \langle [f(x)-\langle f\rangle]^2\rangle = \int [f(x) - \langle f\rangle]^2 \, d\mu_o(x) \quad . \tag{4}$$

For measurements which take a certain amount of time to perform, the appropriate dynamical functions are of the form

$$\bar{f}_T(x) = \frac{1}{T}\int_0^T f(x_t)dt \quad . \tag{5}$$

and

$$\langle \bar{f}_T(x)\rangle = \frac{1}{T}\int_0^T \langle f(x_t)\rangle \, dt = \langle f\rangle \quad , \tag{6}$$

where x_t is the point in the phase space or dynamical state of the system at time t if x is its dynamical state at $t=0$, and in (6) we have used the fact that $d\mu_o(x)$ is time invariant, $d\mu_o(x_t) = d\mu_o(x)$. x_t is obtained from x through the solution of the Hamiltonian equations of motion

$$\dot{\underline{r}}_i = \frac{\partial H}{\partial \underline{p}_i} = m^{-1}\underline{p}_i = \underline{v}_i \quad , \quad \dot{\underline{p}}_i = -\frac{\partial H}{\partial \underline{r}_i} \quad , \tag{7}$$

together with 'boundary conditions' on the surface of the region V in R^3. (The boundaries are usually taken to be reflective, i.e., the component of the velocity $\underline{v}_i$ is reversed when the ith particle hits the boundary.)

In particular, thermodynamic quantities such as the pressure are equated in statistical mechanics with the ensemble average of the corresponding dynamical observables.

The two aspect of equilibrium statistical mechanics are then (1) the investigation, for a given Hamiltonian, of the actual form of the thermodynamic functions and other expectation values, such as those which enter in the scattering of X-rays by fluids or crystals and (2) an understanding of why the predictions of statistical mechanics work as well as they do in relating the observed properties of an equilibrium macroscopic system to the corresponding ensemble averages.

Considerable success has been achieved in the first aspect, and there is no doubt at all that statistical mechanics works for macroscopic systems. The fact that the systems are macroscopic is very important here. Indeed it is only in the so-called thermodynamic limit in which the size of the system formally becomes infinite, $N \to \infty$, $V \to \infty$, $N/V \to \rho$, $E/V \to e$, that the statistical mechanically computed entropy per unit volume (or free energy per unit volume), $s(e,\rho)$ (or $a(\beta,\rho)$) have the right thermodynamic stability properties. It is also only in this limit that the most striking aspects of equilibrium phenomena, phase transitions, have an unambiguous qualitative meaning; they correspond to (mathematical) singularities of $s(e,\rho)$ or $a(\beta,\rho)$. One can also show, in the thermodynamic limit, the equivalence of various ensembles in predicting the results of macroscopic measurements.

The justification of the use of ensembles is in a much less satisfactory state at the present time. Some of the elements entering into an explanation are (1) the special nature of the dynamical functions which are accessible to measurement in macroscopic systems: they all have small dispersions, e.g., $\{[\langle f^2\rangle - \langle f\rangle^2]/\langle f\rangle^2\}$ is very small, (the dispersion actually goes to zero in the thermodynamic limit when $\langle f\rangle$ can be identified with a thermodynamic quantity) and (2) the ergodic hypothesis: $\lim_{T\to\infty} \bar{f}_T(x) \equiv \lim_{T\to\infty} 1/T \int_0^T f(x_t)dt \equiv \bar{f}$ exists and is independent of x for almost all x, $x \in S_E$ and $f(x) \in L_1$, i.e., $\langle |f(x)|\rangle = \int |f(x)| d\mu_0(x) < \infty$.

It is easy to show that when the system is ergodic, $\bar{f} = \langle f(x)\rangle$. Hence, if the 'effective time' T, in macroscopic observables $\bar{f}_T(x)$, were comparable with the time involved in the ergodic statement, i.e., with the T necessary for $\bar{f}_T(x)$ to become approximately equal to $\bar{f}$ for almost all dynamical states x, we would have an "explanation." This is, however, clearly not the case, since if it were true we would never observe non-equilibrium phenomena (except possibly for dynamical states x which lie on trajectories whose measure is zero, in which case the ergodicity property would be irrelevant). It is true, however, that if a macroscopic system is observed for a 'very long time,' then the time average of the macroscopic observables $f(x)$ agrees with the predictions of equilibrium statistical mechanics. Hence some ergodicity property for macroscopic observables seems necessary but not sufficient to justify the Gibbs formalism.

The property of small fluctuations of macroscopic observables also seems more of a necessary than a sufficient condition for justifying the Gibbs assumptions. It helps us in understanding how a macroscopic system whose dynamical state keeps on changing all the time can have macroscopic observables $f(x)$ or $\bar{f}_T(x)$ which appear to be constant in time, i.e., the system appears to reach and remain in an equilibrium state.

The property of small fluctuations, which can actually be proved quite generally, may also have some relevance to the question of ergodicity. If the fluctuations in f are small, then $f(x) \sim \langle f \rangle$ for <u>most</u> $x \in S_E$, which implies that starting with a 'typical' x, $f(x_t) \sim \langle f \rangle$ for most of the time t, which in turn makes it plausible that $T^{-1}\int_0^T f(x,t)dt \to \langle f \rangle$ as $T \to \infty$ for almost all x, i.e., that the system is ergodic with respect to the 'relevant' $f(x)$. (Ergodicity itself however, does not require large systems and does not imply small fluctuations. Consider a simple one-dimensional oscillator $H = (2m)^{-1}p^2 + 1/2\, m\omega^2 q^2$. This system is ergodic on each energy surface with $\langle p^2 \rangle = mE$, $\langle p^4 \rangle = 3/2 \langle p^2 \rangle^2$. Ergodicity is thus not incompatible with large fluctuations.)

I feel therefore that much further work is necessary to explain the applicability of the ensemble method to equilibrium phenomena. This is even more so when we come to a discussion of non-equilibrium phenomena where we shall again adopt the ensemble, or probability, method with or without a good 'explanation.'

III. NON-EQUILIBRIUM ENSEMBLES

There are two entirely equivalent ways in which to proceed. The first approach which builds directly upon what we have already discussed for equilibrium systems is as follows: We consider what Penrose calls a 'compound observation,' measuring first the value of a dynamical function $f(x)$ and then <u>at a time t later</u>, measuring the value of a dynamical function $g(x)$. An appropriate dynamical function for measuring the correlations in this compound observation is the explicitly time-dependent dynamical function $f(x)g(x_t)$ whose expectation value, in an equilibrium ensemble, should be given by

$$\langle g(t)f\rangle = \int g(x_t)f(x)\mu_0(x)dx = \int g(x)[f(x_{-t})\mu_0(x_{-t})]dx_{-t}$$
$$= \int g(x)f(x_{-t})\mu_0(x)dx \quad . \tag{8}$$

You will recognize $\langle g(t)f\rangle$ as a time dependent correlation function in an equilibrium ensemble.

An alternative way of looking at equation (8) is to think of $[f(x)\mu_0(x)]$ as a non-equilibrium ensemble density at time $t=0$, i.e., we set $\mu(x,t=0) = K\mu_0(x)f(x)$ where K is a normalization constant ($f(x)$ can 'always' be made positive by adding a suitable constant). The ensemble density at time t, $\mu(x,t)$, then satisfies the Liouville equation

$$\frac{\partial\mu(x,t)}{\partial t} = (H,\mu) \equiv -iL\mu \quad , \tag{9}$$

where (H,μ) is the Poisson bracket with the Hamiltonian and L is the 'Liouville operator'. The solution of (9) is

$$\mu(x,t) = \exp[-itL]\mu(x,0) = U_{-t}\mu(x,0) = \mu(x_{-t},0) = Kf(x_{-t})\mu_0(x) \quad . \tag{10}$$

U_t is a unitary operator $U_t\varphi(x) = \varphi(x_t)$ on functions in $L_2(\Gamma)$, $\int|\varphi(x)|^2 d\mu_0 < \infty$. $\langle g(t)f\rangle$ in (8) is then the expectation value at time t of the dynamical function $K^{-1}g(x)$ in a system represented by an ensemble density $\mu(x,t)$. The interpretation of $\langle g(t)f\rangle$ in terms of non-equilibrium ensemble densities is particularly clear if $f(x)$ and $g(x)$ are characteristic functions of some sets

$$A, B \subset S_E, \; f(x) = \begin{cases} 1, x \in A \\ 0, x \notin A \end{cases}, \; g(x) = \begin{cases} 1, x \in B \\ 0, x \notin B \end{cases} \quad , \tag{11}$$

(A and B could correspond to regions in S_E where some dynamical functions $F(x)$ and $G(x)$ have values a and b, respectively, or are in the ranges (a_1,a_2), (b_1,b_2).) Calling $\mu_0(A)$ the measure (volume) of A with respect to the equilibrium measure $d\mu_0$, we get

$$\mu_0(A) = \int_A d\mu_0(x) \; , \; \mu_0(B) = \int_B d\mu_0(x) \quad ;$$

then $\mu(x,0) = f(x)\mu_o(x)/\mu_o(A)$ represents a normalized ensemble density which is concentrated 'uniformly' in A. Calling $U_t A = A_t$, with

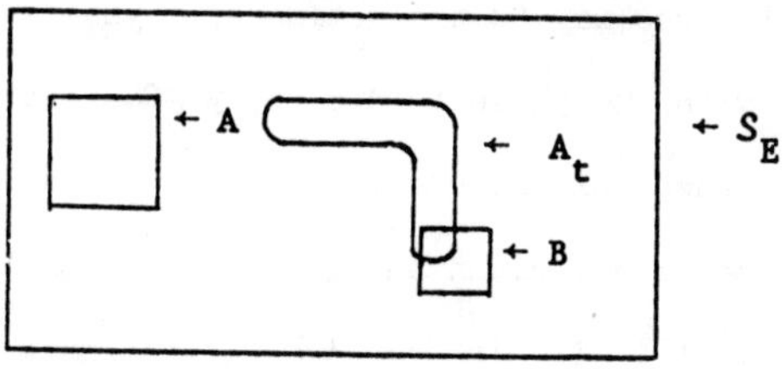

A_t the set of points $\{x: x_{-t} \in A\}$, we then note that $\mu(x,t) = f(x_{-t})\mu_o(x)/\mu_o(A)$ is an ensemble density concentrated uniformly in A_t. The expectation value of $g(x)$ in this non-equilibrium ensemble is then

$$\int g(x)\mu(x,t)dx = [\mu_o(A)]^{-1}\int g(x_t)f(x)d\mu_o(x) = \mu_o(A_t \cap B)/\mu_o(A) \quad , \quad (12)$$

which is simply the measure of the overlap of the set A_t with B, i.e., it is the 'fraction' of systems in the ensemble (all of which were originally in A) that are in B at time t.

A. MIXING FLOW

Pursuing this analysis further, it seems reasonable to say that a necessary condition for a system to 'approach equilibrium' is that, after a 'long time,' the ensemble density $\mu(x,t)$ becomes 'spread out' over the whole energy surface S_E, i.e.,

$$\frac{\mu_o(A_t \cap B)}{\mu_o(A)} \xrightarrow[t\to\infty]{} \mu_o(B) \quad , \quad \text{for all sets } A, B, \text{ with finite measure.} \quad (13a)$$

When the Hamiltonian flow has the property (13a), then the system is called mixing, a notion introduced by Hopf in 1934. The property of mixing may also be expressed directly in terms of the time dependent correlation functions; it can be shown that a system is mixing if and only if

$$\langle g(t)f\rangle \xrightarrow[t\to\infty]{} \langle f\rangle\langle g\rangle \quad , \quad (13b)$$

whenever f and g are in L_2,

$$\int|f|^2 d\mu_0 < \infty \ , \ \int|g|^2 d\mu_0 < \infty \quad .$$

Mixing is related in an inverse way to the notion of 'dynamical stability' of a flow. Suppose we start out with two phase points $x, y \in S_E$ and follow their trajectories x_t, y_t. We can then ask the question of 'how close' x_t and y_t will be if x and y are close, i.e., given an $\epsilon > 0$ does there exist a $\delta > 0$ such that $|x_t - y_t| \leq \epsilon$ for all t, if $|x - y| \leq \delta$? If the answer is yes, then the system is said to be dynamically stable. Clearly, a system which is mixing is dynamically unstable since an M-1 dimensional ball of radius δ centered on x will spread over the whole energy surface S_E as $t \to \infty$.

The property of mixing is stronger than, and implies, ergodicity. (It follows from (13a) that $A_t = A \Longrightarrow \mu_0(A) = [\mu_0(A)]^2$ or $\mu_0(A) = 0$ or 1 which is true iff the system is ergodic.) It represents the kind of 'loss of memory' which Gibbs expected that 'coarse graining' in some way or another would bring about. Note however that a Hamiltonian system which is mixing in the forward time direction is also mixing in the backward time direction, i.e., (13) also holds with $t \to -t$. Mixing, or still stronger properties on the flow, which we shall discuss later, may thus be necessary but are (by themselves) certainly not sufficient to give time a direction. What these mixing properties show is that <u>initial</u> non-equilibrium ensemble densities of a certain type (those which are 'smooth' with respect to $d\mu_0$) will approach, in a certain well-defined sense, the equilibrium ensemble density. (More on this later.)

Until recently there was no example of a dynamical system which is mixing. Recently, however, Sinai was able to prove that a system consisting of a <u>finite</u> number N, $N \geq 2$ of hard spheres (or hard disks in two dimensions), confined to a cubical box, is a mixing system. (Only part of the proof has been published so far.) It follows, therefore, in particular from Sinai's work, that for a <u>finite</u> system of two or more hard spheres in a box in two or three dimensions, the velocity auto-correlation function of any particle, say particle one, approaches zero as $t \to \infty$,

$$\langle v_1(t) v_1 \rangle \xrightarrow[t \to \infty]{} \langle v_1 \rangle \langle v_1 \rangle = 0 \quad . \qquad (14)$$

This is indeed remarkable and contrary to some folklore opinion which holds that it is necessary to go to an infinite size system in order to obtain a true decay of the correlation functions when $t \to \infty$. Here, on the other hand, this is shown to be true for a system consisting of N, $N \geq 2$, particles, as long as N is finite. (It is presumably true also for an infinite system, but this is far from proved.) The usual reason for the belief in the necessity of going to an infinite system is that for a finite system, one always has a finite Poincaré recurrence time for each dynamical state x, and hence it is thought that the time correlation functions too will have such 'recurrences.' Note however that the mixing definition is meaningful only for sets of positive measure and that the equivalent decay of correlation functions definition applies only to square integrable functions (with respect to $d\mu_0$). What mixing therefore implies is that the times when different systems, which were initially together in the same set A (which can be as small as desired as long as $\mu_0(A) > 0$), return to the 'neighborhood of A' are so different from each other that eventually the set A_t spreads out uniformly over all of S_E. This is the important property of mixing flows which does not follow at all from ergodicity alone and is in particularly striking contrast to what occurs in assemblies of oscillators such as harmonic crystals where all the phase points have the same periodicities. (Finite quantum systems behave as oscillators; c.f., remark at the end of Section C.)

B. TRANSPORT COEFFICIENTS

The study of time correlation functions, such as the velocity auto-correlation function in (14), plays a central role in the statistical mechanical theory of non-equilibrium phenomena. Their importance stems from the fact that linear transport coefficients, such as heat conductivity, viscosity, etc., may be expressed as integrals over time (from $t = 0$ to $t = \infty$) of the time correlation of appropriate dynamical functions (Einstein-Green-Kubo). These functions represent the 'fluxes' associated with the transport processes in question. A well known example of such a 'formula' is the Einstein relation between the self-diffusion constant D and the integral of the velocity auto-correlation function.

It might appear from (13b) that for mixing systems these transport coefficients could be defined meaningfully, without going to the thermodynamic limit of an infinite size system, as long as $[\langle g(t)f\rangle - \langle g\rangle\langle f\rangle]$ approached zero sufficiently rapidly to be integrable. Unfortunately, such is not the case since the flux functions, whose time correlations are of interest for transport coefficients, can generally be written as Poisson brackets with the Hamiltonian H, i.e., $f = (F,H)$, $g = (G,H)$; and for mixing systems it can be shown that when f, g, F, G are all square integrable with respect to $d\mu_o$, then

$$\lim_{T\to\infty}\int_o^T \langle f(t)g\rangle dt = \langle (F,H)G\rangle \quad ,$$
$$\lim_{T\to\infty}\int_o^T \langle f(t)f\rangle dt = 0 \quad . \tag{15}$$

Thus for a <u>finite</u> mixing system confined by a wall,

$$\lim_{T\to\infty}\int_o^T \langle v_1(t)v\rangle dt = \lim_{T\to\infty}\langle q_1(T)v_1\rangle = 0 \quad , \tag{16}$$

since

$$v_1 = (q_1,H) \quad \text{and} \quad \int v_1^2 d\mu_o < \infty \quad , \quad \int q_1^2 d\mu_o < \infty \quad .$$

Note that when q_1 is an angle variable, e.g., in the case of periodic boundary conditions, then $v_1 \neq (q_1,H)$ and (16) need not hold. We would still have, however, $\langle v_1(t)v_1\rangle \to 0$ <u>if</u> the system is mixing.

When the system is not mixing, the limit $T\to\infty$ in the above integrals need not exist. It is still true however that for <u>any finite system</u>

$$\lim_{T\to\infty}\int \langle f(t)f\rangle dt = \lim_{T\to\infty} -\frac{d}{dT}\langle F(T)F\rangle = 0 \quad , \text{ \underline{if} \underline{it} exists.} \tag{17}$$

This is so since

$$\langle F(T)F\rangle \leq \langle F(T)F(T)\rangle^{\frac{1}{2}}\langle F^2\rangle^{\frac{1}{2}} = \langle F^2\rangle \quad , \tag{18}$$

so that when F is square integrable, $d/dT\langle F(T)F\rangle$ can either oscillate or approach zero.

These time correlation integrals if they exist at all, will therefore, be equal to zero in any <u>finite</u> system. (The interesting fact is that they do exist for mixing systems.) The Einstein type formulae for transport coefficients can therefore be mathematically meaningful only in the thermodynamic limit.

Therefore one of the most important problems in 'rigorous' statistical mechanics, at the present time, is to investigate, and hopefully establish, the existence in the thermodynamic limit of the time integrals used in the Kubo formulae. Unfortunately, it seems impossible to even tackle this problem before on proves the existence of a <u>time evolution</u> in the thermodynamic limit. This has been established so far, for a general class of systems, only in one dimension (Lanford).

C. SPECTRUM OF LIOUVILLE OPERATOR

There is an intimate relation between the mixing properties of the flow and the spectrum of the Liouville operator L (iL is the generator of the unitary operator U_t). It can readily be shown that if the spectrum of L is absolutely continuous (except for the eigenvalue zero), then the flow is mixing. The space on which L acts here is the Hilbert space of complex valued square integrable functions with the measure $d\mu_0(x)$.

Ergodicity, mixing and absolutely continuous spectrum are members of a hierarchy of increasingly stronger conditions on the <u>flow</u> (or the Hamiltonian H which generates it). It has been shown that absolutely continuous spectrum $\Rightarrow$ mixing $\Rightarrow$ ergodicity, but not the converse. Mixing <u>does</u> however imply that L has no discrete eigenvalues other than zero which is a simple eigenvalue. Such a property of L implies, in turn, that the system is at least <u>weakly mixing</u> (the converse is also true). A weakly mixing system is one in which

$$\lim_{T\to\infty} \frac{1}{T}\int_0^T |\mu_0(A_t \cap B) - \mu_0(A)\mu_0(B)|\,dt = 0 \quad . \tag{19}$$

An eigenfunction corresponding to the eigenvalue zero is, of course, any constant on S_E. It already follows from ergodicity that the eigenvalue zero is <u>simple</u>, i.e., constants are the only eigenfunctions of L with eigenvalue zero. The converse is also true, i.e., if zero is a simple eigenvalue of L, then the flow is ergodic.

This may be a good place to note that, due to the discrete nature of the energy spectrum for finite quantum systems confined to a bounded domain V, there will be no mixing (decay of correlations) in such a system. For such quantum systems we do not therefore gain anything from the use of ensembles, and we are forced to look at the infinite volume limit for signs of long time irreversibility. The remarkable thing about Sinai's result is that if shows that finite classical systems can and do have purely continuous spectra. (Note that when Planck's constant $h \to 0$, the number of energy levels between some fixed E and $E + \Delta E$ becomes infinite.)

D. K-SYSTEMS

Sinai's method of proof that a system of hard spheres is mixing is based not on the study of the spectrum of the Liouville operator but on showing that the 'flow' of the hard sphere system on S_E is similar to the 'flow' of a free particle (geodesic flow) on a surface of negative curvature. Such flows are known to be very unstable and were shown by Sinai to be K-flows. Here K stands for Kolmogoroff who, in the mid-fifties, introduced the notion of a K-flow or a K-system. It can be shown that a K-system is also mixing. Indeed, K-systems seem in some way to have the right kind of 'randomness' which <u>might</u> lead to irreversible kinetic equations like the Boltzmann Equation or the hydrodynamic equations. I shall therefore try to explain what they are.

Imagine the energy surface S_E divided up into K disjoint cells A_i, $i = 1, \ldots, k$,

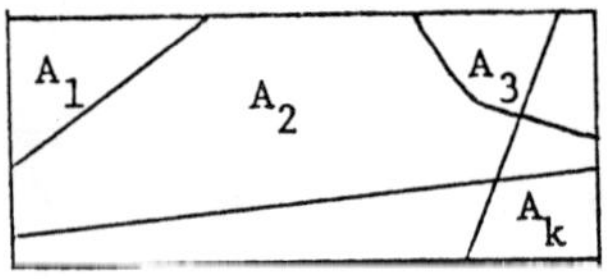

$$\Sigma \mu_0(A_i) = 1 \; , \; \mu_0(A_i \cap A_j) = 0 \quad .$$

This collection of sets $\{A_i\}$ is called a <u>partition</u> $\mathcal{C}$, $\mathcal{C} = \{A_i\}$; the A_i are the 'atoms' of $\mathcal{C}$. Since $\mu_0(A_i)$ is the probability (in the micro-canonical ensemble) of finding the system in A_i, Kolmogoroff defined the 'entropy' (not to be confused with the thermodynamic entropy) of this partition $h(\mathcal{C})$, in analogy with information theory entropy, as

$$h(\mathcal{A}) = -\Sigma\mu_o(A_i)\ln\mu_o(A_i) \tag{20}$$

Clearly, $h(\mathcal{A}) \geq 0$, with the equality holding if and only if $\mu_o(A_j) = 1$, for some j, i.e., there is complete certainty that $x \in A_j$. (We shall generally ignore sets of measure zero, setting $\mu_o(C)\ln\mu_o(C) = 0$ if $\mu_o(C) = 0$ and writing $A_j = S_E$ when $\mu_o(A_j) = 1$.) The maximum value which $h(\mathcal{A})$ can take is $\ln k$, corresponding to $\mu_o(A_i) = k^{-1}$ for all $i = 1, \ldots, k$.

Given two partitions $\mathcal{A} = \{A_i\}$, $i = 1, \ldots, k$ and $\mathcal{B} = \{B_j\}$, $j = 1, \ldots, m$, we denote the 'sum' of the partitions $\mathcal{A}$ and $\mathcal{B}$ by $\mathcal{A} \vee B$; $\mathcal{A} \vee \mathcal{B}$ is the partition whose atoms are all (non-zero measure) sets $A_i \cap B_j$. The entropy of $\mathcal{A} \vee \mathcal{B}$ is

$$h(\mathcal{A} \vee \mathcal{B}) = \sum_{i,j} \mu_o(A_i \cap B_j)\ln\mu_o(A_i \cap B_j) \quad . \tag{21}$$

The 'conditional entropy' of a partition $\mathcal{A}$, relative to a partition $\mathcal{B}$, is defined as

$$h(\mathcal{A}/\mathcal{B}) = \sum_j \mu_o(B_j)\{\sum_i \mu_o(A_i/B_j)\ln\mu_o(A_i/B_j)\} \quad . \tag{22}$$

where

$$\mu_o(A_i/B_j) \equiv \mu_o(A_i \cap B_j)/\mu_o(B_j) \quad .$$

For a given flow operator U_t and some fixed time interval τ, we construct the sets $U_\tau A_i$, $U_{2\tau}A_i$, ... and define $U_\tau A$ as the partition whose atoms are the sets $\{U_\tau A_i\}$. Kolmogoroff then sets

$$h(\mathcal{A}, U_\tau) = \lim_{n\to\infty} \frac{1}{n} H(\bigvee_{j=0}^{n-1} U_{j\tau}\mathcal{A}) \quad . \tag{23}$$

It is readily shown that $h(\mathcal{A}, U_{j\tau}) = jh(\mathcal{A}, U_\tau)$. The K-S entropy (S for Sinai) of the flow U_τ is defined as

$$h(U_\tau) = \sup_{\mathcal{A}} h(\mathcal{A}, U_\tau) = h\tau \quad ,$$

where h is now an intrinsic property of the flow. It was shown by Sinai that a system is a K-system iff $h(\mathcal{A}, U_\tau) > 0$ for all nontrivial partitions $\mathcal{A}$, i.e., for partitions whose atoms are not all of measure zero or one.

K-systems are in some sense 'random' even when the flow is entirely deterministic. To see this suppose that the atoms of the partition $\mathcal{C}, \{A_i\}$, $i = 1, \ldots, k$, correspond to different possible outcomes of the measurement of some dynamical function $f(x)$, i.e., if $x \in A_i$ then the result of the measurement will be α_i , etc. (Since the set of outcomes of the measurement is finite, being equal to k, $k < \infty$, the measurement is a 'gross' one. It need not however be restricted to measuring just one property of the system; we can replace $f(x)$ by a finite set of functions.) The probability (in the micro-canonical ensemble) of an outcome α_i is $\mu_o(A_i) \equiv p(\alpha_i)$. Now if these dynamical functions were measured first at $t = -\tau$ and then at $t = 0$, the joint probability that the result of the first measurement is α_j and the result of the second is α_i is equal to the probability that the dynamical state of the system x at the time of the present measurement $t = 0$ is in the set $A_i \cap U_\tau A_j$, i.e., $p(\alpha_i, \alpha_j) = \mu_o(A_i \cap U_\tau A_j)$. The conditional probability of finding the value α_i , is the result of the previous measurement was α_j , is $p(\alpha_i/\alpha_j) = \mu_o(A_i \cap U_\tau A_j)/\mu_o(U_\tau A_j) = \mu_o(A_i \cap U_\tau A_j)/\mu_o(A_j)$. In a similar way the probability of finding the result α_i at $t = 0$, given that the results of the previous measurements at times $-\tau, -2\tau, \ldots, -n\tau$ were $\alpha_{i_1}, \alpha_{i_2}, \ldots, \alpha_{i_n}$, is

$$p(\alpha_i/\alpha_{i_1}, \ldots, \alpha_{i_n}) = \mu_o(A_i \cap U_\tau A_{i_1} \ldots \cap U_{n\tau} A_{i_n})/\mu_o(U_\tau A_{i_2} \ldots \cap U_{n\tau} A_{i_n}). \quad (24)$$

It can readily be shown that

$$h(\mathcal{C}, U_\tau) = \lim_{n\to\infty} h(\mathcal{C}/\bigvee_{k=1}^{n} U_{k\tau}\mathcal{C}) = \lim_{n\to\infty} \{-\Sigma p(\alpha_{i_1}, \alpha_{i_2}, \ldots, \alpha_{i_n}) \times [\sum_{i=1}^{k} p(\alpha_i/\alpha_{i_1}, \ldots, \alpha_{i_n}) \ell n p(\alpha_i/\alpha_{i_1}, \ldots, \alpha_{i_n})]\} \quad (25)$$

Hence $h(\mathcal{C}, U_\tau) > 0$ for all non-trivial partitions implies that no matter how many measurements of the values of $f(x)$ we make on a system at times, $-\tau, \ldots, -n\tau$, the outcome of the next measurement is still uncertain. (N.B. the measurements are 'coarse' since $\mu_o(A_i) > 0$) .

To see the kind of 'irreversibility' associated with K-systems, I shall use as an example one of the simplest kinds of K-systems: the transformation of a two-dimensional square known as the baker's transformation. This is a discrete transformation, repeated at time intervals τ, on the points $x = (p,q)$ in the unit square (p and q are just labels which are meant to suggest, but have nothing to do with, momentum and coordinates),

$$U_\tau x = U_\tau(p,q) = \begin{cases} (2p, \frac{1}{2} q) & \text{if } 0 \le p < \frac{1}{2} \\ (2p-1, \frac{1}{2}(q+1)) , & \text{if } \frac{1}{2} \le p \le 1 \end{cases} = x_\tau . \tag{26a}$$

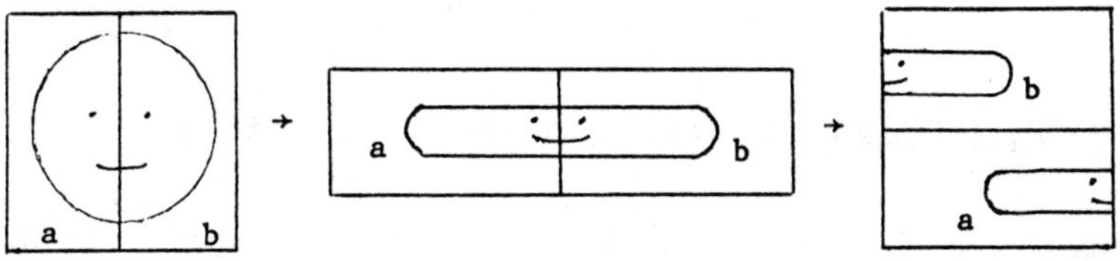

The Baker's Transformation

Using as our equilibrium ensemble density $d\mu_0(x) = dp\, dq$, we note that U_τ is a measure preserving, reversible, one to one transformation

$$U_{-\tau} x = U_{-\tau}(p,q) = \begin{cases} (\frac{1}{2} p, 2q) & , 0 \le q < \frac{1}{2} \\ (\frac{1+p}{2}, 2q-1) & , \frac{1}{2} \le q \le 1 \end{cases} = x_{-\tau} . \tag{26b}$$

The K-S entropy for this transformation is $h(U_\tau) = \ln 2$.

Following Penrose let us consider now a Gibbs' ensemble density which at $t = 0$ is $\mu(x,0)$; $\int_0^1\int_0^1 \mu(p,q;0)dp\, dq = 1$. We have, as in the dynamical flow case, $\mu(x;n\tau) = \mu(x_{-n\tau};0)$, or

$$\mu(p,q;(n+1)\tau) = \mu(U_{-\tau}(p,q);n\tau) = \begin{cases} \mu(\frac{1}{2} p, 2q; n\tau) & , \text{if } 0 \le q < \frac{1}{2} \\ \mu(\frac{1+p}{2}, 2q-1; n\tau) , & \text{if } \frac{1}{2} \le q \le 1 \end{cases} . \tag{27}$$

We now define a <u>reduced</u> ensemble density (mementum distribution function) $W(p;j\tau) \equiv \int_0^1 \mu(p,q;j\tau)dq$. It is readily verified that W satisfies the relation

$$W(p;(n+1)\tau) = \frac{1}{2}[W(\frac{1}{2} p; n\tau) + W(\frac{1}{2} p + \frac{1}{2}; n\tau)] . \tag{28}$$

Equation (28) is an exact, irreversible, 'kinetic' equation for the momentum distribution function, i.e., $W(p;t)$ determines $W(p;t')$ for $t' > t$ but not for $t' < t$, coming from a reversible Liouville equation for the Gibbs ensemble density $\mu(p,q;t)$. As time progresses the momentum distribution function approaches its equilibrium value; $\lim_{n\to\infty} W(p;n\tau) = 1$, in measure (i.e., almost everywhere) <u>if</u> $W(p;0)$ is smooth. This can be seen from (28) by considering the derivative of W,

$$\frac{d}{dp} W(p;\ (n+1)\tau) = \frac{1}{4}\left[\frac{d}{d\eta} W(\eta;n\tau)\Big|_{\eta=\frac{1}{2}p} + \frac{d}{d\eta} W(\eta;n\tau)\Big|_{\eta=\frac{1}{2}p+\frac{1}{2}}\right],$$

so that $\sup_p \frac{d}{dp} W(p;j\tau)$ is a monotonically decreasing function of j.

If we consider, on the other hand, the projection of $\mu(p,q;n\tau)$ on the q-axis, $K(q;n\tau) = \int_0^1 \mu(p,q;n\tau)dp$, then $K(q;n\tau)$ will be determined by $K(q;(n+1)\tau)$ but not by $K(q;(n-1)\tau)$ and $\sup dK(p;n\tau)/dp$ will be <u>non-decreasing</u> as $n\to\infty$. The roles of q and p are reversed if one looks at negative values of n for a given $\mu(x;0)$. It is therefore clearly necessary to choose properly the <u>reduced</u> description of the system if one is to get irreversible behavior. (Alternatively, one has to consider properly restricted initial distributions; Penrose, private communication.)

It should also be pointed out here that any kinetic equation for some projection of the full ensemble density which, like equation (28), holds for arbitrary initial densities $\mu(x;0)$, even when these and the corresponding measures $\mu(dx;0)$ are singular, must have the property of leaving singular parts of the initial reduced distribution (here $W(dp;0)$) singular. This means that it is impossible to have a kinetic equation which is valid for <u>all</u> initial distributions and at the same time acts (like the Boltzmann or diffusion equation) to smooth out initial singular distributions.

<u>Acknowledgments</u>. I would like to thank Oliver Penrose and Oscar Lanford for valuable discussions.

This research was supported by the U.S.A.F.O.S.R. under contract No. F44620-71-C-0013, modification Pool.

BIBLIOGRAPHY

1. Arnold, V. I., and Avez, A., Ergodic Problems of Classical Mechanics, Benjamin, 1968 .

2. Billingsley, P., Ergodic Theory and Information, Wiley, 1965 .

3. Lebowitz, J. L., Hamiltonian Flows and Rigorous Results in Non-Equilibrium Statistical Mechanics, to appear in the Proceedings of I.U.P.A.P. Conference on Statistical Mechanics, Chicago, 1971.

4. Ornstein, D. S., Measure-Preserving Transformations and Random Processes, Amer. Math. Monthly, V78, (1971).

5. Penrose, O., Foundations of Statistical Mechanics, Pergamon, 1970.

6. Sinai, Ja., in Statistical Mechanics Foundations and Applications; Proceedings of the I.U.P.A.P. Meeting, Copenhagen, 1966, T.A. Bak, Editor, Benjamin, 1967, p. 559; Russian Math. Sur., 25, 137 (1970).

7. Wightman, A. S., in Statistical Mechanics at the Turn of the Decade, E. D. Cohen, Ed, Dekker, 1971.

8. Ford, J., The Transition from Analytic Mechanics to Statistical Mechanics, to appear in Advances in Chem. Physics, 1972.

MATHEMATICAL PROBLEMS IN THEORETICAL BIOLOGY

Grégoire Nicolis
Université Libre de Bruxelles
Belgium

1. INTRODUCTION

In this seminar we shall first try to present the conceptual and mathematical difficulties one encounters in trying to formulate and solve certain general problems of theoretical biology. We shall then discuss some results obtained recently on these problems, and, finally, we shall list a number of questions which remain open to future investigations.

Our principal goal will be to understand the physico-chemical basis of certain aspects of biological order, related both to the origin and evolution of prebiological matter as well as to the problem of maintenance of this order in actual living organisms. We all know that biological systems are highly complex and ordered objects. They have the unique ability to store information, i.e., to carry with them structures and functions acquired in the past during a long evolution. On the other hand, the maintenance of life, even in its simplest bacterial form, requires metabolism, synthesis and regulation. These can only be achieved by a highly heterogeneous distribution of matter inside the cell which is the result of active transport and of a few thousand chemical reactions going on simultaneously in a crowded space of the order of a few μ^3. Finally, the process of development of a higher organism from a fertilized egg requires transmission of the genetic information over macroscopic distances and during macroscopic time intervals. If we had to use a single term to describe all these characteristic features of biological systems, coherent behavior would probably be the most appropriate one.

In contrast to this, the physicist feels at home with the idea that the evolution of a physico-chemical system leads to an equilibrium state of maximum disorder. In an isolated system, which cannot exchange energy and matter with the surroundings, this tendency is expressed by the second law of thermodynamics. It amounts to saying that there exists a function of the macroscopic state of the system, the entropy S , which increases monotonically until it becomes a maximum at a state known as thermodynamic equilibrium. Thus, in an isolated system, the formation of ordered structures, i.e., structures of low entropy, is ruled out.

Consider now an open system which is still at equilibrium, but which can now exchange energy and matter with the external world. The situation is then different. As you have seen in the previous talks, there exists a possibility for the formation of ordered structures of low entropy at sufficiently low temperatures. This ordering principle is responsible for the appearance of structures such as crystals. Unfortunately, it cannot explain the formation of biological structures. The probability that at ordinary temperatures a macroscopic number of molecules is assembled to give rise to the highly ordered structures and to the coordinated functions characterizing living organisms is vanishingly small. Thus we must regard the idea of spontaneous genesis of life in its present form as highly impossible, even in the scale of the billions of years of prebiological evolution.

This failure of equilibrium theory to explain biological order suggests that one should examine the feasibility of extending the concept of order to non-equilibrium situations for systems for which the appearance of ordered structures would be very unlikely at thermodynamic equilibrium. In recent years this question has attracted a great deal of attention. The principal results are as follows [1]. For simplicity, we only present them for the following restricted class of systems: Open mixtures of chemically reacting components at mechanical equilibrium, at constant temperature and subject to time-independent boundary conditions. The motivation for focusing on this class of systems is obvious. The behavior of living cells is determined largely by chemical transformations and by transport processes such as diffusion inside the cytoplasm or passage across membranes.

(a) The first result is that linear systems, in particular, systems close to equilibrium, always evolve to a disordered regime corresponding to a steady state which is asymptotically stable with respect to all disturbances. As in Dr. Coleman's talk, stability can be expressed in terms of a variational principle. For the class of systems we defined above, this principle is expressed by the minimum entropy production theorem [1].

(b) Creation of structures may occur spontaneously in non-linear systems maintained, in addition, beyond a critical distance from equilibrium. Let us recall from Dr. Walter's talk that non-linearity seems to be a general rule in biology appearing in various forms, e.g., through auto-catalysis, cross-catalysis, inhibition, etc.

(c) Whenever (b) happens, the steady state corresponding to the extrapolation of the close-to-equilibrium behavior becomes unstable. The system then evolves to a new regime which may correspond to a spatially or temporally organized state. We have called these regimes dissipative structures to indicate that, contrarily to what common intuition would suggest, they are created and maintained by the dissipative, entropy-producing processes inside the system [1].

(d) In spite of the diversity of the situations which may arise in non-linear systems, one can demonstrate a few general thermodynamic theorems underlying these non-equilibrium order phenomena. As we would expect, there seems to be no variational principle in this domain. Nevertheless, it is possible to show that when a non-equilibrium state becomes unstable, the additional amount of dissipation (measured by the produced entropy) introduced by the disturbance (which is assumed small) becomes negative [1]. In this way the stability properties are connected with thermodynamic functions of direct experimental interest.

Our object now will be to see how far one can go, starting from these general ideas, in understanding certain aspects of biological order. We shall carry out this program in three steps. First, we shall illustrate on abstract models the type of non-equilibrium order phenomena that can be expected in chemical kinetics and compare them with the properties of biological systems that we listed earlier in this section. Second, we shall show that the theoretical results are confirmed by several laboratory experiments. And finally we shall indicate several representative biological problems where non-linearity and ordered behavior play a prominent role.

2. SIMPLE MODELS

Consider a reacting mixture of n species. The system is open to the flow of chemicals from the outside world and conversions are allowed between these chemicals and the reacting species inside the reaction volume. Again, we assume that the system is maintained isothermal and in mechanical equilibrium. Moreover, if the concentration gradients are not very high, diffusion could be approximated by Fick's law. Finally, we take a diagonal diffusion coefficient matrix. Under these conditions the instantaneous state of the system is described by the composition variables $\{x_i\}$ $(i = 1,\dots,n)$, denoting, e.g., the average mole functions or the average partial densities of the chemicals. The time evolution of these variables is given by the well known conservation of mass equations:

$$\frac{\partial x_i}{\partial t} = D_i \nabla^2 x_i + V_i(\{x_j\}) \; ; \qquad (1)$$

D_i are the constant diffusion coefficients. The source terms V_i describe the effect of chemical reactions. It will be assumed that the rates of the various chemical transformations are given by the law of mass action. Thus $V_i(x_j)$ will be non-linear (usually algebraic and thus unbounded) functions of x_j's . We shall briefly summarize here the results obtained from the following model chemical reaction chain 1):

$$A \rightleftarrows X$$
$$B + X \rightleftarrows Y + D$$
$$2X + Y \rightleftarrows 3X$$
$$X \rightleftarrows E$$

or

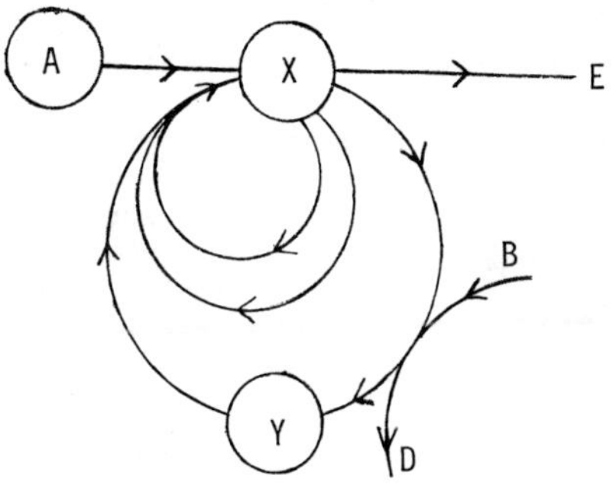

(2)

Notice the presence of the autocatalytic step in the third reaction. This step introduces in the problem the non-linear element which will be largely responsible for the unusual behavior of the system. Otherwise model (2) is unphysical and does not represent any known biochemical reaction chain.

(a) At or near thermodynamic equilibrium, system (2) admits a unique steady state solution which is always stable with respect to arbitrary perturbations (in agreement with our general criteria discussed previously).

(b) Suppose that the system is driven far from equilibrium. In the extreme case that the concentration of D and E is maintained vanishingly small in the system, the distance from equilibrium becomes very large. The rate equations describing the system in this limit are (all rate constants are set equal to unity):

$$\frac{\partial A}{\partial t} = -A + D_A \frac{\partial^2 A}{\partial r^2} ,$$

$$\frac{\partial X}{\partial t} = A - (B+1)X + X^2Y + D_X \frac{\partial^2 X}{\partial r^2} , \qquad (3)$$

$$\frac{\partial Y}{\partial t} = BX - X^2Y + D_Y \frac{\partial^2 Y}{\partial r^2} .$$

We have only considered diffusion along a single spatial dimension. For time independent boundary conditions, one can define from these equations a steady state corresponding to the continuation of the equilibrium-like behavior. The branch of states defined in this fashion will be referred to as a thermodynamic branch and is described in Fig. 1. This time, however, the study of the conservation equations (3) reveals that, because of the non-linearity and the non-equilibrium constraints, states on this branch are not necessarily stable. The different situations which will be realized depend on the type of perturbations to which the system is subject, on the values of the parameters, such as A, B, and on the diffusion coefficients D_A, D_B, D_X, and D_Y. This is illustrated by the non-equilibrium phase diagram of Fig. 2 which is obtained from a conventional linearized stability analysis. For simplicity, the parameters A, D_A, D_X, were held constant, and D_B was taken so large that the distribution of B was maintained uniform in the system.

(c) In the domain II of the diagram, the thermodynamic branch becomes unstable with respect to perturbations in the chemical composition. Beyond the transition threshold, the initial deviations from the steady state are amplified. They finally drive the system to a new steady state corresponding to a regular spatial distribution of X and Y . The latter is represented in Fig. 3.

We obtain a low entropy spatial dissipative structure arising beyond a non-equilibrium transition which spontaneously breaks the symmetry of the initial system. Its behavior is quite different from the properties of systems on the thermodynamic branch. In addition to the coherent character exhibited by such states, one can show that the final configuration depends, to some extent, on the type of initial perturbation. This primitive memory effect makes these structures capable of storing information accumulated in a remote past. We also notice the solution

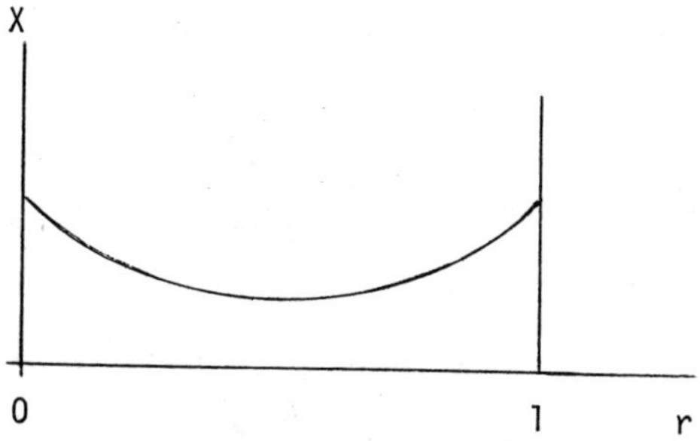

FIGURE 1.

Thermodynamic branch for system (2)

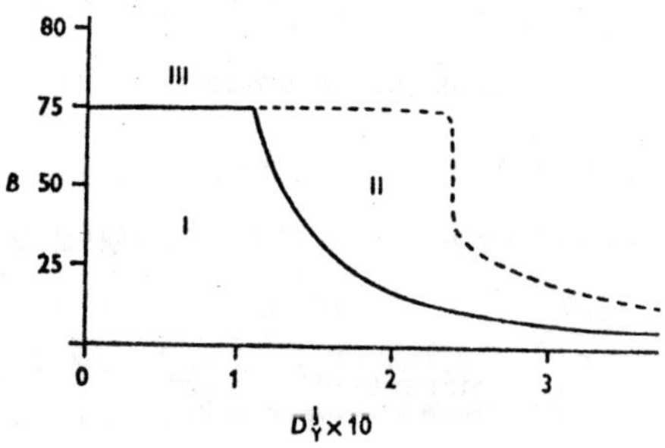

FIGURE 2.

Stability diagram in the B-D_Y plane for $D_A = 197 \times 10^{-3}$, $D_X = 1.05 \times 10^{-3}$, $D_B \to \infty$. The boundary values of A are chosen $\bar{A} = 14.0$.

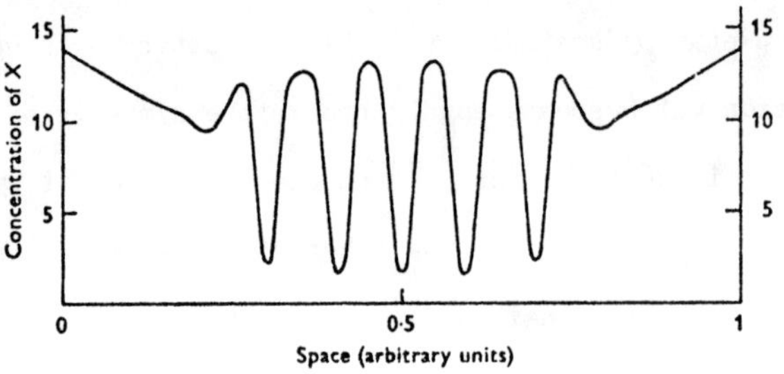

FIGURE 3.

Localized steady state dissipative structure for model (2). The following values have been chosen:
$D_X = 1.05 \times 10^{-3}$, $D_Y = 5.25 \times 10^{-3}$, $D_A = 197 \times 10^{-3}$, $D_B \to \infty$, $B = 26.0$ and $A = \bar{X} = 14$, $\bar{Y} = 1.86$ at both boundaries.

described by Fig. 3 clearly corresponds to an internal transition between two types of solutions of the equations of evolution as described in Dr. Fife's lecture.

(d) For $D_B \gg D_A \gg D_Y \simeq D_X$, the thermodynamic branch may again become unstable. But this time the system is driven beyond instability to a new state which depends not only on space, but is also periodic in time. In the course of one period (see Figs. 4-7), there appear wavefronts of composition which propagate in the reaction volume, first outwards and then, after reflecting on the boundaries, to the interior. Moreover, at each point in space the system performs quasi-discontinuous oscillations in time (Fig. 8). This space-time dissipative structure provides, therefore, a primitive mechanism for propagating information over macroscopic distances, in the form of chemical signals.

(e) For D_A finite, both structures discussed in (c) and (d) are localized inside the reaction volume. Their boundaries, as well as the periods and other characteristics, are determined from the values of the parameters A, B, D_X, etc., independently of the initial conditions. For D_A, $D_B \to \infty$, the structures tend to occupy the whole reaction volume. Finally, when D_X, D_Y are very large (with respect to the chemical rates) the space dependencies disappear in case (d). One obtains a system which, beyond instability, oscillates with the same phase everywhere. The amplitude and period of oscillations are determined by the system itself. Moreover, the periodic motion is an asymptotically stable <u>limit cycle</u>. The existence

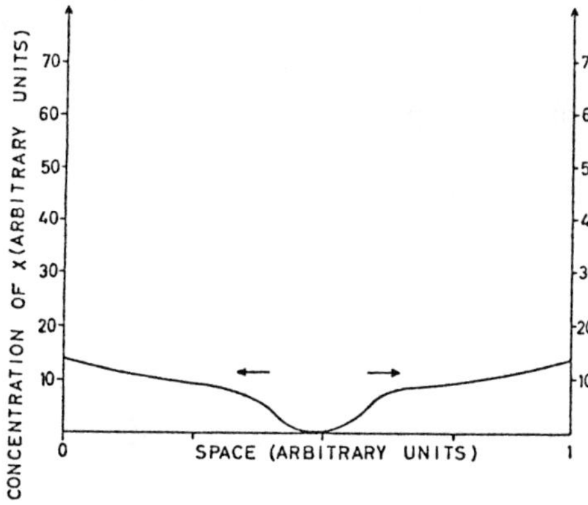

FIG. 4 Spatial distribution of X at a certain stage of its periodic evolution. A well is formed and propagates outwards. The following numerical values of the parameters have been chosen: $D_X = 1.05 \times 10^{-3}$, $D_Y = 0.66 \times 10^{-3}$, $B = 77.0$, $\bar{X} = 14$, $\bar{Y} = 5.50$.

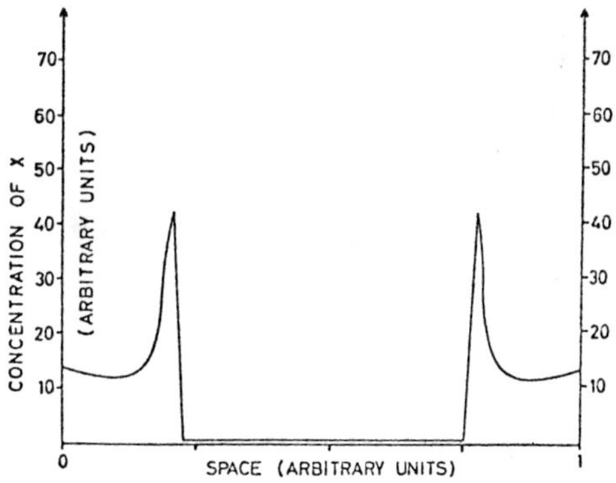

FIG. 5. Spatial distribution of X during the part of the period corresponding to a slow build-up of X at points 9 and 31.

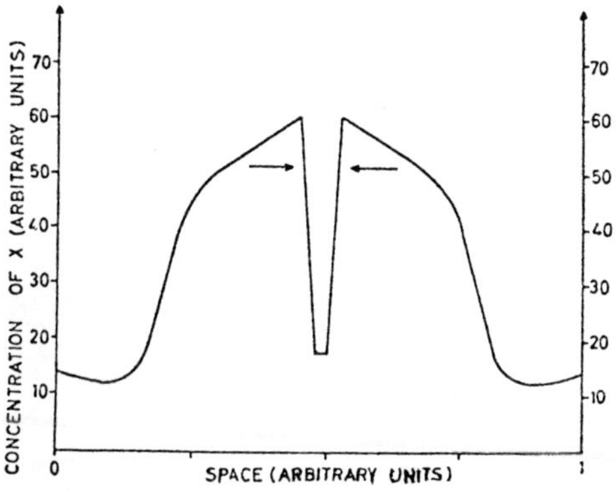

FIG. 6. Spatial distribution of X during the rapid propagation of the two wavefronts toward the middle point.

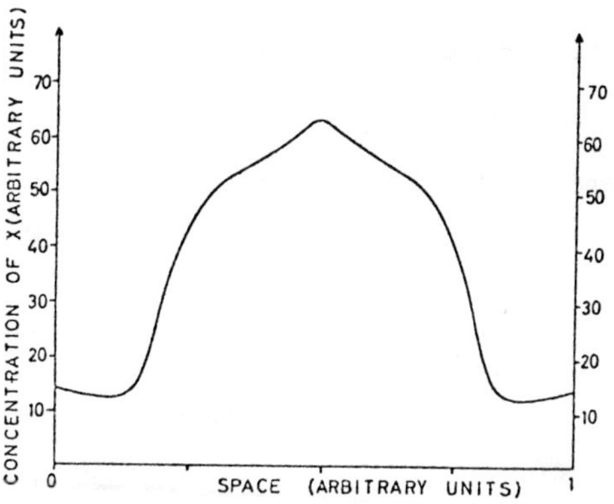

FIG. 7. Spatial distribution of X during the slow overall decrease of concentration to the initial profile.

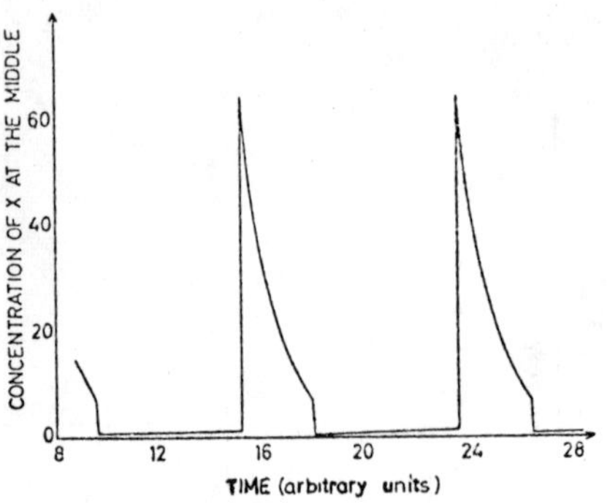

FIG. 8. Time variation of the concentration of X at the middle point (point 20).

of localized states and of wave-like solutions shown in (c) and (d) raises, we believe, a number of interesting mathematical problems related to the existence and stability of periodic solutions for non-linear parabolic partial differential systems with constant diffusion coefficients.

Obviously, the occurence of instabilities far from equilibrium is not a universal phenomenon in chemical kinetics. Coherent behavior requires some very particular conditions on the reaction mechanism, whereas the equilibrium order principle is always valid (for short range forces). But this variety of behavior in chemistry is welcome if we want to account for the variety of situations observed in systems driven far from equilibrium.

In conclusion, we see that non-linear chemical kinetics may give rise to states showing a striking similarity to the properties of biological systems listed in the Introduction. Regulation, storage of information, and transmission of information appear to be natural consequences of two important conditions: Non-linearity and finite distance from thermodynamic equilibrium.

3. EXPERIMENTAL EVIDENCE

So far our discussion has been rather abstract, and the models which we had been able to solve referred to artificial systems. Let us now make a first contact with the real world and discuss the experimental aspects of dissipative structures. There are at present two sources of experimental evidence for dissipative structure formation coming from organic and biochemical reactions, respectively. We shall first deal, briefly, with the non-biological case.

The best known example is an organic reaction in which Ce or Fe ions catalyze the oxidation of analogs of malonic acid by bromate [2]. According to

Degn [2], the process seems to involve the following three overall reactions:

(a) Oxidation of malonic acid

$$CH_2(COOH)_2 + 6Ce^{4+} + 2H_2O \rightarrow 2CO_2 + HCOOH + 6Ce^{3+} + 6H^+$$

(b) Oxidation of cerium ions

$$10Ce^{3+} + 2HBrO_3 + 10H^+ \rightarrow 10Ce^{4+} + Br_2 + 6H_2O$$

(c) Transformation into bromomalonic acid

$$CH_2(COOH)_2 + Br_2 \rightarrow CHBr(COOH)_2 .$$

This reaction chain is known as the Zhabotinski-Belusov reaction. When it occurs in a continuously stirred medium, sustained oscillations in the concentration of intermediates are observed for certain ranges of values of product concentration and temperature [2]. The stability, spontaneous emergence, and perfect reproducibility of the results suggest that they correspond to limit cycle type of oscillations, i.e., that they really occur beyond a dissipative instability. So far, this conjecture cannot be supported by an analytic calculation because the detailed mechanism of the reaction is only beginning to be known [3]. What is certain is that the reaction involves at least one autocatalytic step. The nonlinearity premise, which we stressed repeatedly before, is again found here.

For slightly different values of concentrations, an evolution to a spatially organized state is observed [4]. Of special interest, from the mathematical point of view, is the fact that this state is attained only after the thermodynamic branch becomes unstable with respect to oscillations, which in turn become unstable with respect to diffusion.

Finally, when the same reaction occurs in a thin layer of unstirred solution, a propagation of wavefronts initiated from different centers is observed [5]. At every point in space the system undergoes relaxation oscillations involving a short oxidation phase and a longer reduction phase. For long times the waves coming from different centers collide and are annihilated, with a subsequent synchronization to the highest frequency. Qualitatively, these observations are very similar to the results obtained from the model we discussed a few minutes back in this talk (and, the fact that the same system gives rise to several structures with only slight changes).

The experiments we have just described have been carried out in a closed system. As a result the different structures are maintained for a limited time (of the order of hours). However, the sharpness of the phenomena, the reproducibility of the results and the rapid emergence or disappearance of the patterns in comparison with their lifetime suggest that the effects are really dissipative structures arising beyond symmetry-breaking instabilities. We have here concrete, experimental evidence of a coherent behavior which is perfectly compatible with the laws of thermodynamics and chemical kinetics. Let us point out again that the separation of matter in the Zhabotinski reaction is quite different from the usual phase separation or partial missibility which is observed on certain mixtures in thermal equilibrium. What we have here is rather a striking illustration of the non-equilibrium order principle that we have been referring to in the introductory part of this talk.

Quite recently, Thomas has discovered another striking example of spatial order formation in the distribution of pH across an artificial membrane [6]. His experiment can be described briefly as follows. Two enzymes, the glucose oxidase and the urease are attached homogeneously in an artificial membrane of a thickness of $\sim 100\ \mu$. One can test quite conclusively both that the fixation is homogeneous and that it does not change the catalytic activity of these enzymes.

A peculiar feature of these enzymes is that their activity as a function of pH is different. While V_{ur} is increasing around a pH ~ 7.3, $V_{g\ell}$ is optimum in the region of pH ~ 6.3 as shown in the curves in Figs. 9-11.

One now performs the following experiment. The membrane is immersed in a <u>uniform</u> medium containing glucose and urea where identical concentrations of both substrates are maintained on both sides. One then varies these concentrations and observes the pH distribution across the membrane. For values of ur and $g\ell$ far from saturation, one observes a uniform distribution of pH. But for values corresponding to a large excess of substrates (which corresponds to a far from equilibrium stiuation), one finds a distribution of the type shown in Fig. 12.

We see that, in spite of the initial homogeneity, the system spontaneously broke its symmetry by dividing itself into two parts: one in which

122

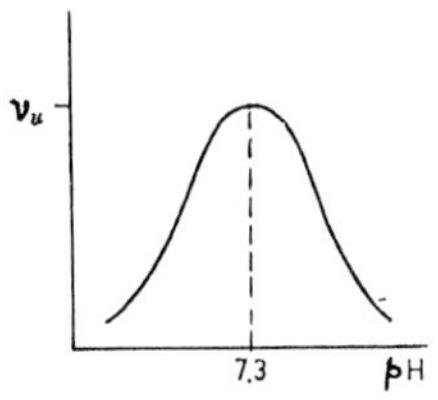

FIG. 9. Urease activity as a function of pH .

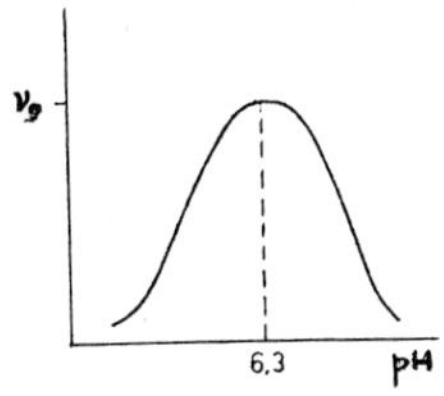

FIG. 10. Glucose oxidase activity as a function of pH .

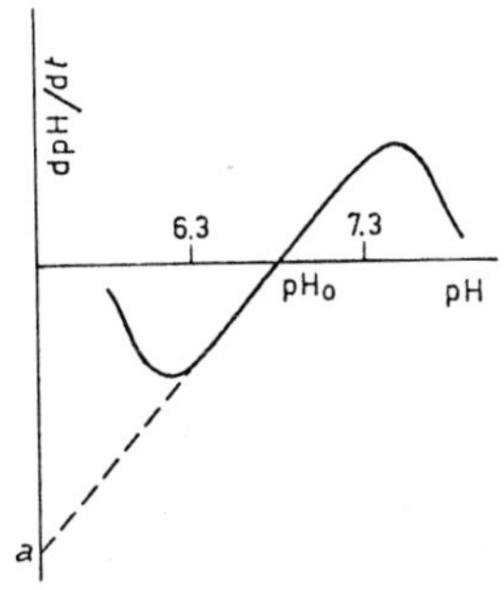

FIG. 11. Rate of pH variation as a function of pH . One sees that around pH_0 there exists a linear region corresponding to a positive feedback.

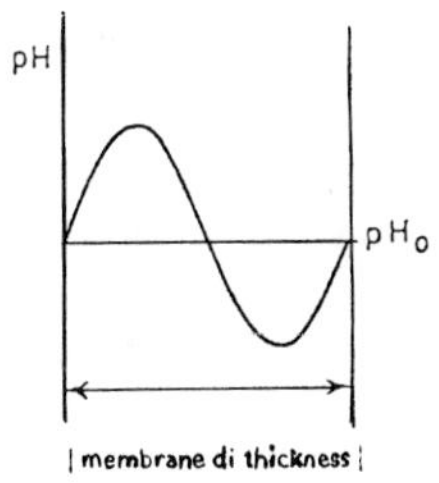

FIG. 12. Spatial distribution of pH inside the membrane.

the urease reaction proceeds preferentially and another in which glucose oxidation is taking place. Notice that one obtains this shape as well as the one where the slope of this profile is inverted. This lack of uniqueness corresponds to the possibility that the initial disturbance, acting, say, on the right half, could increase or decrease the local pH value. Again, this multiplicity of solutions is in agreement with the results of our model calculation of the previous section.

4. SOME REPRESENTATIVE BIOLOGICAL EXAMPLES

Being now convinced, from our general theory, the analysis of models and experimental evidence, that the spontaneous emergence of order is possible provided the system is sufficiently far from equilibrium, we come to the question which concerns us here primarily: What is the usefulness of the results I have just outlined in the understanding of biological phenomena? It will be convenient to discuss separately two types of problems.

- Is it possible to understand the functional order observed in actual living systems? This question refers, therefore, to the physicochemical basis of maintenance of life.

- How did the structures observed in actual living beings (nucleic acids, proteins, cells as a whole) arise from an inert prebiotic mixture of simple molecules? This is the problem of prebiotic evolution or of the origin of life.

We shall first discuss briefly the first point. We shall list a number of typical phenomena which have been analyzed in terms of the concepts we outlined in the first part of this talk.

(a) Regulatory processes. As Dr. Walter emphasized in his talk, there exist elaborate control mechanisms which ensure that the various chemical reactions in living cells happen at the proper rate and at the right time. The first type of control mechanism makes sure that there is no excessive synthesis or lack of small metabolites, e.g., of energy-rich molecules such as ATP. The usual way this mechanism operates is to affect the rate at which a particular protein (enzyme) catalyzing one reaction step acts. One of the best studied biochemical chains from this point of view is glycolysis, a process of great importance for the energetics of living cells which transforms glucose into products useful for the metabolism. Experiments carried both *in vivo* and in the test tube show that the concentrations of the chemicals participating in the reaction present undamped oscillations in time with perfectly reproducible periods ($\sim$ 1 min. *in vitro* and $\sim$ 10 sec. *in vivo*) and amplitudes. On the other hand, starting from known data on the elementary reaction steps, one can construct mathematical models for glycolysis [7]. A detailed study of the rate equations shows that the experimental results may be interpreted quantitatively as oscillations of the limit cycle type arising beyond the instability of a time-independent solution which belongs to the thermodynamic branch. In other terms, a glycolysis is a temporal dissipative structure. This result is expected to extend to a whole series of regulatory processes at the metabolic level.

To give an idea of the mathematics involved in this type of problem, let us write some rate equations referring to a particular series of intermediate steps involving the enzyme phosphofructokinase. This peculiar enzyme is activated by one of the products of the reaction which is ADP (in contrast with the examples mentioned by Dr. Walter where products inhibit) and inhibited by the substrate which happens to be the ATP. When one works out the kinetics and makes the quasi-steady approximation for the enzymes, one finds equations of the form:

$$\frac{\partial(ATP)}{\partial t} = v_1 - F(ATP,ADP) \quad ,$$
$$\frac{\partial(ADP)}{\partial t} = G(ATP,ADP) - k(ADP) \quad , \qquad (5)$$

where F and G are non-linear functions involving fractional powers $(ATP)^\gamma$, $(ADP)^\gamma$ and where γ is some factor >1. Although one can carry out analytically the linearized stability analysis for (5), the subsequent evolution to the limit cycle has, up to now, been studied only numerically.

A second type of control mechanism in living cells affects the rate of synthesis of the various protein molecules which exist in a cell. Usually this mechanism works on a group of more than one type of enzyme molecule. Jacob and Monod have proposed several ingenious models wherein either the products of the metabolic action of the enzymes act on the genetic material to inhibit the synthesis, or the initial metabolites added to the medium have the effect of switching on the action of a part of the genetic material responsible for the synthesis of these enzymes. Again, one can construct mathematical models for this process [8]. The study of rate equations reveals that the activated and inactivated regimes belong to two different branches of solutions which, under certain conditions, are separated by an instability.

(b) Biosynthesis of Proteins. Gibbs [9] has recently proposed a probabilistic treatment of the process of protein biosynthesis. His model can be summarized as follows:

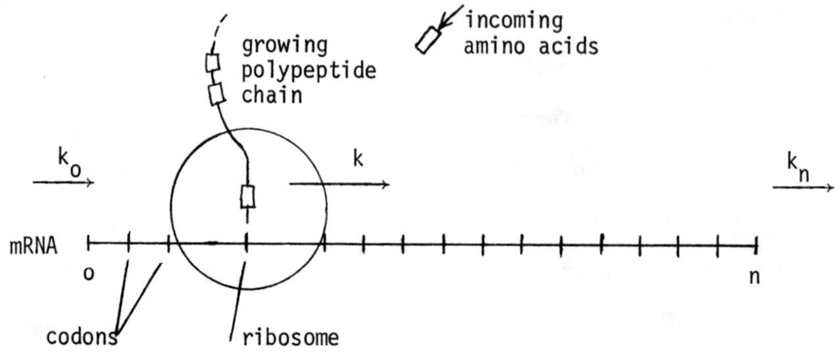

FIGURE 13.

Let P_j = probability to have a growing chain in site j,

k_o = "initiation" rate constant,

k_n = "termination" rate constant,

k = ribosome "propagation" rate constant: $k_1 = \dots = k_{n-1} = k$,

L = number of codons covered by a ribosome.

P_j satisfies equations of the form:

$$\frac{\partial P_j}{\partial t} = F_j(\{P_j\}, k_o, k_n, k, L) \quad , \tag{6}$$

where F is non-linear in P_j.

The (numerical) analysis of (6) shows the following results:

1. In a whole range of values of the parameters, there exist two j-independent solutions (termed "uniform"), one "high density" solution p^I and another "low density" solution p^{II}.
2. For $k_o \simeq k_n$ the system may exhibit internal transitions between these two states or even spatial oscillations around one of them as shown in Fig. 14.

The analogy to Dr. Fife's results and to the discussion of Section 2 is striking. At present there exist no conclusive biological data for deciding which type of solution occurs in living cells.

(c) Cell Aggregation. Certain unicellular organisms attain, during their life, a level of organization wherein the individual cells aggregate in colonies. A primitive form of differentiation between cells is also observed in these colonies. One of the best studied families showing this behavior are the slime molds. Their aggregation is mediated by a cyclic AMP which can be secreted by the cells. It has been shown [10] that the initiation of this aggregation can be interpreted as an instability of the uniform distribution (corresponding to the absence of aggregation)

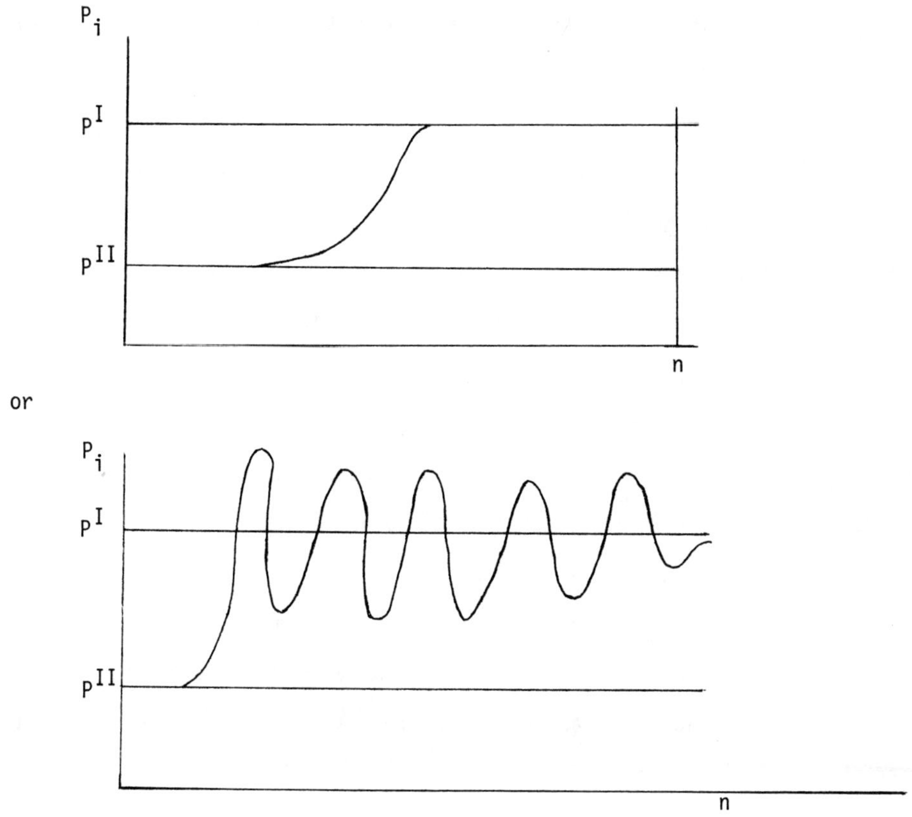

FIGURE 14.

of the individual cells, which again belongs to the thermodynamic branch. The equations describing these processes are of the form (neglecting the effect of cell division which is, anyway, small during aggregation):

$$\frac{\partial a}{\partial t} = F(a,A) + D_a \nabla^2 a \quad ,$$

$$\frac{\partial A}{\partial t} = \nabla [D_A \nabla A - f_{Aa}(A,a) \nabla a] \quad , \tag{7}$$

where A = concentration of aggregating cells,

a = concentration of AMP in the extracellular medium,

D_a = diffusion coefficient of AMP,

D_A = mobility coefficient of the cells,

f_{Aa} = coefficient (depending on A and a) which measures the effectiveness of the mediation effect by the AMP.

One is tempted to hope that these aggregation phenomena will provide valuable indications about the mechanisms of development of higher organisms. In this case the interpretation in terms of dissipative structures would provide a much needed unifying principle for all these extremely diverse and complex processes.

(d) Development and Morphogenesis. The preceding example brings us to one of the most puzzling problems of biology, the phenomenon of pattern formation and differentiation in developing organisms. The complexity of the problem is such that several hours of discussion are needed before one can even formulate properly the various questions. We shall try instead to focus on a few general aspects and illustrate some mathematical problems involved therein.

It is believed that what a cell, or a family of cells, will become in a developing organism depends on the intercellular concentration of one or several characteristic substances known as morphogens. Let X , Y be two such morphogens whose time evolution is given by

$$\frac{\partial X}{\partial t} = F(X,Y) + D_X(r)\frac{\partial^2 X}{\partial r^2} \quad ,$$

$$\frac{\partial Y}{\partial t} = G(X,Y) + D_Y(r)\frac{\partial^2 Y}{\partial r^2} \tag{8}$$

where r is an appropriate coordinate. This system, which looks very much like the model of Section 2, was investigated for the first time by Turing in 1952 [11]. He showed that when the concentration of X or Y exceeds some critical value, the uniform steady solution of equation (8) may become unstable and evolve to a standing wave of concentration of the morphogens. It is reasonable to expect that this should explain, somehow, the development of regular patterns of repeated parts, such as bristles in drosophila.

Incidentally, similar phenomena may arise in ecology. One may ask, what are the consequences, e.g., for Volterra-Lotka dynamics (see Dr. Thomson's lectures), of the fact that the individuals of the two populations move about? To what extent will this motion synchronize the frequencies of the different trajectories? Could one obtain the spontaneous formation of domains corresponding to some sort of ecological niche?

Consider now the more realistic case where the concentration of the morphogens is triggered by a biochemical oscillator. Goodwin, Cohen and Robertson have studied this type of situation in great detail [12]. They conjecture the existence of organizing waves of morphogen concentration, whose phase is controlled via a wave-like propagation mechanism. Although it is still too early to judge the generality of this conjecture, it does seem that their idea is compatible with recent observations on aggregating slime molds [12].

5. EVOLUTION

In the remaining part of this talk we should like to confront the idea of order through non-equilibrium instabilities with the major problem in biology, namely, how did biological systems arise? We shall discuss more specifically the thermodynamic meaning of pre-biological evolution.

The first general principle characterizing evolution was enunciated by Darwin more than 100 years ago. However, his principle of "survival of the fittest" through mutations and natural selection can only apply in the later stage where the pre-biological evolution has led to the formation of some primitive living beings. A new evolutionary principle which would replace Darwin's idea in the context of pre-biotic evolution has been proposed recently by Eigen in an impressive paper which Dr. Thomson is discussing in his lectures [13]. It amounts to optimizing a quantity measuring the faithfulness, or quality of the macromolecules arising from mutations in reproducing themselves via autocatalytic action (known as template action). We have recently proposed an alternative description of pre-biological evolution [15]. The main idea is to look for the possibility that a pre-biological system may evolve through a whole succession of transitions leading to a hierarchy of more and more complex and organized states. As we have seen before, such transitions can only arise in non-linear systems which are maintained far-from-equilibrium. There exists then a critical threshold beyond which the steady state regime becomes unstable and the system evolves to a new configuration. As a result, if the system is to keep the ability to evolve through successive instabilities, a mechanism has to be developed whereby each new transition favors further evolution by increasing the non-linearity and the distance from equilibrium. An obvious mechanism for this is that each transition enables the system to increase the entropy production. We may visualize this evolutionary feedback as follows:

Threshold → instability through fluctuations → increased dissipation .
↑ (feedback from increased dissipation back to threshold)

We have shown that there exists a large class of systems of interest that may exhibit this evolutionary feedback.

It is important to insist on the complementarity between this increasing dissipation behavior and the commonly observed tendency of physico-chemical systems near equilibrium towards a state of minimum dissipation or entropy production. One is tempted to argue that it is only after having synthesized the key substances necessary for its survival (which implies an increase in dissipation), that a pre-biological system tends to adjust its entropy production to a low value compatible with the external constraints. Thus if we follow, for such a system, the time evolution of the specific rate of dissipation p , we should expect to find a curve as in Fig. 15.

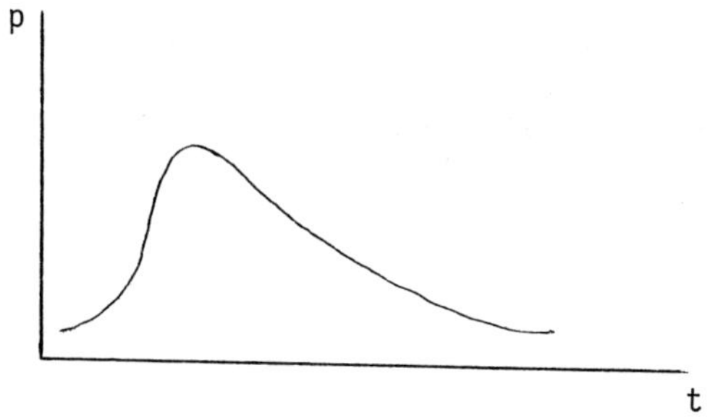

FIGURE 15.

We would like to close this lecture with the following comments. Many of the problems we mentioned earlier cannot be treated in a really satisfactory manner by means of the deterministic equations of mass conservation.

One obvious example is pre-biological evolution. As we have seen evolution is a result of mutations. Now on the level of a macroscopic description, mutations are random events obeying some unknown distribution. Thus, the equations describing evolution (see also Dr. Thomson's lectures) are subject to random driving forces. Similarly, in an ecological problem there are inevitable perturbations coming from the surroundings which continuously modify the instantaneous rates of the various processes and cause fluctuations in the composition of the population.

If the steady state of the deterministic equations (i.e., of the equations which neglect fluctuations) lacks asymptotic stability, one expects that the stochastic effects will "mix" the deterministic solutions in a complicated fashion. In the Lotka-Volterra model this would give rise to a highly irregular, quasi-periodic evolution of the populations. This conjecture has been verified recently by Prigogine and myself [15]. We have proposed the term generalized turbulence to describe these peculiar situations, wherein the macroscopic averages are driven by the fluctuations in some sort of random walk. One of the consequences of this observation is that a strongly stochastic behavior in chemical kinetic systems does not necessarily imply the bifurcation of quasi-periodic solutions in the deterministic equations of evolution. This latter conjecture has been made recently by several authors, among others, by Ruelle [16].

REFERENCES

1. For recent reviews, see Glansdorff, P., and Prigogine, I., Thermodynamic Theory of Structure, Stability and Fluctuations. Wiley-Interscience, New York (1971), or Prigogine, I., and Nicolis, G., Quart. Rev. Biophys. 4, 107 (1971).

2. Zhabotinski, A., Biophysics (USSR) 9, 306 (1964); Degn, H., Nature 213, 589 (1967).

3. Noyes, et al. (J. Am. Chem. Soc. since (1971) have recently elucidated most of the elementary steps involved in the reaction.

4. Herschkowitz-Kaufman, M., Comptes Rendus Acad. Sci. (Paris) 270c, 1049 (1970).

5. Winfree, A. T., Sciences 175, 634 (1971).

6. Thomas, D., University of Rouen (to be published).

7. Sel'kov, E. E., Europ. J. of Biochem. 4, 79 (1968); Goldbetter, A., and Lefever, R., Biophys. Journal (in press (1972)).

8. Babloyantz, A., Nicolis, G., J. Theor. Biology 34, 185 (1972). Mac Donald, C., and

9. Gibbs, J., Biopolymers, 6, 1 (1969).

10. Keller, E., and Segel, L., J. Theor. Biology 26, 399 (1970).

11. Turing, A. M., Philos. Trans. Roy. Soc. (London) B234, 37 (1952).

12. Goodwin, B., Cohen, M., and Robertson, A., series of papers in J. Theor. Biology since 1969.

13. Eigen, M., Naturwiss 58, 965 (1971).

14. Prigogine, I., Nicolis, G., and Babloyantz, A., to appear in Physics Today.

15. Nicolis, G., and Prigogine, I., Proc. Nat. Acad. Sci. (U.S.A.) 68, 2102 (1971).

16. Ruelle, D., Some Comments on Chemical Oscillators, preprint (1972).

ON PREDATOR-PREY EQUATIONS SIMULATING AN IMMUNE RESPONSE

George H. Pimbley
University of California
Los Alamos Scientific Laboratory

1.

To gain insight into the process of the immune response of an organism to invasion by active self-replicating antigens such as viruses, bacteria, or foreign cells, Dr. George I. Bell of the Los Alamos Scientific Laboratory has proposed a "simplest possible" model which is described below.

The substance manufactured by the organism to fight the antigen is known as antibody. For simplicity it is assumed that each unit of antigen has only a single site for binding antibody, and that each antibody can bind only a single antigen site. While this is not true for most antigens, by treating binding sites rather than cells or molecules as the fundamental units, a fairly reasonable and simple representation is obtained [1].

Let A_g denote the concentration of antigen, and let A_b denote the concentration of antibody. It is assumed that, in the absence of antibody, antigens multiply with a rate constant λ_1 but that if antigen is bound to antibody, it is eliminated with the rate constant α_1. If $(A_g)_b$ denotes the concentration of bound antigen, the first equation is

$$\frac{dA_g}{dt} = \lambda_1 A_g - \alpha_1 (A_g)_b \quad . \tag{1a}$$

Equation (1a) could represent antigen behavior even in a more complicated model. Antibody behavior is less clear since antibodies are produced by cells of the organism that have been stimulated by antigen. A suggested mechanism is that antibody production is stimulated by the binding of antigen to antibodies which are stuck on the cell surfaces [1].

It is assumed however that, in the absence of antigen, the antibody concentration decays with rate constant λ_2, while binding of antigen to antibody stimulates the production of antibody with rate constant α_2. Also taken into account is the fact that the capacity of an organism to produce antibodies is limited. A saturation factor is introduced namely $1 - \frac{A_b}{\theta}$, where θ is an antibody concentration level that cannot be exceeded by the organism. There results the second equation

$$\frac{dA_b}{dt} = -\lambda_2 A_b + \alpha_2 (A_b)_b \left(1 - \frac{A_b}{\theta}\right) \tag{1b}$$

where $(A_b)_b$ denotes the concentration of bound antibodies.

Since one antibody binds one antigen, $(A_g)_b = (A_b)_b$. Moreover, if we assume chemical equilibrium between bound and free antigen and antibodies and that the binding has an association constant k, then the law of mass action states that

$$(A_b)_b = (A_g)_b = k[A_g - (A_g)_b][A_b - (A_b)_b] \tag{2}$$

where $A_g - (A_g)_b$ and $A_b - (A_b)_b$ denote the concentrations of unbound antigen and antibodies respectively [2].

Equation (2) is a quadratic to be solved for $(A_b)_b$ in terms of k, A_g, and A_b. With equations (1), the model is thus specified. Solving a quadratic equation, however, results in inconvenient radicals. Instead, we can write equation (2) in the form

$$(A_b)_b = \frac{k A_b A_g}{1 + k(A_b + A_g) - k(A_b)_b} \quad .$$

Then the last term in the denominator is neglected. The resulted expression underestimates $(A_b)_b$ but never by more than a factor of two [2]. It will suffice for a qualitative investigation.

With the use of a common denominator, we arrive at the pair of ordinary differential equations which govern the model:

$$\frac{dA_g}{dt} = A_g \frac{\lambda_1 + k\lambda_1 A_g - k(\alpha_1 - \lambda_1)A_b}{1 + k(A_b + A_g)}$$

$$\frac{dA_b}{dt} = A_b \frac{-\lambda_2 + k(\alpha_2 - \lambda_2)A_g - k\lambda_2 A_b - \frac{k\alpha_2}{\theta} A_b A_g}{1 + k(A_b + A_g)} \quad . \tag{3}$$

The above material has been excerpted from a forthcoming paper by George I. Bell [2].

We introduce new notation now, setting $u = A_b$ and $v = A_g$, and to remove the denominators common to both equations (3), we introduce the new independent variable $s = \int_0^t [1 + k(u + v)]^{-1} dt'$. There results

$$\frac{du}{ds} = u[-\lambda_2 - k\lambda_2 u + k(\alpha_2 - \lambda_2)v - \frac{k\alpha_2}{\theta} u\, v]$$

$$\frac{dv}{ds} = v[\lambda_1 - k(\alpha_1 - \lambda_1)u + k\lambda_1 v] \tag{4}$$

where we have five adjustable parameters λ_1 $\lambda_2, \alpha_1, \alpha_2, \theta$.

A phase plane investigation is useful in discussing equations (4), (see Fig. 1). The singular points are the intersections of the two curves

$$v = \frac{\lambda_2}{k} \frac{1 + ku}{\alpha_2 - \lambda_2 - \frac{\alpha_2}{\theta} u} \tag{a}$$

$$v = \frac{1}{k\lambda_1} [-\lambda_1 + k(\alpha_1 - \lambda_1)u] \quad . \tag{b}$$

We assume that $\alpha_1 > \lambda_1 + \lambda_2$ and $\alpha_2 > \lambda_1 + \lambda_2$, this being sufficient for the existence of these intersections.

Equations (4) define a direction field which can be traced on the positive quadrant of the phase plane as shown by arrows in Fig. 1. In particular curve (a) is crossed vertically by the direction field, downward to the right of (u_f, v_f) and upward to the left of (u_f, v_f) . Curve (b) is crossed horizontally, rightward to the right of (u_f, v_f) and leftward to the left of (u_f, v_f) . Thus (u_f, v_v) is easily seen to be a focus, center or node while from the same considerations, $(0,0)$ and (u_s, v_s) are saddle points.

If curves (a) and (b) in Fig. 1 do not intersect in two points all integral curves in the positive quadrant tend to infinite v for finite values of u . The

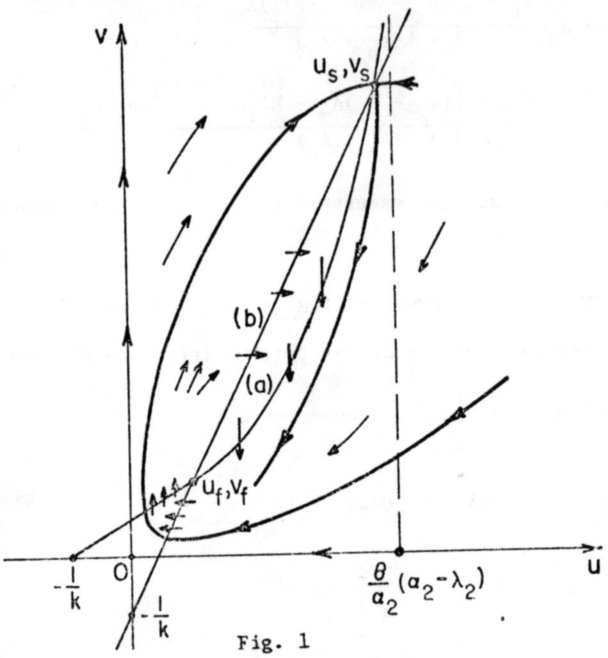

Fig. 1

resulting proliferation of antigen presumably kills the host.

The two exceptional integral curves leading into, and away from the saddle point (u_s, v_s) form two separatrixes which for most sets of parameters, wind one inside the other. Presumably there must be a set, or sets, of parameters for which the two unite to form one closed separatrix.

The phase plane portrait changes, at least quantitatively, as we change parameters. For purposes we have in mind, it is desirable to introduce an auxiliary parameter β as follows:

$$\frac{du}{ds} = \beta u[-\lambda_2 - k\lambda_2 u + k(\alpha_2 - \lambda_2)v - \frac{k\alpha_2}{\theta} u\, v] \tag{5}$$
$$\frac{dv}{ds} = v[\lambda_1 - k(\alpha_1 - \lambda_1)u + k\lambda_1\, v] \quad .$$

Clearly the singular points (u_f, v_f) and (u_s, v_s) remain fixed when β is varied, α_1, α_2, λ_1, λ_2, θ being held fixed.

2.

We are interested in periodic solutions for equations (5). Stable periodic solutions would represent, in the process of immunity, a situation where the antigen invasion is under control and the organism is healthy. From the phase plane heuristics of Fig. 1, these are likely to be found as limit cycles around the singular point u_f, v_f . Thus we represent the differential equations in terms of coordinates centered at u_f, v_f . We substitute

$$u(s) = u_f + \varepsilon \bar{u}(s)$$

$$v(s) = v_f + \varepsilon \bar{v}(s)$$

into equations (5), with parameter ε whose significance will be seen later. We obtain the system

$$\frac{d\bar{u}}{ds} + \beta u_f k[(\lambda_2 + \frac{\alpha_2}{\theta} v_f)\bar{u} - (\alpha_2 - \lambda_2 - \frac{\alpha_2}{\theta} u_f)\bar{v}]$$

$$= -\beta k\, \varepsilon[(\lambda_2 + \frac{\alpha_2}{\theta} v_f)\bar{u}^2 - (\alpha_2 - \lambda_2 - 2\frac{\alpha_2}{\theta} u_f)\bar{u}\bar{v}] - \beta k\, \varepsilon^2 \frac{\alpha_2}{\theta}\bar{u}^2\bar{v} \quad (6)$$

$$\frac{d\bar{v}}{ds} + v_f k[(\alpha_1 - \lambda_1)\bar{u} - \lambda_1 \bar{v}] = -k\varepsilon[(\alpha_1 - \lambda_1)\bar{u}\bar{v} - \lambda_1 \bar{v}^2 \quad .$$

First it is necessary that we study the linearized problem.

Linearized Problem at u_f, v_f .

$$\frac{dh_0}{ds} = -\beta u_f k[(\lambda_2 + \frac{\alpha_2}{\theta} v_f)h_0 - (\alpha_2 - \lambda_2 - \frac{\alpha_2}{\theta} u_f)k_0]$$

$$\frac{dk_0}{ds} = -v_f k[(\alpha_1 - \lambda_1)h_0 - \lambda_1 k_0] \quad ; \quad (7)$$

h_0, k_0 are dependent variables for the linearized system.

This is a linear system with constant coefficients, and we seek periodic solutions. The characteristic equation is $\nu^2 + kb\nu + k^2 u_f v_f \beta c = 0$ where

$$b = \beta u_f \lambda_2 - v_f \lambda_1 + \frac{\beta\alpha_2}{\theta} u_f v_f$$

$$c = (\alpha_1 - \lambda_1)(\alpha_2 - \lambda_2 - \frac{\alpha_2}{\theta} u_f) - \lambda_1(\lambda_2 + \frac{\alpha_2}{\theta} v_f) \quad . \quad (8)$$

The characteristic roots are $\nu = -1/2\, kb \pm 1/2\sqrt{k^2 b^2 - 4\beta u_f v_f k^2 c}$. Since periodic solutions can only result when the characteristic roots are conjugate imaginary [4, p. 53], we have the requirement that $b = 0$, $c > 0$ for periodic solutions of the linearized problem. In fact we are able to state the following result.

Theorem 1. If $b < 0$ in equations (8) and $c > 0$, the solutions of linear equations (7) are outward tending spirals. If $b = 0$ and $c > 0$, the solutions of linear equations (7) are periodic. If $b > 0$ and $c > 0$, the solutions of linear equations (7) are inward tending spirals.

We note that positive c is assured by the assumptions

$$\alpha_1 > \lambda_1 + \lambda_2 \qquad \alpha_2 > \lambda_1 + \lambda_2 \tag{9}$$

made earlier, for sufficiently large θ .

With reference to equations (8), let

$$\beta_0 = \frac{\lambda_1 v_f}{\lambda_2 u_f + \frac{\alpha_2}{\theta} u_f v_f} \quad . \tag{10}$$

Corollary 1. Only at $\beta = \beta_0$ do we have $b = 0$ and the possibility of bifurcated periodic solutions for equations (5) and (6), assuming that $c > 0$. For $\beta > \beta_0$, the constant solution (u_f, v_f) for equations (5) is stable, while for $\beta > \beta_0$ it is unstable (by the principle of linear stability [4, p. 57]).

We shall refer to β_0 as a critical value for β .

A periodic solution of the linear problem is represented by

$$\begin{aligned} h_0 &= A\,L\,e^{ki\sqrt{\beta_0 u_f v_f c}\,s} + \bar{A}\,\bar{L}\,e^{-ki\sqrt{\beta_0 u_f v_f c}\,s} \\ k_0 &= B\,L\,e^{ki\sqrt{\beta_0 u_f v_f c}\,s} + \bar{B}\,\bar{L}\,e^{-ki\sqrt{\beta_0 u_f v_f c}\,s} \end{aligned} \tag{11}$$

where $A = -\lambda_1 v_f + i\sqrt{\beta_0 u_f v_f c} = a_1 + ia_2$, $B = -v_f(\alpha_1 - \lambda_1)$,

and $L = \frac{b_2}{2B} + i\left[\frac{a_1}{a_2}\frac{b_2}{2B} - \frac{b_1}{2a_2}\right]$.

Here $b_1 = h_0(0)$, $b_2 = k_0(0)$ represent a pair of initial conditions. We can write

$$\begin{bmatrix} h_0 \\ k_0 \end{bmatrix} = Y(s) \begin{bmatrix} b_1 \\ b_2 \end{bmatrix} \tag{12}$$

where $Y(s)$ is the fundamental matrix [4, p. 100] of the linear system (7) with $Y(0) = 1$. If P_0 represents the period of the periodic solutions, then

$$Y(P_0) = 1 \quad \text{also and} \quad P_0 = \frac{2\pi}{k\sqrt{\beta_0 u_f v_f c}} .$$

Next, we discuss the case where $\theta = \infty$ in equations (6).

Case $\theta = \infty$. One now observes a considerable simplification in equations (5) and (6). In Fig. 1, curves (a) and (b) are both straight lines, and there is only only one finite intersection, namely u_f, v_f .

Here, when $\beta = \beta_0 = \dfrac{\lambda_1 v_f}{\lambda_2 u_f} = \dfrac{\alpha_1}{\alpha_2}$, it is possible to obtain periodic solutions of the nonlinear problem. From equation (5), we can write

$$\frac{du}{dv} = \beta_0 \frac{u}{v} \frac{-\lambda_2 - k\lambda_2 u + k(\alpha_2 - \lambda_2) v}{\lambda_1 - k(\alpha_1 - \lambda_1)u + k\lambda_1 v} \tag{13}$$

or

$$v[\lambda_1 - k(\alpha_1 - \lambda_1) u + k\lambda_1 v] \frac{du}{dv} + \beta_0 u[\lambda_2 + k\lambda_2 u - k(\alpha_2 - \lambda_2) v] = 0 .$$

An integrating factor for equation (13), when $\alpha_1 = \beta_0, \alpha_2 = \alpha$, is

$$\frac{u^{\lambda_1 - 1} v^{\beta_0\lambda_2 - 1}}{[1 + k(u + v)]^{\alpha+1}} ;$$

this was pointed out by George I. Bell. In fact

$$\frac{v[\lambda_1 - k(\alpha_1 - \lambda_1)u + k\lambda_1 v] u^{\lambda_1 - 1} v^{\beta_0\lambda_2 - 1}}{[1 + k(u + v)]^{\alpha_1 + 1}} \frac{du}{dv}$$

$$+ \frac{\beta_0 u[\lambda_2 + k\lambda_2 u - k(\alpha_2 - \lambda_2)v] u^{\lambda_1 - 1} v^{\beta_0\lambda_2 - 1}}{[1 + k(u + v)]^{\beta_0\alpha_2 + 1}} = 0$$

is a perfect differential, since

$$\frac{\partial}{\partial u} \frac{u^{\lambda_1} v^{\beta_0\lambda_2}}{[1 + k(u + v)]^{\alpha_1}} = \frac{v[\lambda_1 - k(\alpha_1 - \lambda_1) u + k\lambda_1 v]}{[1 + k(u + v)]^{\alpha_1 + 1}} u^{\lambda_1 - 1} v^{\beta_0\lambda_2 - 1}$$

and

$$\frac{\partial}{\partial v} \frac{u^{\lambda_1} v^{\beta_0\lambda_2}}{[1 + k(u + v)]^{\beta_0\alpha_2}} = \frac{\beta_0 u[\lambda_2 + k\lambda_2 u - k(\alpha_2 - \lambda_2)v] u^{\lambda_1 - 1} v^{\beta_0\lambda_2 - 1}}{[1 + k(u + v)]^{\beta_0\alpha_2 + 1}} .$$

Thus we have

$$\frac{d}{dv}\,\frac{u^{\lambda_1}v^{\beta_0\lambda_2}}{[1 + k(u + v)]^{\alpha_1}} = 0$$

and so

$$\frac{u^{\lambda_1}v^{\beta_0\lambda_2}}{[1 + k(u + v)]^{\alpha_1}} = \text{const.}$$

Thus we obtain a family of closed curves in the first quadrant of the phase plane nested around the point u_f, v_f, provided conditions (9) are fulfilled.

With reference to the linear problem and the principle of linear stability, when $\beta > \beta_0$ we have an inward tending spiral, and when $\beta < \beta_0$, an outward tending spiral, in some neighborhood of u_f, v_f. Approximate methods devised by George I. Bell [2] indicate the same tendencies for spiralling behavior at large amplitudes. This would lead to the belief that for $\beta \neq \beta_0$, all solutions are spirals.

We can summarize this case in the following theorem.

Theorem 2. When $\theta = \infty$ in equations (5) and $\beta = \beta_0$ (the critical value), the singular point u_f, v_f is a center, and there exists a dense family of closed integral curves.

One could call this a case of vertical bifurcation of periodic solutions.

Now we come to the general case of equations (6).

Case $\theta < \infty$. Of course we assume θ to be large enough so that $c > 0$ under conditions (9). Were this not so, there could be no periodic solutions. We refer to Corollary 1 for the behavior of solutions for $\beta \neq \beta_0$ in a neighborhood of u_f, v_f. By the principle of linearized stability [4, p. 57], we have inward tending spirals for $\beta > \beta_0$ and outward tending spirals for $\beta < \beta_0$.

We note that β_0 is a value of β where the complex conjugate characteristic exponents $\nu = -1/2\ kb \pm i\sqrt{\beta u_f v_f k^2 c - 1/4\ k^2 b^2}$, which are the two eigenvalues of the matrix in linear system (7), cross the imaginary axis as β decreases through β_0 and as b in equations (8) decreases through zero. That this is a bifurcation situation is clear from the following known result [4, p. 122].

Friedrichs' Bifurcation Theorem. Suppose the equation (where $F(x,\beta)$ satisfies given differentiability hypotheses)

$$\frac{dx}{dt} = F(x,\beta) \ , \quad (x,F \text{ are 2-vectors}) \tag{14}$$

has a constant solution $x = a(\beta)$, i.e., $F(a(\beta),\beta) = 0$, such that for the value β_0 of the parameter, the Jacobian matrix

$$A(\beta_0) = F_x(a(\beta_0), \beta_0)$$

has purely imaginary eigenvalues $\pm i\omega_0, \ \omega_0 \neq 0$. Suppose further that the trace of the matrix

$$B(\beta_0) = \lim_{\beta \to \beta_0} \frac{A(\beta) - A(\beta_0)}{\beta - \beta_0} = \frac{d}{d\beta} A(\beta) \Bigg|_{\beta=\beta_0}$$

$$= F_{xx}(a(\beta_0), \beta_0) \, a_\beta(\beta_0) + F_{x\beta}(a(\beta_0), \beta_0)$$

does not vanish. Here $A(\beta) = F_x(a(\beta), \beta)$. Then there exist functions $\beta(\epsilon)$ and $P(\epsilon)$ of an additional parameter ϵ with $\beta(0) = \beta_0, \ P(0) = P_0 = \frac{2\pi}{\omega_0}$, and derivatives $\beta_\epsilon(0) = 0, \ P_\epsilon(0) = 0$, and further a function $y(s,\epsilon)$ with the period P_0 in s , assuming a prescribed value $y(0,\epsilon) = b^0$ for $s = 0$, such that the function

$$x = a(\beta(\epsilon)) + \epsilon \, y\left(\frac{P_0}{P(\epsilon)} t, \ \epsilon\right)$$

is a solution of the differential equation

$$\frac{dx}{dt} = F(x, \beta(\epsilon)) \quad . \tag{15}$$

In applying Friedrichs' Theorem, we note that for $\beta = \beta_0$ we already have seen that there exists a pair of imaginary conjugate eigenvalues, $\pm i\sqrt{\beta_0 u_f v_f k^2 c}$. Since $u_f, \ v_f$, our constant solution, does not shift when we vary β , we have $a_\beta(\beta_0) = 0$ and $B(\beta_0) = F_{x\beta}(a(\beta_0), \beta_0)$. Thus we have only to check the trace of the matrix

$$F_{x\beta}(a(\beta_0), \beta_0) = \begin{bmatrix} -u_f k(\lambda_2 + \frac{\alpha_2}{\theta} v_f) & u_f k(\alpha_2 - \lambda_2 - \frac{\alpha_2}{\theta} u_f \\ 0 & 0 \end{bmatrix} .$$

By inspection, the trace does not vanish. Thus Friedrichs' Bifurcation Theorem applies to our problem.

We use the Friedrichs' Bifurcation Theorem because of its long availability in print. There is a theorem of E. Hopf which could be employed [5]. Moreover D. H. Sattinger has recently proved a bifurcation theorem of this type, [6, 8]. D. Joseph and Sattinger have shown that the branch of bifurcating periodic solutions is unique up to a phase shift and exists on one side only of the critical value β_0 [8].

We can summarize the situation in the following result.

Theorem 3. There exists a one-parameter family, or branch, of periodic solutions of equations (5), which bifurcates from the constant solution u_f, v_f at the critical value $\beta = \beta_0$. This branch of periodic solutions is unique up to a phase shift, and exists on one side of the critical value only.

3.

In order to determine on which side of the critical value β_0 the bifurcated periodic solutions of Theorem 3 lie, we substitute $\epsilon = \frac{P_0}{P(\epsilon)} s$ into equations (6) as a new independent variable. Then we introduce the following perturbation series.

$$\begin{aligned} \beta &= \beta_0 + \epsilon\beta_1 + \epsilon^2\beta_2 + \dots \\ P &= P_0 + \epsilon P_1 + \epsilon^2 P_2 + \dots \\ \bar{u} &= u_0 + \epsilon u_1 + \epsilon^2 u_2 + \dots \\ \bar{v} &= v_0 + \epsilon v_1 + \epsilon^2 v_2 + \dots \quad . \end{aligned} \tag{16}$$

Here, ϵ is the additional parameter mentioned in Friedrichs' Theorem and which was also introduced in equations (6). There follows the usual hierarchy of linear problems, the first of which is linear problem (7) with the arbitrary initial conditions $b^0 = (b_1, b_2)$ mentioned in the theorem. (That there is no undue restriction in imposing an arbitrary non-trivial initial condition is discussed by

Friedrichs' [4, p. 116]). Thus (u_0, v_0) is given by equations (11) or (12), and the parameter ϵ is the magnitude of the projection of $(\bar{u}, \bar{v})$ along the null space.

By Friedrichs' Theorem, $\beta_1 = \beta_\epsilon(0) = 0$ and $P_1 = P_\epsilon(0) = 0$.

The constants β_2 and P_2 are found from the hierarchy equations for (u_2, v_2) , solved with conditions $u_2(0) = u_2(P_0) = 0$, $v_2(0) = v_2(P_0) = 0$. Very lengthy expressions result for β_2 and P_2 which afford no visual insight regarding the sign of β_2 .

The functions $\beta(\epsilon)$ and $P(\epsilon)$ can be presumed to be analytic, since in Friedrichs' Theorem they result from use of the Implicit Function Theorem. Thus expansions (16) converge in some neighborhood. If $\beta_2 < 0$, β-bifurcation would be to the left, and according to Sattinger [8] the bifurcating periodic solutions would be stable with an exchange of stabilities. On the other hand if $\beta_2 > 0$, the bifurcating periodic solutions would be unstable.

Added Note. The following result has been obtained, subsequent to the submission of this paper, by means of a study of the direction field at $\beta = \beta_0$.

Theorem 4. If $\theta < \infty$ in equations (5), β-bifurcation is to the left, and the bifurcated periodic solutions are stable.

4.

In Section 1, equations (5), we introduced an auxilliary parameter β for convenience, choosing to hold fixed the five natural parameters $\alpha_1, \alpha_2, \lambda_1, \lambda_2, \theta$ and, therefore, the phase plane portrait (Fig. 1). Now we must interpret the results obtained in terms of β for the natural parameters.

Let us first prove a preliminary result. We assume, as we have been doing, that $\alpha_1 > \lambda_1 + \lambda_2$, $\alpha_2 > \lambda_1 + \lambda_2$.

Theorem 5. There exists a value $\alpha_1 = \alpha_{10} > \alpha_2$ such that $\lambda_2 u_f - \lambda_1 v_f + \frac{\alpha_2}{\theta} u_f v_f = 0$. For $\alpha_1 \gtrless \alpha_{10}$ we have $\lambda_2 u_f - \lambda_1 v_f + \frac{\alpha_2}{\theta} u_f v_f \gtrless 0$.

Proof. Indeed from the definition of u_f, v_f in Section 1, we have

$\lambda_1[\lambda_2 + k\lambda_2 u_f - k(\alpha_2 - \lambda_2)v_f + \frac{k\alpha_2}{\theta} u_f v_f] + \lambda_2[-\lambda_1 + k(\alpha_1 - \lambda_1)u_f - k\lambda_1 v_f] = 0$,

whence $\lambda_2\alpha_1 u_f - \lambda_1\alpha_2 v_f + \lambda_1 \frac{\alpha_2}{\theta} u_f v_f = 0$. Now when $\alpha_1 \leq \alpha_2$, $0 = \lambda_2\alpha_1 u_f - \lambda_1\alpha_2 v_f + \lambda_1 \frac{\alpha_2}{\theta} u_f v_f \leq \lambda_2\alpha_2 u_f - \lambda_1\alpha_2 v_f + \lambda_1 \frac{\alpha_2}{\theta} u_f v_f = \alpha_2[\lambda_2 u_f - \lambda_1 v_f + \frac{\lambda_1}{\theta} u_f v_f] < \alpha_2[\lambda_2 u_f - \lambda_1 v_f + \frac{\alpha_2}{\theta} u_f v_f]$. Thus α_{10} , if it exists, is such that $\alpha_2 < \alpha_{10}$. However in Fig. 1, u_f, v_f is the intersection of curves (a) and (b). As $\alpha_1 \to \infty$ with α_2, λ_1, λ_2 , and θ fixed, curve (a) remains fixed so that $u_f \to 0$ and $v_f \to \frac{\lambda_2}{k(\alpha_2-\lambda_2)}$. Hence $\lambda_2 u_f - \lambda_1 v_f + \frac{\alpha_2}{\theta} u_f v_f \to -\frac{\lambda_1\lambda_2}{k(\alpha_2-\lambda_2)} < 0$. Since for $\alpha_1 = \alpha_2$, $\lambda_2 u_f - \lambda_1 v_f + \frac{\alpha_2}{\theta} u_f v_f$ is positive but becomes negative when $\alpha_1 \to \infty$, there exists a zero intercept α_{10} , where $\lambda_2 u_f - \lambda_1 v_f + \frac{\alpha_2}{\theta} u_f v_f \lessgtr 0$ for $\alpha_1 \gtrless \alpha_{10}$. This proves the theorem.

Remembering the expression for β_0 in equation (10), we see that if $\alpha_1 > \alpha_{10}$, then $\beta_0 > 1$, while if $\alpha_1 < \alpha_{10}$, then $\beta_0 < 1$. Assuming continuity as is reasonable, and assuming that α_2, λ_1, λ_2 and θ are fixed, we find that β_0 increases through the value unity as α_1 increases through the value α_{10} . Now unity is that value of β appearing in equations (4) prior to the introduction of β . These remarks indicate the following theorem.

Theorem 6. If, as in Theorem 4, we have left β-bifurcation at β_0 , then the bifurcation for equations (4) in terms of parameter α_1 at α_{10} is to the right. If β-bifurcation is vertical, as in Theorem 2, so also is α_1-bifurcation.

George I. Bell [2] gives the following case where computer results indicate a stable limit cycle when α_1 is on the right of a critical value α_{10} , namely, the case where $\lambda_1 = 1$, $\lambda_2 = 1$, $\alpha_2 = 4$, $k = 1$, $\theta = 10$.

This work was performed under the auspices of the U. S. Atomic Energy Commission.

REFERENCES

1. Bell, George I., "Mathematical Model of Clonal Selection and Antibody Production," Journal of Theoretical Biology, 33, 339 (1971).

2. Bell, George I., 'Predator-Prey Equations Simulating an Immune Response," to be published in Mathematical Biosciences: An International Journal.

3. Coddington, E. A. and Levinson, N., Theory of Ordinary Differential Equations (McGraw-Hill Publ. Co., New York, 1955).

4. Friedrichs, K. O., Lectures on Advanced Ordinary Differential Equations, New York University Lecture Notes, 1948-1949.

5. Hopf, E., "Abzweigung einer periodischen Lösung eines Differentialsystems," Aus den Berichten der Mathematisch-Physikalischen Klasse der Sächsischen Akademie der Wissen-schaften zu Leipzig XCIV, 1-22 (1942).

6. Joseph, D. D. and Sattinger, D. H., "Bifurcating Time Periodic Solutions and Their Stability," Arch. for Rational Mech. and Anal., 45, No. 2, 79-109 (1972).

7. Sattinger, D. H., "Stability of Bifurcating Solutions by Leray-Schauder Degree," Arch. for Rational Mech. and Anal., 43, 154-166 (1971).

8. Sattinger D. H., Topics in Stability and Bifurcation Theory, Springer Lecture Note Series, No. 309, 1973.

9. Waltman, P. E., "The Equations of Growth," Bulletin of Mathematical Biophysics, 26, 39-43 (1964).

BIFURCATION THEORY FOR GRADIENT SYSTEMS

Michael Reeken
Battelle Advanced Studies Center
Geneva

INTRODUCTION

We shall describe new results about bifurcation theory for gradient systems; detailed proofs are given in [1]. The ultimate aim is to establish existence theorems for small solutions of equations of the form

(1) $$A(u) = \lambda u, \ A(0) = 0 ,$$

where $A: H_1 \to H_2$ is a map from one Hilbert space into another and H_1 must be thought of as a dense subspace of H_2. For simplicity of exposition we simply assume, for the purpose of this article, that $H_1=H_2=H$. We attack the problem in the spirit of perturbation theory; that is, we assume that A can be linearized and investigate to what extent solutions of the linearized problem indicate the existence of small solutions of the full equation (1). It is well known that, in general, the answer is negative. Solutions of the linearized equation do not always imply small solutions of (1). But there is one class of problems where a satisfactory theory is possible. This is the case if A is a gradient, that is, there exists a function $G: H \to \mathbb{R}$ such that $d_uG(h)=(A(u),h)$. d_uG denotes the Frechet derivative of G at u.

Then, instead of looking for solutions of (1), one looks for critical points of G on the sphere $S_R=\{u \mid \|u\| = R\}$. These critical points are in one to one correspondence with solutions of (1) with norm R .

If S_R is compact (that is, for finite dimensional H) the reason for the satisfactory nature of these problems is purely topological, and it suffices to suppose G to be C^1 . In infinite dimensions there are troubles with topological arguments. Either a suitable substitute for the compactness of S_R in the form of a certain compactness property of the map A itself is assumed, or, (as we shall do) a global version of the implicit function theorem, is developed to reduce an infinite dimensional problem to a finite dimensional one. In that case differentiability of G has to be strengthened to C^2 .

II. TOPOLOGICAL THEORY

We study a C^1-function $f : M \longrightarrow \mathbb{R}$ where M is a C^2-Finsler manifold. As usual we call a value c of f critical or regular depending whether there are or there are not critical points in $f^{-1}(c)$. This set $f^{-1}(c)$ we shall call the c-level. We shall need one more dichotomy. We call c a normal value if there is an open interval I containing c such that condition C is valid on $f^{-1}(I)$; that is, for any $V \subset f^{-1}(I)$ on which df is not bounded away from zero, there is a critical point of f in $\overline{V}$. If this condition is violated, we call c a limit value. The nomenclature is inspired by quadratic forms. Let $f(u) = (u,Bu)$ be the quadratic form associated to the bounded self-adjoint operator B. Let S be the unit sphere of the Hilbert space. Then any limit point of the spectrum of B is a limit value of f on S.

It is this condition which ensures that the situation is the same as in finite dimensions where the condition is automatically fulfilled.

We shall also use the following notation. The set $f^{-1}(I)$ where $I = [a,b]$ will be called a slice and be denoted by f_a^b. If $a = -\infty$ or $b = +\infty$ we shall simply suppress the corresponding index, for example, $f^{-1}((-\infty,b]) = f^b$.

The essential idea of our approach to studying critical points of f is to imitate Morse theory. There it is shown that f^a changes in a very definite way whenever a crosses a critical level which contains only nondegenerate critical points, that is critical points where the Hessian defines an invertible linear operator. If there is exactly one nondegenerate critical point on the c level, then $f^{c+\varepsilon}$ is diffeomorphic to $f^{c-\varepsilon}$ with a handle added to it. For degenerate critical points, this is not true. The topological changes occuring in f^a when crossing a critical level are, intuitively speaking, more diffuse. A much broader concept than addition of handles is needed to describe what happens when a critical value is crossed. It is this subject that we turn to now.

Given a topological space X and a subset X_1 of X, the category of X_1 in X (written $\mathrm{cat}_X X_1$), is the minimal number of contractible sets by which X_1 can be covered. For example, a ball in $\mathbb{R}^n$ is contractible and therefore of category 1 in $\mathbb{R}^n$. A sphere in $\mathbb{R}^{n+1}$ is not contractible in itself. One needs two contractible sets to cover it; so $\mathrm{cat}_{S^n} S^n = 2$. But if the ambient space is $\mathbb{R}^{n+1}$, then clearly S^n is contractible; that is, $\mathrm{cat}_{\mathbb{R}^{n+1}} S^n = 1$. This shows that the ambient space is essential, since the notion of contractibility depends upon it. Lusternik-Shnirelman theory proceeds by using this concept. One looks at $\mathrm{cat}_M f^a$ as a function of the real variable a. This function is integer valued and monotone, so it can only have jumps. It turns out that any value c where this function has a jump is critical, and the set of critical points on that level has category at least equal to the size of the jump. Since $f^{-1}(\mathbb{R}) = M$ for a bounded function, it is clear that the jumps must add up to $\mathrm{cat}_M M$.

So there are at least $\mathrm{cat}_M M$ critical points. All this is true provided that condition C is fulfilled on the whole of M. A counter-example is given by the quadratic form on the unit sphere of any bounded self-adjoint operator with continuous spectrum only. This function has no critical points at all, but condition C is violated, of course.

As said before, we want to describe critical points locally without reference to what happens globally on M. So we define the relative category of one set over another as follows. Let $X_1 \subset X_2 \subset X$. Then the relative category of X_2 over X_1 in X is the minimal number of contractible sets U_i such that $X_2 - \bigcup U_i$ is strongly retractable in X onto X_1. We shall write $\mathrm{cat}_X(X_2/X_1)$ for the relative category. It will be convenient to introduce also the symmetric relative category

$$\overline{\mathrm{cat}}_X(X_1,X_2) = \max\{\mathrm{cat}_X(X_2/X_1),\mathrm{cat}_X(X_1^c/X_2^c)\} ,$$

where the superscript means complement in X.

The invariant $\mathrm{cat}_X(X_2/X_1)$ has the following essential properties :

1.) $\mathrm{cat}_X(X_2/X_1) = 0$ whenever X_1 is a strong deformation retract of X_2,
2.) $\mathrm{cat}_X(X_3/X_1) \leq \mathrm{cat}_X(X_3/X_2) + \mathrm{cat}_X(X_2/X_1)$ where $X_1 \subset X_2 \subset X_3 \subset X$,
3.) $X_1 \subset X_2 \subset X_3$ implies $\mathrm{cat}_X(X_3/X_1) \geq \mathrm{cat}_X(X_2/X_1)$.

Having introduced this topological notion, we can turn now to the study of the kind of topological change occuring when degenerate critical points are present.

The situation can be described as follows:

THEOREM : Assume that c is a normal value of f, and let $K_{f,c}$ be the set of critical points on the c level. Then there is an $\varepsilon_0 > 0$ and a set $\tilde{U}$ containing $K_{f,c}$ and having the same category, $\mathrm{cat}_M\tilde{U} = \mathrm{cat}_M K_{f,c}$ such that

$$f_{c-\varepsilon}^{c+\varepsilon} - \tilde{U} \approxeq (f^{-1}(c) - f^{-1}(c) \cap \tilde{U}) \times [c-\varepsilon, c+\varepsilon] .$$

Here $\approxeq$ means a locally Lipschitzian homeomorphism. That is, after removing a set of the same category as the set of critical points,the slice has the structure of a cylinder.

This means that, for any $\varepsilon \leqslant \varepsilon_o$, $f^{c+\varepsilon} - \tilde{U}$ is strongly retractable onto $f^{c-\varepsilon}$, and similarly $f_{c-\varepsilon} - \tilde{U}$ is strongly retractable onto $f^{c+\varepsilon}$. So it follows that

$$\overline{\text{cat}}_M(f^{c+\varepsilon}, f^{c-\varepsilon}) \leqslant \text{cat}_M K_{f,c} \quad \text{for} \quad \varepsilon \leqslant \varepsilon_o .$$

We now define the indicator-function δ_f as follows:

$$\delta_f(c) = \underline{\lim}_{\varepsilon > 0} \overline{\text{cat}}_M(f^{c+\varepsilon}, f^{c-\varepsilon}) .$$

From what we said above it is clear that $\delta_f(c) \leqslant \text{cat}_M K_{f,c}$ if c is normal, that is, if condition C holds locally at the c-level. Therefore whenever $\delta_f(c)$ is not zero, there must be sufficiently many critical points ,since the category of $K_{f,c}$ is bounded below by $\delta_f(c)$.

If there is a whole interval $I = [a,b]$ of normal value,then we can extend these considerations over the whole interval. To do this we define two functions Σ_f and Δ_f . If there are infinitely many critical values in I, then Σ_f is infinite,and if there are infinitely many critical values with $\delta_f(c) \neq 0$, then Δ_f is infinite. If there are only finitely many critical values,then

$\Sigma_f(a,b) = \sum_i \mathrm{cat}_M K_{f,c_i}$ where c_i are the critical values in $[a,b]$. Similarly we define $\Delta_f(a,b) = \sum_i \delta_f(c_i)$. From our results and property 2.) of the relative category, we get

$$\overline{\mathrm{cat}}_M(f^a,f^b) \leqslant \Delta_f(a,b) \leqslant \Sigma_f(a,b) .$$

This is a generalization of the result of Lusternik-Shnirelman theory.

What have we gained by this? It is certainly, in general, a hopeless task to calculate the quantity $\overline{\mathrm{cat}}_M(f^a,f^b)$. But since this is a topological quantity related to the sets f^a and f^b, we expect that, for deformations of f which are small in a proper sense, the above quantity should stay constant. This is exactly what happens. The essential continuity properties of the relative category used are expressed by properties 2.) and 3.). We prove that given a f, there is an $\varepsilon > 0$ such that for all g with $\sup_M |f-g| \leqslant \varepsilon$, $\overline{\mathrm{cat}}_M(f^a,f^b) = \overline{\mathrm{cat}}_M(g^a,g^b)$.

So, if we can show that normal values of f are normal values of g for sufficiently small ε, then we are done, since then our above equalities hold for all g as well as for f . This, of course, seems to need restrictions on g . Later, when applying our results to a special case, we shall give a hypothesis ensuring that normal values remain normal.

III. ANALYTICAL THEORY

In order to motivate our approach we recall the following theorem which is a consequence of the inverse function theorem.

THEOREM : *Let* $f : (U_1 \times U_2) \times U_3 \longrightarrow U_4$ *be a differentiable map and* $f((u_1',u_2'),u_3') = u_4'$. *The* U_i *are open sets of Banach spaces* B_i. *Assume that the Frechet derivative of* f *with respect to* (u_1,u_2) *at* (u_1',u_2') *is surjective and has kernel* B_1. *Then there exist local submanifolds* $N_{u_3} \subset U_1 \times U_2$ *which are diffeomorphic to an open subset* U_1' *of* U_1 *and such that* $f((u_1,u_2),u_3) = u_4'$ *for all* $(u_1,u_2) \in N_{u_3}$ *and* $(u_1',u_2') \in N_{u_3'}$.

Our aim is to find a global version of this theorem for a cross-section ω_ε of a vector bundle $E \xrightarrow{\pi} M$. That is,we assume that ω_o is zero on a certain submanifold N of M, and we suspect, in analogy to the above theorem,that there will be manifolds N_ε diffeomorphic to N on which ω_ε vanishes.

This generalization is not quite trivial because none of the manipulations in the proof of the above theorem has an invariant meaning. Also the notion of range and kernel for the derivative of a cross-section has to be defined carefully. There is an invariant meaning to these notions only at points where the cross-section vanishes. This is due to the fact that the tangent space to a vector bundle has a complicated transformation behavior.

We introduce the following notion:

DEFINITION : Given a cross-section ω of a vector bundle $E \xrightarrow{\pi} M$, we call a closed submanifold N of M a nondegenerate zero-manifold for ω if ω vanishes on N, and if the kernel of T_ω at any point of N equals the tangent space to N at this point and if T_ω is surjective at all points of N.

We prove the following:

THEOREM : Given two cross-sections ω and τ of $E \xrightarrow{\pi} M$ (where M admits smooth partitions of unity) and a compact nondegenerate zero manifold N for ω, there exists an $\varepsilon_0 > 0$ such that for all $|\varepsilon| < \varepsilon_0$ there is an unique nondegenerate zero-manifold N_ε for $\omega_\varepsilon = \omega + \varepsilon\tau$ and a diffeomorphism $\phi_\varepsilon : N \longrightarrow N_\varepsilon$ depending differentiably upon ε.

In order to apply this theorem to gradient systems, we need some more notions.

DEFINITION : Given a vector bundle $E \xrightarrow{\pi} M$, two subbundles E_1 and E_2 are said to provide a complementary splitting of E if $E = E_1 \oplus E_2$.

DEFINITION : Given three subbundles E_1, E_2 and E_3 of a vector bundle $E \xrightarrow{\pi} M$. We say that $E_1 \approx E_2 \bmod E_3$ if $E = E_1 \oplus E_3 = E_2 \oplus E_3$.

DEFINITION : Given a manifold M and a function $f : M \longrightarrow \mathbb{R}$, we call a compact n-dimensional submanifold N of M a subcritical manifold for f if there is a complementary splitting $E_1 \oplus E_2$ of $T^*(M)$ on a neighbourhood of N such that E_1 has n-dimensional fiber, N is a nondegenerate zero-manifold for the projection of df onto E_2 along E_1 and $E_2/N \approx {}^{\perp}T^*(N) \bmod E_1/N$. Here ${}^{\perp}T(N)$ is the orthogonal complement bundle of $T(N)$ in $T^*(M)/N$ and $/N$ means restriction to N. A manifold is critical if it is subcritical and consists of critical points.

The importance of this notion lies in the fact that, on a subcritical manifold N the critical points of f are exactly those of f/N. With this we can apply the theorem about nondegenerate zero manifolds to get the following result.

THEOREM : Given a C^{r+4}-manifold M $(r \geq 0)$ admitting smooth partitions of unity and given two C^{r+2}-functions $f : M \longrightarrow \mathbb{R}$ and $g : M \longrightarrow \mathbb{R}$ as well as a subcritical manifold N for f, there is a neighbourhood $U(N)$ of N in M and a C^{r+1}-diffeomorphism $\phi_\varepsilon : N \longrightarrow N_\varepsilon$ for all $|\varepsilon| < \varepsilon_0$ such that the critical points of $f_\varepsilon = f + \varepsilon g$ in $U(N)$ are exactly the critical points of $f_\varepsilon/N_\varepsilon$.

This is the important result which enables us to reduce the problem of finding critical points on the infinite dimensional manifold M to that of finding critical points of the function $f_\varepsilon / N_\varepsilon$ defined on the finite dimensional manifold N_ε.

Before applying these results to bifurcation theory, we define what we mean by bifurcation.

DEFINITION : A point $m \in M$ is called a point of bifurcation for $f_\varepsilon = f + \varepsilon g$ if for any neighbourhood $U(m)$ of m there is an $\varepsilon_o > 0$ such that for all $|\varepsilon| \leqslant \varepsilon_o$ there are critical points of f_ε in $U(m)$.

If we now assume that we are given a critical manifold N of f on M, then we can ask for the bifurcation points in N for $f_\varepsilon = f + \varepsilon g$. It turns out that the bifurcation points are contained in the set of critical points of g/N on N. This, together with the results of §II, leads to the following result.

THEOREM : Given M, f and g and a critical manifold for f as before, then there is an open interval J around $c = f(N)$ and an $\varepsilon_o > 0$ such that $\Sigma_{f_\varepsilon}(J) \geqslant \Delta_{g/N}(-\infty,\infty)$ for all $|\varepsilon| \leqslant \varepsilon_o$ and the set of bifurcation points is contained in the set of critical points of g/N. Moreover, there is a "fine structure" of the set of critical points: If g/N has finitely many critical values c_i and if we choose disjoint neighbourhoods V_i of them, then there is an $\varepsilon_o > 0$ and a neighbourhood U of N such that for all $|\varepsilon| \leqslant \varepsilon_o, \Sigma_{f_\varepsilon}(c+\varepsilon V_i) \geqslant \delta_{g/N}(c_i)$ and the corresponding critical points of f_ε are bifurcating from the c_i-level of g/N.

IV. APPLICATION TO NONLINEAR EIGENVALUE PROBLEMS IN HILBERT SPACE

We are interested in equations of the form

$$(*) \qquad Bu + C(u) = \lambda Du$$

where B and D are bounded, linear, self-adjoint operators in a Hilbert space H, and $D \geq \varepsilon_o > 0$. We assume that the relative eigenvalue problem $Bu = \lambda Du$ has at least one isolated eigenvalue of finite multiplicity. We assume further that $C(0) = 0$ and that C is the gradient of a potential $G : H \longrightarrow \mathbb{R}$ such that $d_u G(h) = (C(u),h)$. Then we can formulate the problem of finding solutions of (*) of given value of $\frac{1}{2}(u,Du)$ as a variational problem, that is finding the critical points of $F + G$ on M_R where

$$F(u) = \tfrac{1}{2}(u,Bu) \; ; \; M_R = \{u \,|\, \tfrac{1}{2}(u,Du) = R^2\} \quad .$$

F is a C^∞-functional and M_R is a C^∞-manifold because $\frac{1}{2}(u,Du)$ is a C^∞-functional and $R^2 \neq 0$ is a regular value of it. Since $\frac{1}{2}(u,Du)$ defines a norm equivalent to $\|u\|$, the bifurcation problem for (*) is equivalent to the problem of finding solutions of arbitrary small R.

We introduce now polar coordinates; that, is, we set $u = R\tilde{u}$ where $\frac{1}{2}(\tilde{u},D\tilde{u}) = 1$. The reason is that we want to define our variational problem on a fixed manifold with R as a parameter of the function. It is evident that for $R \neq 0$ the critical points of $F(u) + G(u)$

on M_R are exactly those of $F(\tilde{u}) + (1/R^2)G(R\tilde{u})$ on $M = \{\tilde{u} | \frac{1}{2}(\tilde{u}, D\tilde{u}) = 1\}$. We define $\tilde{G}(R,\tilde{u}) = (1/R^2)G(R\tilde{u})$. Its gradient is $(1/R)C(R\tilde{u})$.

We shall show that the lowest nonvanishing term in the Taylor series of C determines the bifurcation behavior.

Corresponding to the introduction of polar coordinates, we have to introduce the notion of direction of bifurcation.

<u>DEFINITION</u> : Given an equation of the form (*), we call the normalized vector u_o a direction of bifurcation for λ_o if λ_o is an isolated eigenvalue of finite multiplicity of $Bu = \lambda Du$ and if, for every neighbourhood $U(u_o)$ of u_o on $M = \{u | \frac{1}{2}(u,Du) = 1\}$ and any $\varepsilon > 0$, there is an R_o such that for all $R \leqslant R_o$ there are nontrivial solutions (u,λ) of (*) such that $\frac{1}{2}(u,Du) = R^2$, $\frac{u}{\frac{1}{2}(u,Du)} \in U(u_o)$, and $|\lambda - \lambda_o| \leqslant \varepsilon$.

Then we get the following theorem as a consequence of the results of §II and III.

<u>THEOREM</u> : <u>Given an equation of the form</u>

$$(*) \qquad Bu + C(u) = \lambda Du$$

<u>in a Hilbert space</u> H, <u>where</u>

1. B <u>and</u> D <u>are bounded, linear, self-adjoint operators and</u> $D \geqslant \varepsilon_o > 0$. <u>The relative eigenvalue problem</u> $Bu = \lambda Du$ <u>has an isolated eigenvalue</u> λ_o <u>of finite multiplicity</u> n_o.

2. <u>There is a</u> C^{m+2}<u>-function</u> $(m \geq 1)$ $G : H \longrightarrow R$ <u>such that</u> $d_u G(h) = (C(u),h)$, <u>and</u> $d_o^k G = 0$ <u>for</u> $k \leq m+1$ <u>and</u> $G_o(u) = \frac{1}{(m+2)!} d_o^{m+2} G(u,u,\ldots,u) \neq 0$. <u>Then it is true that</u>

A. λ_o <u>is a point of bifurcation for equation (*) and for each topologically critical value</u> c <u>of</u> G_o/N <u>(where</u> N <u>is the submanifold of eigenvectors</u> u <u>with</u> $Bu = \lambda_o Du$ <u>and</u> $\frac{1}{2}(u,Du) = 1$) <u>the corresponding c-level of</u> G_o/N <u>contains at least</u> $\delta_{G_o/N}(C)$ <u>directions of bifucation for</u> λ_o <u>and</u> $\Delta_{G_o/N}(-\infty,\infty) \geq 2$.

B. <u>If</u> G <u>is an even function and if we indicate the manifold resulting from identification of opposite points of any centrally symmetric manifold by a bar,</u> F <u>and</u> G <u>define functions</u> $\overline{F}$, $\overline{G}$ <u>and</u> $\overline{G}_o$ <u>on</u> $\overline{M}$. <u>Then again each topologically critical c-level contains at least</u> $\delta_{\overline{G}_o/\overline{N}}(c)$ <u>directions of bifurcation for</u> λ_o <u>and</u> $\Delta_{\overline{G}_o/\overline{N}}(-\infty,\infty) \geq n_o$.

We can intuitively describe this in the following way. Let λ_o be an isolated eigenvalue of finite multiplicity of $Bu = \lambda Du$ and let N be as above, that is the set of eigenvectors to λ_o , normalized by the condition $\frac{1}{2}(u,Du) = 1$. As defined above G_o comes from the first non-vanishing higher order term of the Taylor series of A around zero (where we define $A(u) \equiv Bu + C(u)$) . The theorem tells us that λ_o is a point of bifurcation for the equation $A(u) = \lambda Du$, that is,any neighbourhood of $(\lambda_o,0) \in \mathbb{R} \times H$

contains nontrivial solutions of the equation. Furthermore if we consider the distribution of all possible vectors $\frac{u}{\frac{1}{2}(u,Du)}$ on N, where u is supposed to be a solution of the equation, then the theorem states that for $(u,Du) \longrightarrow 0$ there are always such vectors clustering around the topologically critical levels of G_o restricted to N and if such a level is characterized by a relative category s then for all sufficiently small R, the categories of the sets of solutions with fixed value of $F + G$ and $\frac{1}{2}(u,Du) = R$ must add up to at least s.

Above we have used the results of §II only for the finite dimensional problem to which we have reduced the problem by using the global implicit function theorem. It is natural to ask why one should not apply §II directly to the full problem. We can define $f = (u,Bu)$ on the unit sphere S. If we identify opposite points on S, we get the infinite dimensional projective space $\overline{S}$, and on it we can define the function $\overline{f}$ in an obvious way since f is even (so it takes the same value on opposite points). Now we could expect that any isolated eigenvalue of finite multiplicity of $\mathcal{B}$ is a topologically critical value of $\overline{f}$ on $\overline{S}$. Then, of course, we could apply §II directly. Unfortunately, this is not true in general. Let us denote the range of the projection operator corresponding to the spectrum of $\mathcal{B}$ from $-\infty$ up to c by $\mathcal{B}^c$ and that corresponding to the spectrum of $\mathcal{B}$ from c to $+\infty$ by $\mathcal{B}_c$.

THEOREM : Let c be an isolated eigenvalue of multiplicity 1 for $\mathcal{B}$ and let $\dim \mathcal{B}^c = \dim \mathcal{B}_c = \infty$; then $\delta_{\bar{f}}(c) = 0$. That is, c is not topologically critical. On the other hand, if c is an isolated eigenvalue of multiplicity m for $\mathcal{B}$ and if $\dim \mathcal{B}^c$ or $\dim \mathcal{B}_c$ is finite, then $\delta_{\bar{f}}(c) = m$.

In the latter case we can therefore apply directly the results of §II, if we can make sure that the perturbation does not destroy normality. This we can ensure by assuming C to be a compact operator.

THEOREM : If the gradient C of G is a compact operator, and if F is the quadratic form of a bounded self-adjoint operator $\mathcal{B}$ and if f and g denote the restrictions of F and G to the unit sphere, then for $[a,b]$ a normal interval for f, there is an $\varepsilon > 0$ such that $[a,b]$ is normal for $f + g$ for all g such that $\|g\|_1 = \sup|g| + \sup\|dg\| < \varepsilon$.

This leads us to the final result.

THEOREM : Let B be a bounded self-adjoint operator and $f = (u,Bu)$ its quadratic form restricted to the unit sphere S , let G be a C^1-function from H into $\mathbb{R}$ with gradient $C(u)$, and let $C(u)$ be a compact odd operator and $\|C(u)\| = o(\|u\|)$. Then any isolated eigenvalue λ of finite multiplicity m of B such that either $\dim \mathcal{B}^\lambda < \infty$ or $\dim \mathcal{B}_\lambda < \infty$ is a point of bifurcation for the equation $Bu + C(u) = \lambda u$, and on any sufficiently small sphere there will be at least $2m$ solutions.

REFERENCES

[1] M. Reeken, Stability of Critical Points Under Small Perturbations
Part I : Topological Theory
Part II: Analytic Theory, to appear in manuscripta mathematica.

SIX LECTURES ON THE TRANSITION TO INSTABILITY

D.H. Sattinger
School of Mathematics
University of Minnesota
Minneapolis, Minnesota

1. ORDINARY DIFFERENTIAL EQUATIONS

In these lectures we are going to consider the two simplest cases of bifurcation which may take place when an equilibrium solution loses its stability. Namely, we shall discuss bifurcation phenomena associated with that class of problems which may be written in the form

$$\frac{\partial u}{\partial t} = L(\lambda)u + N(\lambda,u)$$

where u is an element of a Banach space, L is a linear operator, and N is a non-linear operator. The number λ is a real parameter which is determined by the physical constants of the system. We assume that $N(\lambda,0) = 0$ so that $u = 0$ is an equilibrium solution, and we shall discuss the bifurcation and stability of non-trivial stationary solutions and of time periodic oscillations of the equations when the null solution loses stability.

These lectures are based on Chapters I, IV, and V of the lecture notes "Topics in Stability and Bifurcation Theory," which have been put out by the University of Minnesota. These lectures will be somewhat formal, whereas the technical details and examples are (we hope) fully explained in the lecture notes.

Let us begin by reviewing some simple facts and examples from ordinary differential equations. These provide a guide for what to look for in the case of partial differential equations, although the techniques of partial differential equations are often quite different.

Consider a system of ordinary differential equations

$$\dot{x} = f(x,\mu) \tag{1.1}$$

where $x = (x_1,\dots x_n)$, $f = (f_1,\dots f_n)$, and μ is a parameter.

Suppose $f(x_0,\mu_0) = 0$ for some point x_0 in R^n , $\mu = \mu_0$; then x_0 is called an equilibrium solution.

Definition 1.1. We say that x_0 is a stable equilibrium of (1.1) if, given any $\epsilon > 0$, there is a $\delta > 0$ such that $|v - x_0| < \delta$ implies that $|x(t) - x_0| < \epsilon$ for all $t > 0$, where $x(t)$ is the solution of (1.1) with initial data v . We say that x_0 is asymptotically stable if in addition $|x(t) - x_0| \to 0$ as $t \to \infty$.

Here $|x| = \sqrt{x_1^2 + \dots + x_n^2}$ or any equivalent norm.

Theorem 1.1. (Lyapounov) Let $A = \left(\frac{\partial f_i}{\partial x_j}\right)\Big|_{(x_0,\mu_0)}$ be the Jacobian of f at (x_0,μ_0) . If all eigenvalues of A have negative real parts, then x_0 is a stable equilibrium. If some eigenvalues of A have positive real parts, then x_0 is unstable.

Theorem 1.1. can be called "the principle of linearized stability." The reason for this is the following:

Let the solution of (1.1) be written in the form $x(t) = u(t) + x_0$; thus u is the perturbation from equilibrium. From $\dot{x} = f(x,\mu_0)$ we get

$$\begin{aligned} \dot{u} = \dot{x} &= f(u + x_0,\mu_0) = f(x_0,\mu_0) + \frac{\partial f_i}{\partial x_j} u_j + O(|u|^2) , \\ \dot{u} &= Au + O(|u|^2) \end{aligned} \tag{1.2}$$

where $A = \frac{\partial f_i}{\partial x_j}\Big|_{x_0,\mu_0}$. $O(|u|^2)$ denotes a term $g(u)$ such that $|g(u)| \le c|u|^2$.

If we neglect the second order term in (1.2), we obtain the linear equation

$$\dot{u} = Au \tag{1.3}$$

whose solution is $u(t) = e^{tA}u_0$. As is well known, all solutions of (1.3) decay provided the spectrum of A lies in the left half-plane; some solutions of (1.3) grow exponentially if A has eigenvalues in the right half-plane.

It is argued that the second order term may be neglected when the perturbations are small. This heuristic reasoning is justified by Lyapounov's theorem. In the case of partial differential equations one can prove theorems of an analogous nature [19]. The term $O(|u|^2)$ appearing in (1.2) may in that case, however, be a nonlinear partial differential operator (for example, $\vec{u}\cdot\nabla\vec{u}$ in the case of the Navier-Stokes equations), and the extension of Lyapounov's theorem to this case is not immediately obvious.

At any rate, we call Lyapounov's theorem the principle of linearized stability because it says it is enough to look at the linearized problem in order to determine stability. Incidentally, if some of the eigenvalues of $\frac{\partial f_i}{\partial x_j}$ have zero real parts (critical case), then it is not enough to look at the linearized problem - and one must look at the effects of the nonlinear terms in order to determine stability or instability.

Now let us consider the case where $x_0(\mu)$ is an equilibrium solution for μ in an interval $a \leq \mu \leq b$:

$$f(\mu, x_0(\mu)) = 0 \ .$$

Suppose that as μ crosses μ_0 , $a < \mu_0 < b$, some of the eigenvalues of $A(\mu) = \left.\frac{\partial f_i}{\partial x_j}\right|_{(\mu, x_0(\mu))}$ cross the imaginary axis, as in the diagrams below:

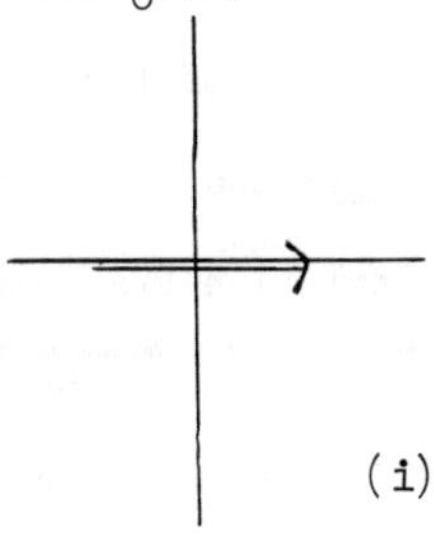

(i)

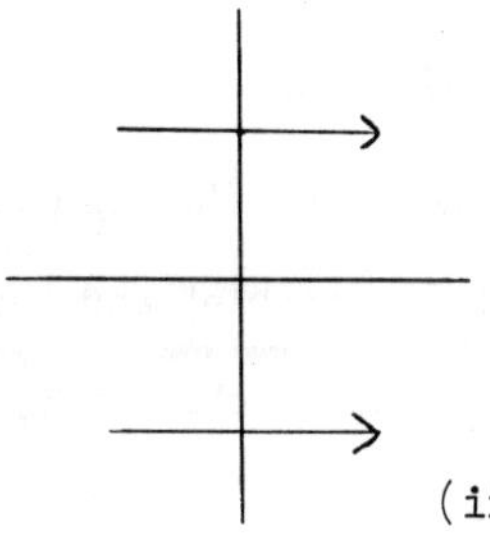

(ii)

Thus, for $\mu < \mu_0$, $x_0(\mu)$ is stable, while $x(\mu)$ becomes unstable when $\mu > \mu_0$. What happens? This is where the phenomenon of bifurcation occurs.

Two important cases are those in which a simple eigenvalue crosses through the origin (i) or in which a pair of simple complex conjugate eigenvalues cross the imaginary axis (ii). (We are restricting ourselves here to the case where $f(x,\mu)$ is real for real x and μ ; this includes a wide range of physical problems. In that case the eigenvalues of $A(\mu)$ always appear in complex conjugate pairs.) In these two cases one can give a fairly complete description of what happens.

In both cases we shall assume

$$\operatorname{Re} a'(\mu_0) > 0 ,$$

where $a(\mu)$ is one of the critical eigenvalues of $A(\mu)$ (the simple real one in case (i), one of the complex ones in case (ii)). Then in case (i) we get (in general) one of the following pictures

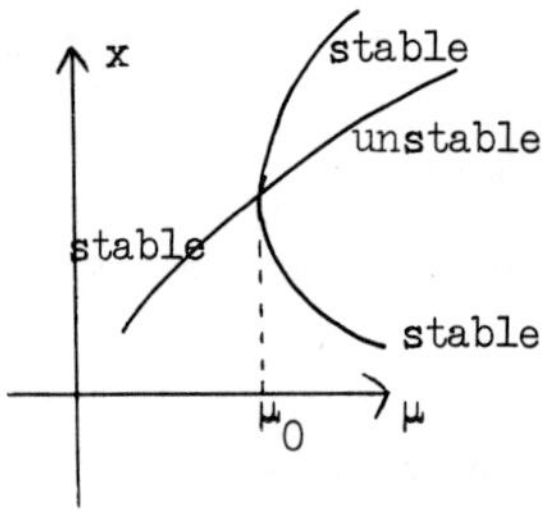

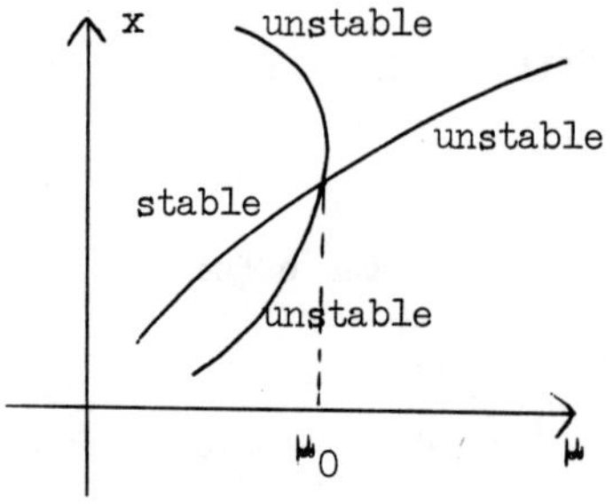

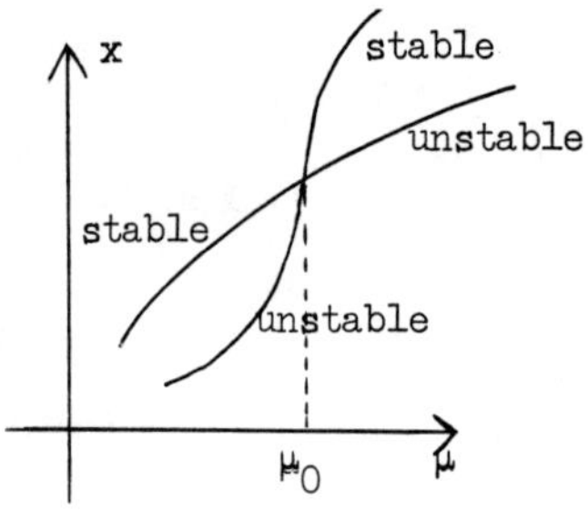

Each curve in the diagrams represents a solution curve $(\mu, x(\mu))$ of the equations $f(x(\mu),\mu) = 0$. The point $(\mu_0, x_0(\mu))$ is a <u>bifurcation point</u> - two solution branches intersect. The labels "stable" and "unstable" denote which solution branches are stable equilibria of the problem $f(\mu, x(\mu)) = 0$. We see that the given branch $x_0(\mu)$ is stable for $\mu < \mu_0$ and unstable for $\mu > \mu_0$, as we assumed. In the case of the bifurcating branch, solutions which bifurcate above criticality are stable while solutions which bifurcate below criticality are unstable.

In case (ii) there occurs a bifurcation of time periodic motions of $\dot{x} = f(x,\mu)$. As opposed to the stationary case, we get only one periodic motion — (note that there were two in the stationary case) — apart from phase shifts. Again, solutions which bifurcate above criticality are stable while solutions which bifurcate below criticality are unstable. The bifurcation of periodic solutions from equilibrium solutions was investigated completely by E. Hopf [5] for ordinary differential equations.

2. PHYSICAL SYSTEMS AND EXAMPLES

Physical systems which exhibit the phenomenon of bifurcation range, literally, from yardsticks to stars. The buckling of a yardstick when the thrust on the end is increased past a critical value is a familiar example of bifurcation. Below the critical thrust the yardstick maintains a vertical position which is stable to displacements. Above the critical thrust the yardstick bows to one side or the other, and the vertical, unbowed, position is no longer stable. Astrophysicists have attempted to explain the formation of binary stars by a process of loss of stability and bifurcation. This problem was the subject of classic papers by Poincaré (1885) and Lyapounov (1908); the ideas of both of these mathematicians concerning bifurcation continue to this day to have a strong influence on the subject. Below we discuss a few specific physical systems which exhibit bifurcation phenomena and mention briefly some of the mathematical models used to describe these systems.

Chemical Kinetics

The general equations governing chemical reactions between various agents are

$$\begin{aligned} \frac{\partial C_i}{\partial t} &= k_i \Delta C_i + f_i(C_1, \dots C_n, T) \\ \frac{\partial T}{\partial t} &= K \Delta T + g(C_1, \dots C_n, T) \end{aligned} \tag{2.1}$$

where $C_1, \dots C_n$ are the concentrations, and T is the temperature. Equations (2.1) are supplemented by boundary conditions, such as

$$\begin{aligned} &\frac{\partial T}{\partial \nu} + \beta T = 0 \qquad \text{or} \qquad T = T_0 \\ &\frac{\partial C_i}{\partial \nu} = 0 \qquad \text{or} \qquad C_i = C_{i,0} \ . \end{aligned} \tag{2.2}$$

Equations (2.1) - (2.2) form a parabolic boundary value problem. Their derivation and specific forms for the reaction rates g_i and f may be found in the book of Gavalas [3].

A special simple case which is often considered in thermal combustion processes is

$$\frac{\partial C}{\partial t} = K\Delta C - ce^{-E/RT}$$
$$\frac{\partial T}{\partial t} = k\Delta T + QCe^{-E/RT} \quad . \tag{2.3}$$

The rate factor $\exp\{-E/RT\}$ is called the <u>Arrenhius</u> rate factor; Q is the heat of reaction, E is called the activation energy, and R is the gas constant. If, as is sometimes done, it is assumed that the concentration remains constant during the initial stages of the reaction, we get the following equation for the temperature

$$\frac{\partial T}{\partial t} = k\Delta T + QC_0 e^{-E/RT} \; ,$$

where C_0 is the initial concentration. Steady state solutions of this equation are governed by

$$0 = \Delta T + \lambda e^{-E/RT} \; , \quad \lambda = \frac{QC_0}{k} \; . \tag{2.4}$$

Numerical analysis of this problem (see Parks) on a sphere of radius r_0 has shown the existence of multiple solutions (with boundary conditions $T = T_0$ on ∂D). By plotting $T(0)$, the center temperature, against λ, one obtains a solution diagram with roughly the shape shown in the figure at the left. We see from the

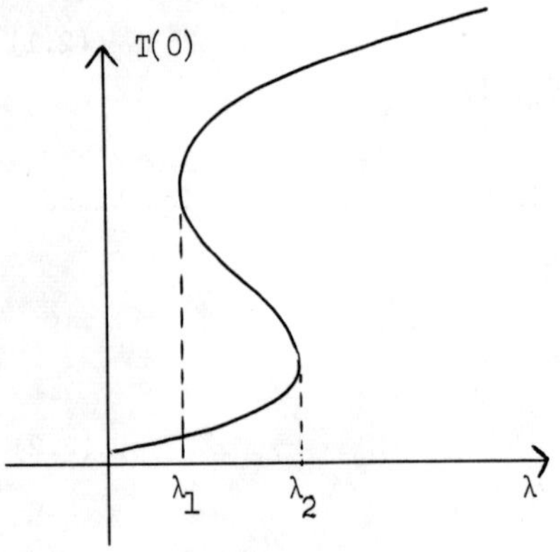

graph that there are three solutions for $\lambda_1 < \lambda < \lambda_2$ while for $\lambda < \lambda_1$ and $\lambda > \lambda_2$ there is only one solution. As λ increases one of these solutions increases while the other decreases until it meets the lower solution as λ reaches λ_2. The lower and middle solution then vanish as λ increases past λ_2. This phenomenon of multiple equilibrium solutions and their stability is of great interest in applications. Time periodic oscillations are also known to occur in other circumstances. (See Lee and Luss [14]).

Fluid Dynamics

The motion of a viscous incompressible fluid is governed by the Navier-Stokes equations

$$\frac{\partial u_i}{\partial t} + u_j \frac{\partial u_i}{\partial x_j} = -\frac{\partial p}{\partial x_i} + \frac{1}{R}\Delta u_i$$
$$\frac{\partial u_i}{\partial x_i} = 0 \,. \tag{2.5}$$

Here u_1, u_2, u_3 are the Cartesian components of the fluid velocity, and p is the pressure. Repeated indices denote summation. The number R, called the Reynolds number, is a pure number which results from introducing non-dimensional variables. Equations (2.5) may be considered on infinite domains with periodic boundary conditions, or they may be considered on a finite domain with boundary conditions

$$u_i \Big|_{\partial D} = \psi_i \,. \tag{2.6}$$

It is known that for small values of R (2.5) - (2.6) has a unique and unconditionally stable solution: all perturbations decay and the system returns to its equilibrium state. As R increases this basic solution loses its stability and new stationary solutions may branch off, or time periodic oscillations may set in. In 1944, L. Landau proposed that turbulence arises from repeated loss of stability and branching. Thus, the secondary solution in turn loses its stability at higher values of R and is replaced by yet another stationary solution or perhaps by a time periodic solution. With the further increase in R successive bifurcations may take place, with periodic solutions replaced by quasi-periodic motions of more and more base periods. E. Hopf [6] constructed an intriguing mathematical model (a system of nonlinear equations which bears a similarity to the Navier-Stokes equations) which exhibited precisely this behavior of repeated loss of stability and branching. Hopf then went on to construct a statistical mechanics for his model.

The situation regarding fluid mechanics is, of course, somewhat more complicated. For example, the bifurcation may be subcritical and the bifurcating solutions unstable. This appears to be the case for plane Poiseuille flow. It is also possible for a flow to lose stability with no bifurcation taking place at all. In such cases as these the fluid would not progress toward turbulence through a continuous branching process, but might make a sudden transition to turbulence when the critical parameter value is crossed. An interesting discussion of other matters related to Landau's ideas may be found in the article by Ruelle and Takens [16]. They point out that true quasi-periodic motion is non-generic in the case of ordinary differential equations. Thus we may expect that turbulence should be more complicated than quasi-periodic motion. Ruelle and Takens investigate the bifurcation of closed invariant manifolds from periodic solutions as these lose stability. Their work is discussed in some detail by Prof. Lanford in his lectures, which appear elsewhere in this volume. D. Joseph gives a purely formal construction of series for quasi-periodic solutions. These formal series may be asymptotic series for the solutions which are valid over a long period of time.

3. BIFURCATION OF STATIONARY SOLUTIONS

For simplicity we consider the bifurcation of solutions of an equation of the form

$$(L_o + \tau L_1)u + N(u,u) = 0 \tag{3.1}$$

where the nonlinear term is quadratic. The computations for the more general case

$$L(\lambda)u + N(\lambda,u) = 0 \;,$$

where N is analytic in λ and u, are a little more complicated, while the final outcome is qualitatively the same. We refer the reader to the first of these lectures, the more complete lecture notes, and his own imagination for particular examples of systems of partial differential equations which may be written in the above form. However, we mention specifically the Navier-Stokes equations and

Boussinesq equations of fluid dynamics, the parabolic systems of chemical reactor kinetics, and the simple model problem

$$\Delta u + \lambda u + u^2 = 0$$

$$u\big|_{\partial D} = 0 .$$

We suppose that for (3.1) L_o, L_1 and N are operators on a Banach space B with the following properties:

(1) L_o has a one-dimensional null space: $L_o\varphi_o = 0$

(2) L_o satisfies a Fredholm alternative: There is an element $\varphi_o^* \in B^*$ such that

(a) $L_o u = f$ is solvable iff $\langle f, \varphi_o^* \rangle = 0$.

(b) $\langle L_o \psi, \varphi_o^* \rangle = 0$ for all ψ in the domain of L_o .

(c) $\langle \varphi_o, \varphi_o^* \rangle = 1$.

Let $M_o = \{u : u \in B, \langle u, \varphi_o^* \rangle = 0\}$. By (2) L_o is invertible on M_o . We denote this inverse by K_o . We denote the projection onto M_o by Q .

(3) K_oQL , and K_oQN are continuous operators on B .

As an important class of examples in applications where the above three assumptions are valid, take the case where L_o is a second order elliptic operator or system. This class also includes the equations of fluid mechanics (see [19]). For B we may take the Banach space $C_{2+\alpha}$. Then K_o picks up two derivatives and L_1 and N may be second order partial differential operators.

(4) We assume that as τ crosses zero a simple real eigenvalue $\mu(\tau)$ of $L_o + \tau L_1$ crosses the origin and that $\mu'(0) \neq 0$.

Theorem 3.1. Under assumptions (1) - (4) there exists an analytic one parameter family $u = u(\epsilon)$, $\tau = \tau(\epsilon)$ of solutions of (3.1) such that $\langle u(\epsilon), \varphi_o^* \rangle = \epsilon$ and $\tau(0) = u(0) = 0$.

Proof. Put $[u] = \langle u, \varphi_o^* \rangle$ for $u \in B$. Then $[\varphi_o] = 1$ and the projection P onto the null space of L_o is given by $Pu = [u]\varphi_o$. $Q = I - P$ is the projection onto the range of L_o .

First let us see the consequences of condition (4). We have

$$(L_o + \tau L_1)\varphi(\tau) = \mu(\tau)\varphi(\tau) \tag{3.2}$$

with $\varphi(0) = \varphi_o$, and $\mu(0) = 0$. Expanding

$$\varphi(\tau) = \varphi_o + \tau\varphi_1 + \ldots$$

$$\mu(\tau) = \quad \tau\mu_1 + \ldots,$$

substituting in, and collecting powers of τ , we get

$$L_o\varphi_o = 0$$

$$L_o\varphi_1 + L_1\varphi_o = \mu_1\varphi_o \ .$$

Taking the bracket of the second equation, we get

$$[L_o\varphi_1] + [L_1\varphi_o] = \mu_1[\varphi_o] \ .$$

Since $[L_o\varphi_1] = \langle L_o\varphi_1, \varphi_o^*\rangle = \langle \varphi_1, L_o^*\varphi_o^*\rangle = 0$ and $[\varphi_o] = 1$, we get

$$\mu_1 = \mu'(0) = [L_1\varphi_o] \neq 0 \ .$$

We now look for a family of non-trivial solutions of (3.1) expressed parametrically in the form

$$z = z(\epsilon) \qquad \tau = \tau(\epsilon)$$

where $z(0) = 0$ and $\tau(0) = 0$. We may write $z = Pz + Qz = [z]\varphi_o + Qz$. We shall set $\epsilon = [z] = \langle z, \varphi_o^*\rangle$. Let us put $z = \epsilon W$, $\tau = \epsilon\sigma$, and write $W = \varphi_o + \xi$ where $\xi = QW$ and $[\xi] = 0$. Substituting in (3.1) we get

$$L_o\xi + \epsilon\sigma L_1(\varphi_o + \xi) + \epsilon N(\varphi_o + \xi, \ \varphi_o + \xi) = 0 \tag{3.3}$$

Now we operate by the projection Q :

$$L_o\xi + \epsilon\sigma QL_1(\varphi_o + \xi) + \epsilon QN(\varphi_o + \xi, \ \varphi_o + \xi) = 0 \ . \tag{3.4}$$

(Note that $QL_o\xi = L_o\xi$) . Taking brackets of (3.3), which is the equivalent of operating by P, we get

$$\sigma[L_1\varphi_o] + \sigma[L_1\,\xi] + [N(\varphi_o + \xi,\ \varphi_o + \xi)] = 0 \ . \tag{3.5}$$

(Recall that $[L_o\,\xi] = 0$.)

Now operate on (3.4) by K_o . We get

$$\xi + \epsilon\sigma K_o QL_1(\varphi_o + \xi) + \epsilon K_o QN(\varphi_o + \xi,\ \varphi_o + \xi) = 0 \ . \tag{3.6}$$

Equations (3.5) and (3.6) are the Lyapounov-Schmidt equations. A solution pair (ξ,σ) of (3.5) - (3.6) gives rise to a solution of (3.1) and vice-versa.

Consider the mapping

$$\mathcal{F}(\xi,\sigma,\epsilon) = (\xi + \epsilon\sigma K_o QL_1(\varphi_o + \xi) + \epsilon K_o QN(\varphi_o + \xi,\ \varphi_o + \xi)\ ,$$

$$\sigma[L_1\varphi_o] + \sigma[L_1, \xi] + [N(\varphi_o + \xi,\ \varphi_o + \xi)]) \ .$$

The zeros of $\mathcal{F}$ are the solutions of (3.1). By assumption (3) $\mathcal{F}$ is a continuous (also analytic) mapping of $M_o \times R$ to $M_o \times R$. When $\epsilon = 0$ we may take $\xi = 0$ and $\sigma = \sigma_o$, where

$$\sigma_o[L_1\varphi_o] + [N(\varphi_o,\varphi_o)] = 0 \ ;$$

then $\mathcal{F}(0,\sigma_o,0) = 0$.

The Frechet derivative of $\mathcal{F}$ at $(0,\sigma_o,0)$ is easily seen to be

$$\begin{bmatrix} \dfrac{\partial\mathcal{F}_1}{\partial\xi} & \dfrac{\partial\mathcal{F}_1}{\partial\sigma} \\ & \\ \dfrac{\partial\mathcal{F}_2}{\partial\xi} & \dfrac{\partial\mathcal{F}_2}{\partial\sigma} \end{bmatrix} = \begin{bmatrix} I & 0 \\ & \\ \sigma_o[L_1\cdot] + [M(\varphi_o,\cdot)] & [L_1\varphi_o] \end{bmatrix} \ .$$

The symbol $[L_1\cdot]$ denotes the linear functional $\xi \to [L_1\xi]$ defined on M_o . The operator $M(\varphi_o,\cdot)$ is the operation

$$\xi \to M(\varphi_o,\xi) = N(\varphi_o,\xi) + N(\xi,\varphi_o) \ .$$

$M(\varphi_o,\cdot)$ is the Frechet derivative of N at φ_o .

The Frechet derivative of $\mathcal{F}$ is invertible provided $[L_1\varphi_0] \neq 0$. As we have seen, this follows from our assumption (4). Therefore we may apply the implicit function theorem to $\mathcal{F}$. Since $\mathcal{F}$ is analytic, we conclude that there exists a one parameter, analytic family

$$\xi = \xi(\epsilon) \ , \quad \sigma = \sigma(\epsilon)$$

of solutions of $\mathcal{F}(\xi, \sigma, \epsilon) = 0$ and hence of (3.1).

4. FORMAL CALCULATION OF THE STABILITY

In this section we calculate the critical eigenvalue by perturbation methods. For simplicity, we consider the equation

$$(L_0 + \tau L_1)u + N(u,u) = 0 \ . \tag{4.1}$$

The eigenvalue problem associated with the linearized stability problem for u is

$$(L_0 + \tau L_1)\varphi + M(u,\varphi) = \sigma\varphi \ . \tag{4.2}$$

We have arrived at the eigenvalue problem (4.2) in a purely formal manner. The derived operator $L(\lambda) + M(u,\cdot)$, where u is a solution of (4.1), is the analogue of the Jacobian $(\frac{\partial f_i}{\partial x_j})$ in the case of ordinary differential equations. (See Theorem 1.1, Lecture 1)

When $\tau = 0$ and $u = 0$ we have $\sigma = 0$ and $\varphi = \varphi_0$ in (4.2) by our assumption that $\tau = 0$, $u = 0$ is a bifurcation point. The stability of the trivial solution $u = 0$ is determined by the eigenvalues of the operator $L_0 + \tau L_1$. We assume that the trivial solution is unstable for $\tau > 0$ and stable for $\tau < 0$. We wish to determine the critical eigenvalue $\sigma(\tau)$ of (4.2) for small τ , when u is a nontrivial solution of (4.1)

As we have seen, the solution branch of (4.1) can be expanded in a power series ($u = \epsilon w$)

$$w = \varphi_0 + \epsilon w_1 + \epsilon^2 w_2 + \dots$$

$$\tau = \epsilon \tau_1 + \epsilon^2 \tau_2 + \dots \ .$$

For now, assume that σ and φ also have power series expansions

$$\sigma = \epsilon\,\sigma_1 + \epsilon^2\,\sigma_2 + \dots$$
$$\varphi = \varphi_0 + \epsilon\,\varphi_1 + \dots\ . \qquad (4.3)$$

The convergence of these series is proved in [19]. We normalize φ so that $[\varphi] \equiv 1$; this means that in the power series for φ, $[\varphi_j] = 0$ for $j = 1,2,\dots$.

Theorem 4.1. If $\tau_1 \neq 0$ then

$$\sigma_1 = -\,\tau_1[L_1\varphi_0]\ ; \qquad (4.4)$$

if $\tau_1 = 0$ then $\sigma_1 = 0$ and

$$\sigma_2 = -\,2\,\tau_2[L_1\varphi_0]\ . \qquad (4.5)$$

Accordingly, subcritical branches are unstable and supercritical branches are stable if $\tau_1 \neq 0$ or if $\tau_1 = 0$ and $\tau_2 \neq 0$. (We assume, of course, that all other eigenvalues of $L_0 + \tau L_1$ have negative real parts for small τ.)

Proof: We have

$$(L_0 + \epsilon\,\tau_1 L_1 + \epsilon^2\,\tau_2 L_1 + \dots)(\varphi_0 + \epsilon\,w_1 + \dots)$$
$$+ N(\varphi_0, \varphi_0) + \epsilon\,M(\varphi_0, w_1) + \dots = 0\ ;$$

hence, equating the coefficient of each power of ϵ to zero, we get

$$L_0\varphi_0 = 0$$
$$\tau_1 L_1\varphi_0 + L_0 w_1 + N(\varphi_0, \varphi_0) = 0\ . \qquad (4.6)$$

Taking brackets of the second equation yields

$$\tau_1[L_1\varphi_0] + [N(\varphi_0, \varphi_0)] = 0\ . \qquad (4.7)$$

(Since $[L_0\psi] = 0$ for any ψ).

If $\tau_1 = 0$ then the third equation obtained is

$$\tau_2 L_1\varphi_0 + L_0 w_2 + M(\varphi_0, w_1) = 0\ .$$

Taking brackets of this equation, we get

$$\tau_2[L_1\varphi_0] + [M(\varphi_0, w_1)] = 0 \quad \text{if} \quad \tau_1 = 0 \ . \tag{4.8}$$

Now we calculate the perturbation terms for σ. By a procedure similar to the one above, we get the equation

$$L_0\varphi_1 + \tau_1 L_1\varphi_0 + M(\varphi_0, \varphi_0) = \sigma_1\varphi_0 \ , \tag{4.9}$$

and, in case $\tau_1 = 0$, the next equation is

$$L_0\varphi_2 + \tau_2 L_1\varphi_0 + M(w_1, \varphi_0) + M(\varphi_0, \varphi_1) = \sigma_2\varphi_0 + \sigma_1\varphi_1 \ . \tag{4.10}$$

If $\tau_1 \neq 0$, then by taking brackets of (4.9) we obtain

$$\tau_1[L_1\varphi_0] + [M(\varphi_0, \varphi_0)] = \sigma_1 \ . \tag{4.11}$$

From (4.7) and the fact that $M(u,u) = 2N(u,u)$, we get equation (4.4).

If $\tau_1 = 0$ then (4.7) and (4.11) show that $\sigma_1 = 0$. Taking brackets of (4.10), we then see that

$$\tau_2[L_1\varphi_0] + [M(w_1, \varphi_0)] + [M(\varphi_0, \varphi_1)] = \sigma_2 \ .$$

From this and (4.8) we then see that $\sigma_2 = [M(\varphi_0, \varphi_1)]$. Since $\tau_1 = \sigma_1 = 0$, φ_1 satisfies

$$L_0\varphi_1 + 2N(\varphi_0, \varphi_0) = 0$$

(see (4.9)), while w_1 satisfies (4.6) with $\tau_1 = 0$. Since $[\varphi_1] = [w_1] = 0$, we see that $\varphi_1 = 2w_1$. Hence

$$\sigma_2 = 2[M(\varphi_0, w_1)] = -2\tau_2[L_1\varphi_0]$$

(see (4.8)).

The final statement of the theorem follows from the fact that $[L_1\varphi_0] > 0$ if the null solution loses stability for $\tau > 0$. The reader can now easily verify the exchange of stability diagrams of Lecture 1.

5. BIFURCATION OF PERIODIC SOLUTIONS*

In this lecture we treat the transition to instability when an equilibrium solution loses stability by virtue of a complex conjugate pair of simple eigenvalues crossing the imaginary axis. We again treat a system of equations in operator form

$$\frac{\partial u}{\partial t} + L(\lambda)u + N(u,u) = 0 \ , \tag{5.1}$$

where, for simplicity, we assume that $L(\lambda) = L_0 + \tau L_1$, where $\tau = \lambda - \lambda_0$ and L_0 is an operator satisfying the assumptions in Lecture 3.

The null solution $u = 0$ is an equilibrium (time independent) solution of (5.1); according to the principle of linearized stability, its stability is determined by the spectrum of the operator $L(\lambda)$. We assume that $L(\lambda)$ has a simple eigenvalue $\gamma(\lambda)$ such that $\gamma(\lambda_0) = i$ and $\operatorname{Re} \gamma'(\lambda_0) \neq 0$. Under these conditions we wish to investigate the bifurcation of periodic solutions of (5.1). In constructing the periodic solutions, the period, as well as the amplitude of the oscillations, must be determined in terms of $\lambda - \lambda_0$. This is accomplished in the following way: put $s = \omega t$ and $u = \epsilon v$ in (5.1); then that equation becomes

$$\omega \frac{\partial v}{\partial s} + L(\lambda)v + \epsilon N(v,v) = 0 \ . \tag{5.2}$$

We are going to construct a one parameter family $v = v(s,\epsilon)$, $\omega = \omega(\epsilon)$, $\lambda = \lambda(\epsilon)$ of solutions of (5.2), where v is 2π periodic in s. This gives a family of $2\pi/\omega$ periodic solutions of (5.1), namely, $u(t,\epsilon) = \epsilon\, v(\omega(\epsilon)t, \epsilon)$. In this way, ω, λ, and u are determined as functions of ϵ, and indirectly, ω and u are given as functions of $\lambda - \lambda_0$.

The procedure for constructing periodic solutions bears some similarities to that for constructing bifurcating stationary solutions, developed in Lecture 3.

Fredholm alternative for $\frac{\partial}{\partial s} + L_0$.

Let $\varphi(\lambda)$ and $\gamma(\lambda)$ denote the critical eigenfunction and eigenvalue of $L(\lambda)$; thus, $L(\lambda)\varphi(\lambda) = \gamma(\lambda)\varphi(\lambda)$, $\gamma(\lambda_0) = i$, and $\operatorname{Re} \gamma'(\lambda_0) \neq 0$. The operator $L(\lambda)$ is

*Based on joint work with Professor D.D. Joseph [8].

assumed to commute with complex conjugation: $\overline{L(\lambda)u} = L(\lambda)\bar{u}$. Consequently the eigenvalues of $L(\lambda)$ appear in complex conjugate pairs. Letting $\varphi_0 = \varphi(\lambda_0)$ we have

$$L_0\varphi_0 = i\,\varphi_0 \quad \text{and} \quad L_0\bar{\varphi}_0 = -i\,\bar{\varphi}_0 \ .$$

From the Fredholm theory for L_0 there exists a vector φ_0^* such that

$$L_0^*\varphi_0^* = -i\,\varphi_0^* \quad \text{and} \quad L_0^*\overline{\varphi_0^*} = i\,\overline{\varphi_0^*} \ .$$

Since i is a simple eigenvalue we may normalize φ_0^* so that $\langle\varphi_0,\varphi_0^*\rangle = 1$. The notation $\langle\ ,\ \rangle$ denotes linear functionals φ^* acting on the Banach space B .

Let us compute $\operatorname{Re}\gamma'(\lambda_0)$. We write

$$\varphi(\lambda) = \varphi_0 + \tau\,\varphi_1 + \dots$$

$$\gamma(\lambda) = i + \tau\,\gamma_1 + \dots \ .$$

Substituting these series into $L(\lambda)\varphi(\lambda) = \gamma(\lambda)\varphi(\lambda)$, we obtain

$$L_0\varphi_0 = i\,\varphi_0 \ .$$

and

$$L_0\varphi_1 + L_1\varphi_0 = i\,\varphi_1 + \gamma_1\varphi_0 \ .$$

Proceeding as in Lecture 3 we obtain

$$\operatorname{Re}\gamma_1 = \operatorname{Re}\gamma'(\lambda_0) = \operatorname{Re}\langle L_1\varphi_0,\varphi_0^*\rangle \ .$$

Again we assume that

$$\operatorname{Re}\gamma'(\lambda_0) = \operatorname{Re}\langle L_1\varphi_0,\varphi_0^*\rangle \neq 0 \ . \tag{5.3}$$

We define the linear space

$$P^{2\pi} = \{u(s) : u(s+2\pi) = u(s),\ u(s) \in B\}$$

and put an appropriate norm on $P^{2\pi}$ to make it a Banach space.

Let $J_0 = \frac{\partial}{\partial s} + L_0$ and let $z_1(s) = e^{-is}\varphi_0$ and $z_2(s) = \bar{z}_1$. Then $J_0z_1 = J_0z_2 = 0$.
We define

$$\langle u,\varphi^*\rangle_{2\pi} = \frac{1}{2\pi}\int_0^{2\pi} \langle u(s),\ \varphi^*(s)\rangle\,ds \ ,$$

where u takes values in B and $\varphi^*(s)$ takes values in B^*. The functional $\langle\ ,\ \rangle$ is to have the property $\langle u, \lambda\varphi^*\rangle = \bar{\lambda}\langle u, \varphi^*\rangle$ for complex λ.

We assume that

(1) z_1 and z_2 are the only null functions of J_0.

(2) The Fredholm alternative is valid for J_0: If φ_0^* is an eigenfunction of L_0^* $(L_0^*\varphi_0^* = i\,\varphi_0^*)$, we set $z_1^* = e^{-is}\varphi_0^*$ and $z_2^* = \overline{z_1^*}$. Then

(a) $J_0 u = f$ is solvable iff $\langle f, z_i^*\rangle_{2\pi} = 0$, $i = 1, 2$.

(b) $\langle J_0 u, z_i^*\rangle_{2\pi} = 0$ for $i = 1, 2$ and all u in the domain of J_0.

Let K_0 denote the inverse of J_0 restricted to

$$P_0 = \{u : u(s) \in P^{2\pi},\ \langle u, z_i^*\rangle_{2\pi} = 0,\quad i = 1, 2,\ u \text{ real}\}.$$

Let Q denote the projection onto the range of J_0.

(3) $K_0 QL$, $K_0 QN$, and $K_0 Q \frac{\partial}{\partial s}$ are continuous operators on P_0.

The Fredholm alternative (properties (1) - (3)) for the operator $\frac{\partial}{\partial s} + L_0$ are carried out in [8] and [19] for the case of the Navier-Stokes equations.

<u>Theorem 5.1.</u> <u>Under assumptions</u> (1) - (3) <u>and</u> (5.3) <u>above there exists a one parameter analytic family</u> $v(s, \epsilon)$, $\omega(\epsilon)$, $\lambda(\epsilon)$ <u>of periodic solutions of</u> (5.2).

It is also true that the family of periodic solutions is unique up to a phase shift (contrary to the case of bifurcation of stationary solutions - where there are two distinct solutions). This is proved in [8].

Proof. It will be somewhat more convenient to introduce the bracket

$$[u] = \langle u, z_1^*\rangle_{2\pi}.$$

If u is real then $[u] = 0$ implies that $\overline{[u]} = 0$, but

$$\overline{[u]} = \overline{\langle u, z_1^*\rangle}_{2\pi} = \langle u, z_2^*\rangle_{2\pi},$$

since $\overline{z_1^*} = z_2^*$. Therefore, for u real, $[u] = 0$ contains two orthogonality conditions.

Then the equation $J_0 u = f$ is solvable iff $[f] = 0$ (provided u and f are real). We let $P_0^{2\pi}$ be the Banach space

$$P_0^{2\pi} = \{u : u \in P^{2\pi},\ u \text{ real},\ [u] = 0\} .$$

The projection P onto the null space of J_0 is given by

$$Pu = 2 \text{ Re } [u] z_1 = [u]\ 2 \text{ Re } z_1 .$$

We now construct the one parameter family ω, λ, v of periodic solutions of (5.2) with v lying in the Banach space $P_0^{2\pi}$. First, note that we can assume that $[v] > 0$. In fact, letting $(T_\delta v)(s, \in) = v(s + \delta, \in)$ we see that $[T_\delta v] = e^{-i\delta}[v]$, so by an appropriate phase shift, we can assume that $[v] > 0$. Furthermore, we can assume that $[v] = 1$ (this in effect normalizes our choice of the parameter $\in$).

For future reference we note the following relationships, both easily derivable:

$$\text{(i)} \quad [z_1] = 1 , \quad [z_2] = 0\ ^* \tag{5.4}$$

$$\text{(ii)} \quad [\frac{\partial w}{\partial s}] = -i[w] \quad \text{for any } 2\pi \text{ periodic vector field } w .$$

We write equation (3.2) in the form

$$Jv + (\omega - 1) \frac{\partial v}{\partial s} + \tau L_1 v + \in N(v, v) = 0 . \tag{5.5}$$

Letting $v = Pv + Qv = Pv + \varphi$ we have $[\varphi] = 0$, since φ is in the range of J, and $Pv = 2\text{Re } [v] z_1 = 2\text{Re } z_1$, since $[v] = 1$. We denote $2\text{Re } z_1$ by u_0 and write $v = u_0 + \varphi$. Substituting this into (5.5) we get

$$J\varphi + (\omega - 1)\frac{\partial u_0}{\partial s} + (\omega - 1) \frac{\partial \varphi}{\partial s} + \tau L_1 u_0 + \tau L_1 \varphi + \in N(u_0 + \varphi,\ u_0 + \varphi) = 0 .$$

We now apply the projections P and Q to this equation to obtain

$$J\varphi + (\omega - 1)\frac{\partial \varphi}{\partial s} + \tau\ QL_1 u_0 + \tau\ QL_1 \varphi + \in QN(u_0 + \varphi,\ u_0 + \varphi) = 0 \tag{5.6}$$

$$(\omega - 1)[\frac{\partial u_0}{\partial s}] + (\omega - 1)[\frac{\partial \varphi}{\partial s}] + \tau[L_1 u_0] + \tau[L_1 \varphi] + \in[N(u_0 + \varphi,\ u_0 - \varphi] = 0 \tag{5.7}$$

---

*(i) is a consequence of the relationships $\langle \varphi_0, \varphi_0^* \rangle = \langle \overline{\varphi_0}, \overline{\varphi_0^*} \rangle = 1$ and $\langle \overline{\varphi_0}, \varphi_0^* \rangle = \langle \varphi_0, \overline{\psi_0^*} \rangle = 0$.

(see the remarks below). By using (5.4), (ii) we can put the second equation in the form

$$-i(\omega-1)+\tau[L_1u_0]+\tau[L_1\varphi]+\epsilon[N(u_0+\varphi,\ u_0+\varphi)]=0 .$$

Taking real and imaginary parts of this equation, we obtain

$$\tau\ \mathrm{Re}[L_1u_0]+\tau\ \mathrm{Re}[L_1\varphi]+\epsilon\ \mathrm{Re}[N(u_0+\varphi,\ u_0+\varphi)]=0 \tag{5.8}$$

$$-(\omega-1)+\tau\ \mathrm{Im}[L_1u_0]+\tau\ \mathrm{Im}[L_1\varphi]+\epsilon\ \mathrm{Im}[N(u_0+\varphi,\ u_0+\varphi)]=0 . \tag{5.9}$$

Remarks. 1. In obtaining (5.6) we have used the facts: $QJw = Jw$ for any w, $Q\frac{\partial\varphi}{\partial s}=\frac{\partial\varphi}{\partial s}$ if φ is in the range of J, and $Q\frac{\partial u_0}{\partial s}=0$ since u_0 is in the null space of J.

2. In obtaining (5.7) we have used the fact that $Pw=0$ implies $[w]=0$.

Now we apply the transformation K_0 given above to equation (5.6) to obtain

$$\varphi+(\omega-1)K\frac{\partial\varphi}{\partial s}+\tau\, KQL_1u_0+\tau\ QL_1\varphi+\epsilon\ KQN(u_0+\varphi,\ u_0+\varphi)=0 . \tag{5.10}$$

By assumption (3), each of the operations in (5.10) is continuous.

The equations (5.8) - (5.10) are the equivalent of the Lyapounov-Schmidt equations of Lecture 3. We define a mapping $\mathcal{F}(\phi,\omega,\tau,\epsilon)=(\mathcal{F}_1,\mathcal{F}_2,\mathcal{F}_3)$ with $\mathcal{F}_1$ the left side of (5.10), $\mathcal{F}_2$ the left side of (5.9) and $\mathcal{F}_3$ the left side of (5.8). Then $\mathcal{F}$ is a mapping from $P_0^{2\pi}\times R\times R\times R$ to $P_0^{2\pi}\times R\times R$. The reader should verify that $\mathcal{F}$ is analytic. We have $\mathcal{F}(0,1,0,0)=0$ while the Frechet derivative of $\mathcal{F}$ at $(0,1,0,0)$ is

$$\begin{bmatrix} \frac{\partial\mathcal{F}_1}{\partial\phi} & \frac{\partial\mathcal{F}_1}{\partial\tau} & \frac{\partial\mathcal{F}_1}{\partial\omega} \\ \frac{\partial\mathcal{F}_2}{\partial\phi} & \frac{\partial\mathcal{F}_2}{\partial\tau} & \frac{\partial\mathcal{F}_2}{\partial\omega} \\ \frac{\partial\mathcal{F}_3}{\partial\phi} & \frac{\partial\mathcal{F}_3}{\partial\tau} & \frac{\partial\mathcal{F}_3}{\partial\omega} \end{bmatrix} = \begin{bmatrix} I & KQL_1u_0 & 0 \\ 0 & \mathrm{Re}[L_1u_0] & 0 \\ 0 & \mathrm{Im}[L_1u_0] & -1 \end{bmatrix} . \tag{5.11}$$

This operator is invertible provided $\mathrm{Re}[L_1u_0]\neq 0$; however, a simple computation shows that $\mathrm{Re}[L_1u_0]=\mathrm{Re}\ \gamma'(\lambda_0)=\mathrm{Re}(L_1\varphi_0,\varphi_0^*)$, which is non-zero by assumption (5.3).

This completes the construction of the bifurcating family of periodic solutions of (5.2)

Since the solutions are power series in ϵ, we may write

$$\omega = 1 + \epsilon\omega_1 + \epsilon^2\omega_2 + \dots$$

$$\tau = \epsilon\tau_1 + \epsilon^2\tau_2 + \dots$$

$$v = u_0 + \epsilon v_1 + \epsilon^2 v_2 + \dots .$$

Substituting into (5.5) we get

$$\epsilon\left(Jv_1 + \omega_1 \frac{\partial u_0}{\partial s} + \tau_1 L_1 u_0 + N(u_0, u_0)\right)$$
$$+ \epsilon^2\left(Jv_2 + \omega_2 \frac{\partial u_0}{\partial s} + \omega_1 \frac{\partial v_1}{\partial s} + \tau_1 L_1 v_1 + \tau_2 L_1 u_0\right.$$
$$\left. + M(u_0, v_1)\right) + O(\epsilon^3) = 0$$

where $M(u,v) = N(u,v) + N(v,u)$.

Since $u_0 = 2 \operatorname{Re} z_1 = e^{-is}\varphi_0 + e^{is}\overline{\varphi_0}$, it is easily seen that $[N(u_0, u_0)] = 0$. In fact,

$$[N(u_0, u_0)] = \frac{1}{2\pi}\int_0^{2\pi} \{\text{odd trigonometric polynomial}\} ds = 0 .$$

Therefore the solvability conditions for the coefficient of ϵ are $\omega_1 = \tau_1 = 0$. For v_1 we take the unique (real) solution of

$$Jv_1 + N(u_0, u_0) = 0 \qquad (5.12)$$

for which the bracket is zero: $[v_1] = 0$. The second equation then is

$$Jv_2 + \omega_2 \frac{\partial u_0}{\partial s} + \tau_2 L_1 u_0 + M(u_0, v_1) = 0 .$$

The solvability conditions are

$$\omega_2\left[\frac{\partial u_0}{\partial s}\right] + \tau_2[L_1 u_0] + [M(u_0, v_1)] = 0 .$$

From (5.4) (ii) we have $\left[\frac{\partial u_0}{\partial s}\right] = -i[u_0] = -i$. Taking real and imaginary parts we obtain

$$\tau_2 \ \mathrm{Re}[L_1 u_0] + \mathrm{Re}[M(u_0, v_1)] = 0$$

$$-\omega_2 + \tau_2 \ \mathrm{Im}[L_1 u_0] + \mathrm{Im}[M(u_0, v_1)] = 0 \ . \tag{5.13}$$

These equations determine ω_2 and τ_2. Similarly, we may determine each of the coefficients ω_n and τ_n successively. Assumption (5.3) guarantees that $\mathrm{Re}[L_1, u_0] \neq 0$. In fact,

$$[L_1 u_0] = \frac{1}{2\pi} \int_0^{2\pi} \langle L_1(e^{-is} \varphi_0 + e^{is} \bar{\varphi}_0), \ e^{-is} \varphi_0^* \rangle \, ds$$

$$= \frac{1}{2\pi} \int_0^{2\pi} \langle L_1 \varphi_0, \varphi_0^* \rangle + e^{2is} \langle L_1 \bar{\varphi}_0, \varphi_0^* \rangle \, ds$$

$$= \langle L_1 \varphi_0, \varphi_0^* \rangle \ .$$

Therefore, from (5.3),

$$\mathrm{Re} \ \gamma'(\lambda_0) = \mathrm{Re}[L_1 u_0] \neq 0 \ .$$

6. STABILITY: FLOQUET EXPONENTS

In the case of ordinary differential equations the stability of a periodic solution of an autonomous system

$$\dot{x} = F(x)$$

can be determined by computing the Floquet exponents of the variation equations

$$\dot{y} = F'(x(t))y \quad \text{or} \quad \dot{y}_i = \frac{\partial F_i}{\partial x_j}(x(t))y_j \ ,$$

where $x(t)$ is the periodic solution in question and $F'(x(t))$ denotes the Frechet derivative (Jacobian) of F at $x(t)$.

The Floquet exponents are determined as follows: Suppose we are given a linear system of ordinary differential equations with periodic coefficients

$$\dot{y} = A(t)y \ , \quad \text{where} \quad A(t) = A(t+T) \ . \tag{6.1}$$

Denote by $\Phi(t)$ the fundamental solution matrix; thus

$$\dot{\Phi}(t) = A(t)\Phi \quad \text{and} \quad \Phi(0) = I .$$

The eigenvalues of $\Phi(T)$ are called the <u>Floquet multipliers</u> of (6.1). If μ is a multiplier, then there is a vector ψ such that $\Phi(T)\psi = \mu\psi$. Equivalently, if $\dot{z}(t) = A(t)z$ and $z(0) = \psi$, then $z(T) = \mu\psi = \mu z(0)$. Put $\mu = e^{-\beta T}$ and $w(t) = e^{\beta t}z(t)$. Then $w(T) = e^{\beta T}z(T) = \mu^{-1}z(T) = z(0) = w(0)$, and

$$\dot{w} - \beta w - A(t)w = 0 .$$

Thus the <u>Floquet exponent</u> β can be considered an eigenvalue of the problem

$$\begin{aligned} &\dot{w} - A(t) - \beta w = 0 \\ &w(0) = w(T) . \end{aligned} \tag{6.2}$$

If $\operatorname{Re}\beta > 0$ then $|\mu| < 1$ and $|z(T)| < |z(0)|$. If all the Floquet exponents have positive real parts, then all the eigenvalues of $\Phi(T)$ are less than one in absolute value, and $\Phi(T)$ is a contraction mapping. In the case of a periodic solution $x(t)$ of $x = F(x)$, however, $y(t) = \dot{x}(t)$ is a periodic solution of the variation equations; hence one of the Floquet multipliers is always unity. Nevertheless, <u>if all the remaining Floquet multipliers of the variational equation are less than one in absolute value, the periodic solution $x(t)$ is orbitally stable</u>. (See Coddington and Levinson.)

To be precise, let $y(t,y_0)$ denote the solution of $\dot{y} = F(y)$, $y(0) = y_0$. Then $x(t)$ is orbitally stable if there is a number $\epsilon > 0$ such that whenever $|y_0 - x(0)| < \epsilon$, then $|y(t,y_0) - x(t+\delta)| \to 0$ as $t \to \infty$ for some number $\delta (0 \le \delta < T)$.

The eigenvalue problem for the Floquet exponents corresponding to (6.2) is

$$\begin{aligned} &\frac{\partial w}{\partial t} + L(\lambda)w + \epsilon M(v,w) - \beta w = 0 \\ &w(0) = w\left(\frac{2\pi}{\omega}\right) . \end{aligned} \tag{6.3}$$

Recall that $\lambda = \lambda_0 + \tau(\epsilon)$, $\omega(\epsilon)$ and $v(\epsilon, s)$ are predetermined. We change variables by putting $s = \omega t$ and replacing β by $\omega\beta$; then we get

$$\omega \frac{\partial w}{\partial s} + (L_0 + \tau L_1) w + \epsilon M(v, w) - \beta\omega w = 0$$

$$w(0) = w(2\pi) \quad . \tag{6.4}$$

One solution of (6.4) is $\beta = 0$ and $w = \frac{\partial v}{\partial s}$; this solution is obtained by differentiating (5.2) with respect to s. When $\epsilon = 0$, $\omega = 1$ and $\lambda = \lambda_0$ we have the eigenvalue problem

$$\frac{\partial w}{\partial s} + L_0 w - \beta w = 0$$

$$w(0) = w(2\pi) \ .$$

Solutions of this problem are of the form $w(x, s) = e^{\beta s}\varphi(x)$, where $\beta = \mu + 2\pi i n$, φ is an eigenfunction of L_0 with eigenvalue μ, and n is an arbitrary integer. Under the bifurcation assumption that L_0 has eigenvalues at $\pm i$ and all other eigenvalues in the right half-plane, we see that when $\epsilon = 0$ (6.4) has double eigenvalues at $\pm 2\pi i n$ and all remaining eigenvalues in the right half-plane.

As we have seen, one Floquet exponent of (6.4) is always $\beta = 0$ with $w = \frac{\partial u}{\partial s}$ (for small $\epsilon \neq 0$). We construct the other Floquet function in the form

$$\varphi = a(\epsilon) \frac{\partial v}{\partial s} + \psi \ . \tag{6.5}$$

Recall that $u = \epsilon v$ and $v = v_0 + \epsilon v_1 + \ldots = u_0 + \varphi$. The function $\frac{\partial v}{\partial s}$ satisfies

$$\omega \frac{\partial^2 v}{\partial s^2} + L(\lambda) \frac{\partial v}{\partial s} + \epsilon M(v, \frac{\partial v}{\partial s}) = 0 \ . \tag{6.6}$$

Substituting the expression (6.5) for φ into (6.4) and using (6.6), we get

$$\omega \frac{\partial \psi}{\partial s} + L(\lambda)\psi + \epsilon M(v, \psi) - \beta\omega\psi - \beta\omega a \frac{\partial v}{\partial s} = 0 \ ,$$

$$\psi(0) = \psi(2\pi) \quad . \tag{6.7}$$

We seek to determine $\psi(s, \epsilon)$, $\beta(\epsilon)$ and $a(\epsilon)$ in the form

$$\psi = u_0 + \epsilon\eta, \quad [\eta] = 0$$

$$\beta(\epsilon) = \epsilon^2\sigma(\epsilon) \ , \quad \sigma = \sigma_0 + \epsilon\, \sigma_1 + \dots$$

$$a(\epsilon) = a_0 + \epsilon\, a_1 + \dots \quad .$$

Substituting these series into (6.7), we obtain

$$\{J\eta + M(u_0, u_0)\} + \epsilon\{\omega_2 \frac{\partial u_0}{\partial s} + \tau_2 L_1 u_0 + J\eta_1$$

$$+ M(u_0, \eta_0) + M(v_1, u_0) - \sigma_0 u_0 - a_0\sigma_0 \frac{\partial u_0}{\partial s}\} + O(\epsilon^2) = 0 \ .$$

Since $M(u_0, u_0) = 2N(u_0, u_0)$ we have $[M(u_0, u_0)] = 0$, and accordingly

$$J\eta_0 + M(u_0, u_0) = 0$$

is solvable. The requirement that $[\eta_0] = 0$ means that η_0 is uniquely determined and we must have $\eta_0 = 2v_1$ (compare with equation (5.12)). The solvability condition $[J\eta_1] = 0$ leads to

$$-i\,\omega_2 + \tau_2[L_1 u_0] + 3[M(u_0, v_1)]$$

$$-\sigma + i\, a_0\sigma_0 = 0 \ .$$

(We have used (5.4) (ii), the relation $\eta_0 = 2v_1$, and the fact that M is symmetric.) Taking real and imaginary parts we get

$$\tau_2 \ \mathrm{Re}[L_1 u_0] + 3 \ \mathrm{Re}\ [M(u_0, v_1)] - \sigma_0 = 0$$

$$-\omega_2 + \tau_2 \ \mathrm{Im}[L_1 u_0] + 3 \ \mathrm{Im}[M(u_0, v_1)] + u_0\sigma_0 = 0 \ .$$

Comparing the first of these equations with the first of (5.13), we get

$$\sigma_0 = 2 \ \mathrm{Re}\ [M(u_0, v_1)] = -2 \ \tau_2 \ \mathrm{Re}\ [L_1 u_0] \ .$$

The perturbation series for the critical Floquet exponent is thus

$$\beta(\epsilon) = -2 \ \tau_2 \ \mathrm{Re}\ [L_1 u_0]\epsilon^2 + \dots \ ,$$

or

$$\beta(\epsilon) = -(2 \ \tau_2 \ \mathrm{Re}\ \gamma'(\lambda))\epsilon^2 + \dots \quad .$$

If $\operatorname{Re} \gamma'(\lambda_0) < 0$ then the null solution loses stability as λ increases past λ_0. Since

$$\lambda = \lambda_0 + \tau = \lambda_0 + \tau_2 \epsilon^2 + \dots ,$$

we see that the periodic solutions bifurcate above criticality if $\tau_2 > 0$. In this case, $\beta < 0$ for small ϵ; accordingly, the Floquet analysis indicates that supercritical periodic solutions are stable and subcritical solutions are unstable.

Notes

The bifurcation of periodic solutions of the Navier-Stokes equations has been investigated independently by Sattinger [17], Joseph and Sattinger [8], V.I. Judovic [11], and G. Iooss [7]. Each of these papers treats the problem by different methods. The validity of Floquet analysis in the determination of the stability of periodic motions has been established by both Judovic [9], [10] and Iooss [7]. Judovic's methods in [9, 10] are particularly elegant and simple.

E. Hopf investigated the bifurcation of periodic solutions from an equilibrium solution for ordinary differential equations in 1942. Hopf gave a complete analysis of the problem in that paper, though because of the unavailability of the paper, many of Hopf's results are not yet widely known. In that paper Hopf made note of the result that supercritical motions are unstable - both for the stationary problem and for the time periodic problem. He also showed that the bifurcating time periodic solution was unique up to phase shifts.

Closely related to the phenomenon of bifurcation of periodic solutions from an equilibrium is that of the bifurcation of traveling waves when the domain is infinite in one direction. The bifurcation of traveling waves has been discussed formally by J.T. Stuart [21] and J. Watson [22].

Bibliography

1. Coddington, E.A., and Levinson, N., Theory of Ordinary Differential Equations. McGraw-Hill, New York, 1955.

2. Frank-Kamenetzky, D.A., Diffusion and Heat Exchange in Chemical Kinetics. Princeton Univ. Press, 1955.

3. Gavalas, G.R., Nonlinear Differential Equations of Chemically Reacting Systems. Springer, New York, 1968.

4. Gelfand, I.M., "Some Problems in the Theory of Quasilinear Equations." AMS Translations Ser. 2 29, 295-381, (1963).

5. Hopf, E., "Abzweigung einer Periodischen Lösung eines Differential Systems." Berichten der Math.-Phys. Klasse der Sächischen Akademie der Wissenschaften zu Leipzig, XCIV, 1-22, (1942).

6. ________, "A Mathematical Example Displaying Features of Turbulence." Comm. Pure Appl. Math., 1, 303, (1948).

7. Iooss, G., "Existence et Stabilité de la Solution Périodique Secondaire Intervenant dans les Problèmes d'Evolution du Type Navier Stokes." Arch. Rational Mech. Anal. 4, 47, 301-329 (1972).

8. Joseph, D., and Sattinger, D.H., "Bifurcating Time Periodic Solutions and Their Stability." Arch. Rational Mech. Anal. 45, 75-109 (1972).

9. Judovic, V.I., "On the Stability of Forced Oscillations of Fluid." Sov. Math. Dokl. 11, 1473-1477 (1970).

10. ___________, "On the Stability of Self-Oscillations of a Liquid." Sov. Math. Dokl. 11, 1543-1546 (1970).

11. ___________, "Appearance of Auto Oscillations in a Fluid." Prikl. Mat. Mek. 35, 638-655 (1971).

12. Kirchgässner, K., and Sorger, P., "Stability Analysis of Branching Solutions of the Navier-Stokes Equations." Proc. 12th Int. Cong. Appl. Mech., Stanford Univ. August, 1968.

13. Landau, L., "On the Problem of Turbulence." C.R. Acad. Sci. USSR 44, 311 (1944).

14. Lee, J.C., and Luss, D., "The Effect of Lewis Number on the Stability of a Catalytic Reaction." AIChE Journal, 16, 620-665, July, 1970.

15. Parks, J.R., "Criticality Criteria ... as a Function of Activation Energy and Temperature of Assembly." Jour. of Chemical Physics 34, 46-50 (1961).

16. Ruelle, D., and Takens, F., "On the Nature of Turbulence." Comm. Math. Phys. 20, 167-192 (1971).

17. Sattinger, D., "Bifurcation of Periodic Solutions of the Navier-Stokes Equations." Arch. Rational Mech. Anal. 41, 66-80 (1971).

18. ___________, "Stability of Bifurcating Solutions by Leray-Schauder Degree." Arch. Rational Mech. Anal. 43, 154-166 (1971).

19. Sattinger, D., "Topics in Stability and Bifurcation Theory," Lecture Notes, University of Minnesota, to appear in the Springer Lecture Note Series.

20. ___________, "Monotone Methods in Nonlinear Elliptic and Parabolic Boundary Value Problems." Indiana Univ. Math. Journal, 21, 979-1000 (1972).

21. Stuart, J.T., "On the Nonlinear Mechanics of Wave Disturbances in Stable and Unstable Parallel Flows." Jour. Fluid Mechanics 9, 353-370 (1960).

22. Watson, J., "On the Nonlinear Mechanics of Wave Disturbances in Stable and Unstable Parallel Flows, Part II." J. Fluid Mech. 14, 336 (1960).

GROUNDWATER FLOW AS A SINGULAR PERTURBATION PROBLEM AND REMARKS ABOUT NUMERICAL METHODS

R. B. Simpson
Department of Applied Analysis and Computer Science
University of Waterloo
Waterloo, Ontario, Canada

There are two objectives for this section of this set of lectures; the first is to review the mathematical formulation of groundwater flow problems, and the second is to make some comments on numerical methods for solving the formulated problem. Some recent mathematical models suggest that some of the older formulations of hydrologic problems are a form of singular perturbation limit of the newer equations. To provide some evidence to support this, I will review some classical formulations of hydrologic problems in §1, and then relate them to a recent model in §2. The equation to be described in §2 forms the basis for a number of numerical solutions to groundwater flow problems. However the equation is sufficiently complex that the physical requirements (time and storage) of standard numerical techniques press on or exceed the resources of large computers. These implementation problems will be discussed in §3.

§1. Background Developments

The equations of groundwater flow are based on two principles (e.g., [1]):

(1.1) Darcy's Law

$$\vec{v} = -K \text{ grad } p$$

(1.2) Continuity Equation

$$(\rho\theta)_t = (-\vec{\nabla})\cdot(\rho\vec{v})$$

In these formulae, v is the seepage velocity of the water; $K = g\rho k/\mu$ is the hydraulic conductivity and is obtained as shown from g, the acceleration of gravity; ρ, the density of water; k, the specific permeability of the soil;

and μ , the viscosity of water. Two quite important variables are $p = \psi + z$, the total hydraulic head, which is the sum of ψ , called the hydraulic or pressure head and z , the vertical coordinate; and θ the moisture content defined as the fraction of a volume of the ground occupied by water.

The ground is regarded as consisting of three phases: water, solid, and air; θ is the fraction of volume taken up by water which can vary from zero (at least in principle) to θ_o , where $1-\theta_o$ is the fraction of volume taken up by solids. θ_o is the saturated moisture content; it is spatially nonuniform but for most soils varies between 1/4 and 1/2 or so. Although θ_o and K are spatially varying, we shall ignore this in the subsequent discussion.

'Classical' hydrology is based on the assumption of a rigid soil matrix (solid component of volume). The ground is regarded as consisting of a relatively dry upper layer (the unsaturated region), and a wet substratum (the saturated region), the interface being the water table. Ponds appear where the water table breaks the surface, and wells must be drilled below the water table to fill with water. About 1856, Darcy examined (1.1) experimentally and used (1.1) and (1.2) to formulate steady, saturated flow problems as boundary value problems for Poisson's equation in ψ . About 1935, transient saturated flow phenomena began to be treated by the expedient of allowing time dependent forcing and boundary terms. Some rationale for this is provided by the rigid soil matrix assumption which implies that θ_o is time independent. These models were mainly the response to questions formulated in terms of the saturated zone, e.g., well recharge rates and dam heights. Also around 1930, nonsaturated flow was investigated, and L. A. Richards observed that (1.1) could be used if K were allowed to vary with θ . $K = K(\theta)$ (Fig. 3, next section) has since been essentially taken as an empirical nonlinearity (although some theoretical bases for it have been proposed). Until recently, this line of investigation has generally been concerned with the rate of infiltration of water into dry soil to attempt to give some basis for empirical algebraic relations used for this rate. The substitution of (1.1) into (1.2) produces a nonlinear parabolic equation in θ (see

(2.3) below).

These formulations lead to hydrologic problems being separated into two flow regimes: a saturated flow regime governed by elliptic equations and a unsaturated one governed by nonlinear parabolic equations. This separation is evident in the pattern of virtually all texts on this subject devoting separate sections to each flow regime.

§2. More Recent Developments

Relatively new to hydrology are techniques for providing highly quantitative models for watershed response to hydrologic phenomena, e.g., rain, snow melting, artificial lake level change. Although the motivating questions are clearly basic, with a long history, these developments have been spurred by computer development and take basically three forms.

A) accounting models - essentially based on a macroscopic view of a watershed as a system of black boxes. Bookkeeping is done with algebraic relation - ships to distribute water between system components and to conserve total moisture. The parameters of the system are determined by analysing a series of hydrologic records for the particular watershed. This approach is well exemplified by the Stanford watershed model [2].

B) discrete versions of 'classical' mathematical hydrology as formulated in the preceding section. In this approach one of the main difficulties is to develop and implement suitable interface conditions at the boundary between the saturated and unsaturated zones. Descriptions of this approach can be found in [3] or [4].

C) development of uniform equations for the entire flow field and their numerical solution. A fine discussion of this approach is given by R. A. Freeze in [5], and a partial description of his work will follow in the next section.

Naturally, a certain amount of controversy exists over the relative merits of competing methods, and I have included this description for completeness only. Our intention here is to indicate that the third case appears to give rise to an equation with the second case as its singular perturbation limit.

For this purpose, a more extensive look at the hydrologic variables is needed. The pressure head, ψ , is usually normalized so that $\psi = 0$ at the saturation interface with $\psi > 0$ (< 0) in the saturated (unsaturated) regions. This is reasonable since ψ has dimensions of length and, with this normalization, ψ can be identified as the height to which water would rise above the open end of a pipe inserted into the saturated soil (in the absence of gravity). Moreover in the unsaturated zone, one would have to reduce the pressure inside a pipe to make water enter it from the 'dry' soil.

Ignored in the discussion of the preceding section is a functional relationship between the basic dependent variables ψ and θ . The relation is empirical, varies for each soil type and is complicated by hysteresis effects which are ignored here. However, they have the general shape of Figures 1 and 2, and $K(\theta)$ and $K(\psi)$ are sketched in Figures 3 and 4, based on [5]. These graphs indicate that once θ reaches saturation no increase in pressure affects it, evidently a consequence of the rigid soil matrix assumption, i.e.,

(2.1) $$\theta(\psi) = \theta_o \quad \text{for} \quad \psi \geq 0 .$$

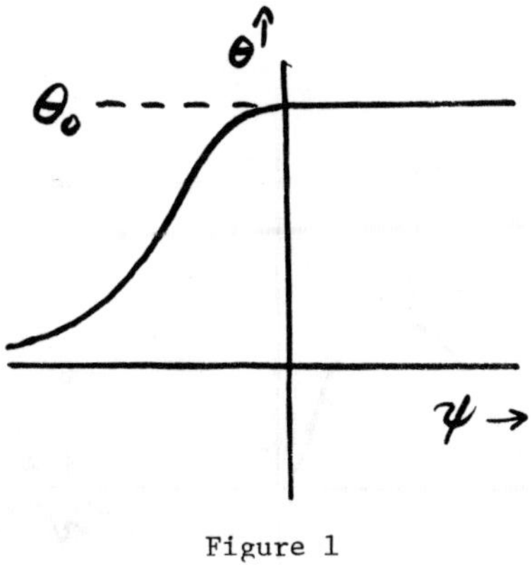

Figure 1

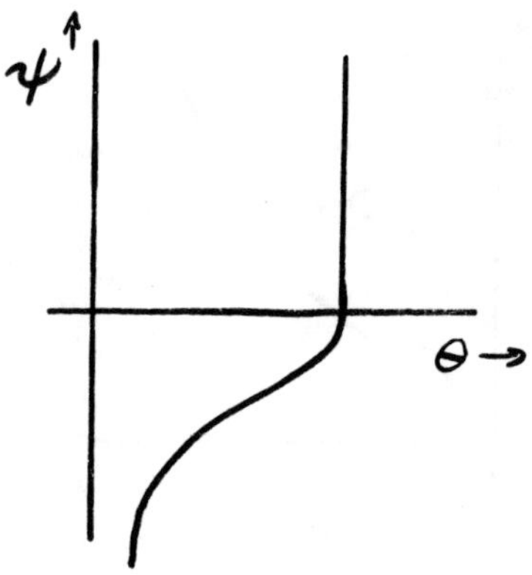

Figure 2

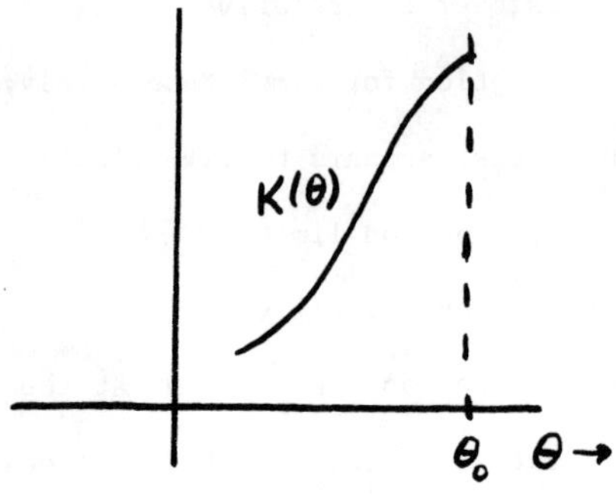

Figure 3

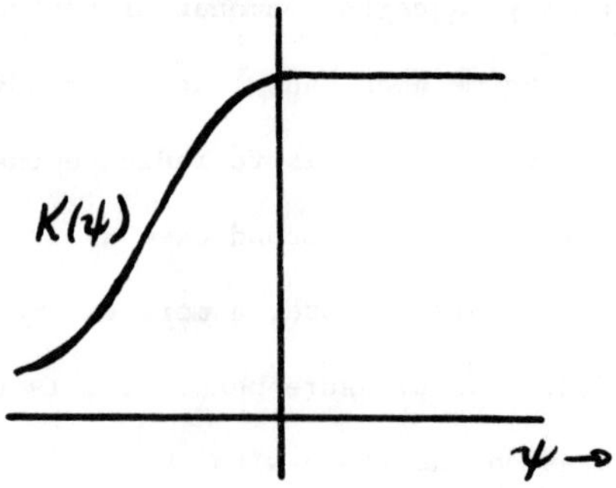

Figure 4

Taking account of these relations when one substitutes (1.1) into (1.2), one gets (for ρ constant)

$$(2.2) \qquad \theta_t = \vec{\nabla}\cdot(K(\psi)\,\mathrm{grad}(\psi + z))$$

which yields an equation in θ

$$(2.3) \qquad \theta_t = \underbrace{\vec{\nabla}\cdot(K(\theta)\psi'(\theta)}_{D(\theta)}\nabla\theta) + (K(\theta))_z$$

or an equation in ψ

$$(2.4) \qquad \theta'(\psi)\psi_t = \vec{\nabla}\cdot(K(\psi)\nabla\psi) + (K(\psi))_z$$

with nonlinear coefficients as graphed in Figures 5 and 6.

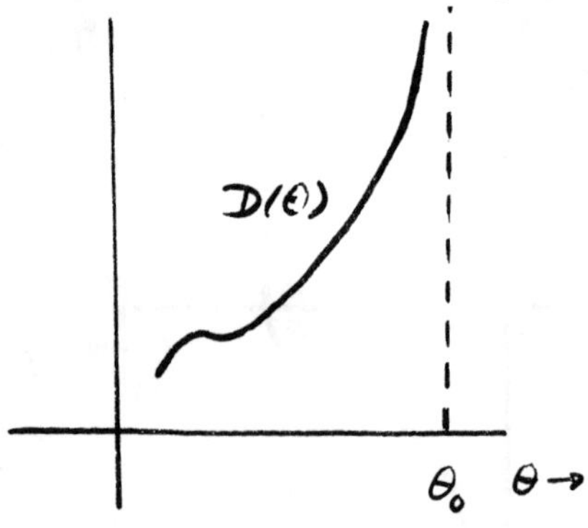

Figure 5

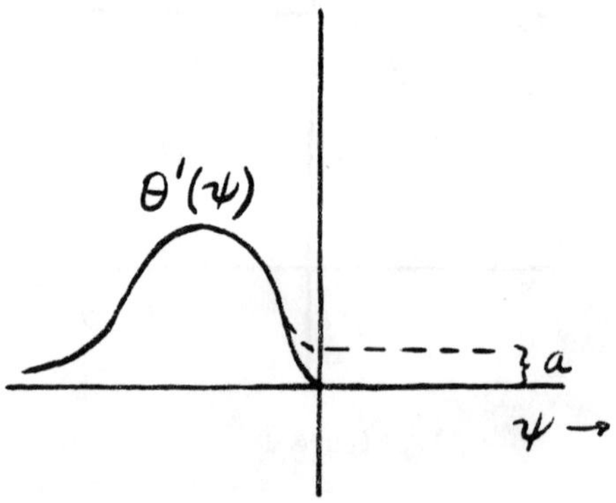

Figure 6

The degeneracies in these coefficients characterize the two cases discussed in §1, i.e., (2.3) is a nonlinear parabolic equation for unsaturated flows, and (2.4) is linear elliptic for saturated flow.

This degenerate behaviour is clearly traceable to the rigid soil matrix assumption, and the development of uniform equations amounts to replacing this assumption by a more realistic one, e.g., a linearly elastic soil in the sense that

$$\theta(\psi) = \theta_o + a\psi \qquad \psi \geq 0 \ . \tag{2.5}$$

This introduces a small parameter, a , the elastic constant, into $\theta'(\psi)$ on the right side of (2.4), resulting in a uniformly parabolic equation for ψ . It appears then, that this new equation is a type of singular perturbation equation whose limits as $a \to 0$ have the classical flow regimes of §1. The idea of a saturated zone is to be interpreted as a region in which elastic effects become important and the classical concept of the water table becomes an internal boundary layer, giving the transition region from unsaturated to saturated zones. In this transition region, large changes in the spatial derivatives of ψ of order one and higher would be expected but not in ψ itself. To give a trivial, but hopefully, suggestive example of this phenomenon, consider setting ϕ_o as a 'saturation' value for a variable $\phi(t)$ which satisfies

$$\{(\phi_o - \phi(t)) + \varepsilon\}\frac{d\phi}{dt} = a\phi(t) + b \tag{2.6}$$

which $0 < \phi(0) < \phi_o$. If we set $y(t) = \phi_o - \phi(t)$ and $b = -a\phi_o$, $y(t)$ satisfies

$$\{y(t) + \varepsilon\}\frac{dy}{dt} = -ay(t) \tag{2.7}$$
$$y(0) > 0 \ ,$$

the solution of which is given implicitly by

$$y(t) + \varepsilon \log y(t) = 1 - at$$

and is graphed in Figure 7 for small positive ε .

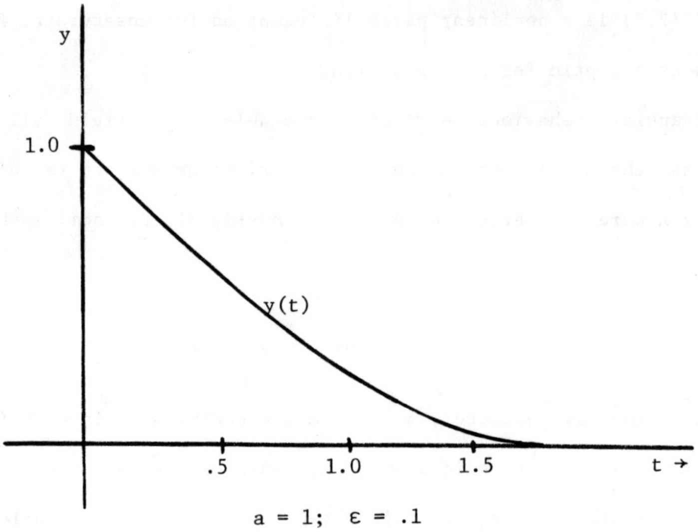

$a = 1; \quad \varepsilon = .1$

Figure 7

The 'reduced' equation for $\varepsilon = 0$ becomes

$$y^{(o)}(t) \frac{dy^{(o)}(t)}{dt} = - ay^{(o)}(t), \text{ i.e.,} \tag{2.8}$$

$$\frac{dy^{(o)}(t)}{dt} = - a \quad \text{or} \quad y^{(o)}(t) = 0 .$$

This degenerate equation has many 'solutions' but only one (i.e., $y^{(o)}(t) = - at + y_o$) which is continuously differentiable and only one (i.e., $y^{(o)}(t) = \max(- at + y_o, 0)$) which is non-negative. Clearly the latter is the correct choice to approximate the solution of (2.7).

§3. Computer Solutions and Problems

A major aim of the uniform description of groundwater flow is to simplify numerical methods for simulating it, as well as, naturally, to provide a more accurate mathematical model. However, numerical difficulties arise which are related to the nonlinear and the singular perturbation aspects of the equation; these difficulties are discussed in this section. For his major numerical study of the use of this equation, R. A. Freeze [5, 6], used a three dimensional finite difference version of a more sophisticated version of (2.4) (taking account of water compressibility and hysteresis effects). The method is based on a nonlinear Crank-Nicolson scheme; for purposes of discussion we shall illustrate some of its features with the simpler equation

$$u_t = D(u)u_{xx} . \tag{3.1}$$

The discrete version in the (x, t) plane on a uniform mesh of mesh spacings h and τ is

$$2\delta_t u(x, t) = D(u(x, t+\tau))\delta_x^2 u(x, t+\tau) + D(u(x, t))\delta_x^2 u(x, t) , \tag{3.2}$$

where $(x, t) = (ih, j\tau)$ for integers i and j , δ_t is the forward divided first time difference, and δ_x^2 is the central divided second x difference. From knowing the solution at time level t , one can (in principle) calculate it at time level $t+\tau$ by solving the nonlinear algebraic system

$$\begin{aligned} 2u(ih, t+\tau) &- \tau D(u(ih, t+\tau))\delta_x^2 u(ih, t+\tau) \\ &= 2u(ih, t) + \tau D(u(ih, t))\delta_x^2 u(ih, t) \end{aligned} \tag{3.3}$$

for $u(ih, t+\tau)$ for a range of i values determined by boundary conditions which have been omitted. Note that if D were independent of u , part of the last term on the left side would be known, and the system becomes a linear system for $u(ih, t+\tau)$.

For the three dimensional version [5], the (in core) storage limitations of a large computer (IBM 360/91) restricts the spatial mesh to about $20 \times 30 \times 50$ nodes. Variable mesh spacings are used, and the occurence of the internal boundary layer requires fine vertical mesh spacings of the order of .5 to 5 ft. apart wherever the boundary layer may appear. For reasonably coarse horizontal mesh spacings then, a $20 \times 30 \times 50$ point grid is restricted to representing watersheds of the order of a few square miles by about 1000 ft. deep. This limitation is rather severe for natural watersheds, and a reasonable objective would be to expand the horizontal capabilities of the model by a factor of 10 to 100.

The mesh referred to above was a static mesh, i.e., fixed with respect to the time parameter. Consequently it had to be finely divided wherever the boundary layer might occur. This suggests the use of a mesh which adapts itself, at least in the vertical direction, to the solution as time goes on, retaining fine mesh spacings only where necessary. Such adaptive meshes have been contemplated recently; however algorithms have not been investigated sufficiently to predict their impact on a problem of this size. Variational techniques (finite element methods, [3]) that have been developing recently seem superior to finite differences in problems with quite a bit of structure in their solution. While such an approach may be helpful, it would appear that the basic difficulty would be the same, i.e., a need for some type of time varying adaptive elements.

There is also a time limitation on the usefulness of the model (from a cost point of view); the τ step sizes required resulted in the larger tests running up to several hours of computing time. The choice of τ is not only determined by the desired accuracy of the discretization, but appears to influence the success of the method in peculiarly nonlinear ways. For example, suppose one tries to solve the simple, nonlinear system (3.3) by a Gauss-Seidel iterative method, evaluating the nonlinear coefficient, $D(u)$, at the preceding iterate (not at the preceding time level). A simple analysis of convergence (linearizing about the solution) shows that a necessary condition for convergence is

$$\tau \leq 2(\max\{|D'(u(x, t+\tau))\delta_x^{\,2}u(x, t+\tau)|, 0\})^{-1}$$

(here $D'(u)$ is $dD(u)/du$). In fact, in [5], it is reported that a similar iterative scheme starts off appearing to converge and then diverges, and that the occurrence of this phenomenon was reduced by decreasing the step size, τ. One would suspect that the use of Newton's method, coupled with a direct method for solving the linear systems would reduce this restriction on τ. Since $D(u)$ is provided by empirical functions or tables, the direct use of Newton's method may not be possible or may not give the benefits anticipated.

However, another perhaps more basic restriction on τ appears to be present.

The Crank-Nicolson scheme for the heat equation (for appropriate boundary conditions) is an unconditionally stable numerical method. However it appears that there may be a stability restriction on (3.2) related to the nonlinear coefficient, D(u) . More precisely, the energy method of analysing stability gives a sufficient condition for stability as

$$\tau \leq (\min |D(u(x, t))|)/\max |\delta_t D(u(x, t))|$$

for the extrema taken over the mesh points used.

I hope that these remarks serve to indicate that there are a number of challenging problems associated with large scale numerical solution of this uniform groundwater flow field equation, in particular, and which, to some extent would be expected in problems of diffusion with boundary layers generally.

BIBLIOGRAPHY

1. Eagleton, P.G. Dynamic Hydrology, McGraw Hill, 1970.

2. Linsley, R.K., and Crawford, N.H. Digital Simulation in Hydrology: Stanford Watershed Model IV, Technical Report #39, Department of Civil Engineering, Stanford University, 1966.

3. Hornberger, G.M., and Remson, I. Numerical Methods in Subsurface Hydrology, with an introduction to the finite element method, Wiley-Interscience, 1971.

4. Hornberger, G.M., and Remson, I. A Moving Boundary Model of a One-Dimensional Saturated-Unsaturated Transient Porous Flow System, Water Resources Research, Vol. 6, 898-905, 1970.

5. Freeze, R.A. Three Dimensional Transient Saturated-Unsaturated Flow in a Groundwater Basin, Water Resources Research, Vol. 7, 347-366, 1971.

6. Freeze, R.A. The Role of Subsurface Flow in the Generation of Surface Runoff, 1. Baseflow Contributions to Channel Flow, IBM Research Report RC 3631, 1971.

NONLINEAR PROBLEMS IN NUCLEAR REACTOR ANALYSIS

I. Stakgold* (Northwestern University) and L. E. Payne** (Cornell University)

1. INTRODUCTION

In the steady-state operation of a bare, homogeneous, monoenergetic reactor, the neutron density $u(x)$ satisfies the boundary value problem

$$(1) \qquad \Delta u + \omega u = 0 \ , \quad x \in D \ ; \quad u = 0 \ , \quad x \in \partial D \ ,$$

where D is the domain (taken in R^n, for generality) occupied by the reactor, ∂D is its boundary, and ω is a positive parameter incorporating the absorption cross section, the multiplication factor and the diffusion constant. Normally ω can be varied by adjusting the position of the control rods within the reactor.

We have previously considered (Payne and Stakgold, [8]) the linear problem where ω is independent of u and the reactor is operating at criticality. This means that $\omega = \lambda_1$, where λ_1 is the fundamental (lowest) eigenvalue of

$$(2) \qquad \Delta\varphi + \lambda\varphi = 0 \ , \quad x \in D \ ; \quad \varphi = 0 \ , \quad x \in \partial D \ .$$

It is well known that λ_1 is simple and positive with an associated eigenfunction $\varphi_1(x)$ which may be chosen positive in the open set D. With this agreement, $\varphi_1(x)$ is determined within a positive multiplicative constant so that the form, but not the size, of the neutron density is specified. This underlines the severe limitation of any linear theory, namely, that many quantities of physical importance such as the operating power of the reactor cannot be calculated within the theory. Nevertheless the linear model predicts accurately the value of ω at criticality and the general form of the neutron density. In our paper, we were able to obtain bounds for the gradient of the neutron density in terms of its maximum value and thus develop isoperimetric inequalities for the mean-to-peak neutron density ratio which depends only on the shape of the reactor.

We shall consider here a simple nonlinear version of (1) with ω a function of the steady temperature, which, when measured with respect to a suitable datum, is itself proportional to the neutron density. We recast (1) in the form

$$(3) \qquad -\Delta u = f(u) \ , \quad x \in D \ ; \quad u = 0 \ , \quad x \in \partial D \ ,$$

---

*Research supported by NSF grant GP 29508
**Research supported by NSF grant GP 33031x

where

$$(4)\qquad \begin{aligned} &f(z) = \lambda z - p(z)\ , \quad \lambda > 0\ , \\ &p(z) \in C^2(-\infty, \infty),\ p(0) = p'(0) = 0,\ p(z) > 0 \text{ for } z > 0\ . \end{aligned}$$

We have written (3) with $-\Delta u$ on the left side to stress the analogy with steady heat conduction for which a positive value of the right hand side characterizes a source of heat, whereas a negative value characterizes a sink. With this physical background in mind, questions of existence and notions of upper and lower solutions discussed in section 2 often acquire a straightforward intuitive meaning. Since $f(0) = 0$, we are dealing with what D. Cohen and H. B. Keller have called the unforced problem. In (4), $p(z)$ is positive for $z > 0$ in keeping with the usual physical requirement of negative feedback. The case $p(z) = z^2$ is particularly important and has been considered by Smets [11] in one dimension (when the solution is an elliptic function), by Kastenberg and Chambré [4], and, in the forced case, by D. Cohen [1] who was concerned with power excursions in a reactor. Outside the specific reactor setting, equation (3) has been widely studied; probably the most general results on positive solutions can be found in H. B. Keller [6] and in D. Cohen and B. Simpson [2].

In section 2, we study the existence and uniqueness of positive solutions for (3) and (4), sharpening slightly some of the results previously known. In section 3, we derive a maximum principle for (3) enabling us to obtain bounds for $|\text{grad } u|$. We then illustrate the use of these bounds in isoperimetric inequalities and norm estimates.

2. EXISTENCE AND UNIQUENESS OF POSITIVE SOLUTIONS

We are interested in positive solutions $u(x)$ of (3). Our first results do not require the specific assumptions (4) on $f(z)$. Instead we shall only demand that $f(0) = 0$ and that $f(z) \in C^2(-\infty, \infty)$. For simplicity in formulating the theorems, we shall exclude the linear case $f(z) \equiv \lambda_1 z$ and the case where $f(z)$ coincides with $\lambda_1 z$ in some finite neighborhood of $z = 0$:

(NL) $\qquad z = 0$ is an isolated zero of the equation $f(z) - \lambda_1 z = 0$.

Our first theorem is a slight generalization of one by H. B. Keller [5].

Theorem 1. If $f(z)$ satisfies (NL) and either

(- +) $\quad f(z) - \lambda_1 z \leq 0 \quad$ for all $z > 0$

or

(+ +) $\quad f(z) - \lambda_1 z \geq 0 \quad$ for all $z > 0$,

then no solution of (3) can be strictly positive in D.

Proof: Let u(x) be a strictly positive solution of (3). Then u also satisfies

$$- \Delta u - \lambda_1 u = f(u) - \lambda_1 u \ , \quad x \in D \ ; \quad u = 0 \ , \quad x \in \partial D \ .$$

Compatibility requires

$$(5) \qquad \int_D [f(u) - \lambda_1 u] \, \varphi_1 \, dx = 0 \ ,$$

where $\varphi_1(x)$ is the strictly positive eigenfunction of (2) corresponding to λ_1. Suppose f(z) satisfies (- +), then (5) is violated unless $f(u(x)) - \lambda_1 u(x) = 0$ for every x in D. Since u(x) vanishes on ∂D, condition (NL) guarantees that the only continuous solution of $f(u(x)) - \lambda_1 u(x) = 0$ is $u(x) \equiv 0$. Thus (3) cannot have a strictly positive solution. Exactly the same argument applies if f(z) satisfies (+ +).

Remark. If f(z) does not satisfy (NL), say $f(z) = \lambda_1 z$ for some neighborhood of $z = 0$, then (3) has strictly positive solutions of the form $c\varphi_1$, where c is a sufficiently small positive constant.

The next theorem shows that condition (- +) actually leads to the much stronger result that u(x) cannot be positive at any point x in D. Condition (+ +) on the other hand still allows for u(x) to be positive at some points in D, negative at other points.

Theorem 2. If f(z) satisfies (NL) and (- +), then u(x) cannot be positive anywhere in D.

Proof. Suppose $u > 0$ at x_0 in D. We can then find a domain $D^* \subset D$ and including x_0 with $u > 0$ in D^* and $u = 0$ on ∂D^*. Let λ_1^* be the fundamental eigenvalue of the operator $- \Delta$ on D^* with vanishing boundary conditions on ∂D^* and let φ_1^* be a corresponding eigenfunction strictly positive in D^*. If follows that u satisfies

$$- \Delta u = f(u), \quad x \in D^* \ ; \quad u = 0 \ , \quad x \in \partial D^* \ .$$

Since $\lambda_1^* \geq \lambda_1$, we infer from (- +) that $f(z) - \lambda_1^* z \leq 0$ for all $z > 0$. Applying Theorem 1 to domain D^*, we obtain a contradiction.

Corollary 1. If f(z) satisfies (NL) and

$$(+\ -) \qquad f(z) - \lambda_1 z \geq 0 \quad \text{for all } z < 0 ,$$

then no solution of (3) can be negative anywhere in D.

Corollary 2. If f(z) satisfies (NL) and

$$(-\ -) \qquad f(z) - \lambda_1 z \leq 0 \quad \text{for all } z < 0 ,$$

then no solution of (3) can be strictly negative in D.

We can now apply these results to the case where f(z) depends on a parameter λ and satisfies conditions (4). If $p(z) < 0$ for $z < 0$, and if $\lambda \leq \lambda_1$, then $f(z) - \lambda_1 z$ is negative for $z > 0$ and positive for $z < 0$; thus f(z) satisfies (- +) and (+ -) so that from theorem 2 and corollary 1, $u \equiv 0$. If $p(z) > 0$ for $z < 0$ or if p(z) changes sign for $z < 0$, and if $\lambda \leq \lambda_1$, then all one can say is that f(z) is (- +) so that u cannot be positive anywhere in D (but negative solutions are possible, as predicted by Lyapunov-Schmidt theory when p is even). In any event, we have

Theorem 3. Under conditions (4), no solution of (3) can be positive anywhere in D if $\lambda \leq \lambda_1$.

Next, we turn to the question of existence of positive solutions for $\lambda > \lambda_1$. We shall use the formalism of upper and lower solutions (see, for instance, Sattinger [9] or [10]).

Definition. An upper solution to the boundary value problem (3) is a function $\bar{v}(x)$ satisfying

$$- \Delta\bar{v} \geq f(\bar{v}) , \quad x \in D ; \quad \bar{v} \geq 0 , \quad x \in \partial D .$$

A lower solution $\underline{v}(x)$ is a function that satisfies

$$- \Delta\underline{v} \leq f(\underline{v}) , \quad x \in D ; \quad \underline{v} \leq 0 , \quad x \in \partial D .$$

Theorem 4. Let $\underline{v}(x)$, $\bar{v}(x)$ be lower and upper solutions, respectively, and suppose

$$\underline{v}(x) \leq \bar{v}(x) , \quad x \in D ,$$

then there exists at least one solution u(x) of (3), satisfying the inequality

$$\underline{v}(x) \leq u(x) \leq \bar{v}(x) , \quad x \in D .$$

The proof is given in Sattinger ([9] or [10]) and will not be repeated here. It is based on constructing 2 monotone approximation sequences, an increasing one starting from $\underline{v}$ and a decreasing one starting from $\bar{v}$. An easy argument then shows that these sequences converge, respectively, to solutions $\underline{u}(x)$ and $\bar{u}(x)$ of (3), where $\underline{u}(x)$ is a minimal, and $\bar{u}(x)$ a maximal solution among those lying between

$\underline{v}(x)$ and $\bar{v}(x)$.

Let us now apply Theorem 4 to problem (3) with f(z) satisfying (4). We try $\bar{v} = C > 0$ so that

$$- \Delta\bar{v} - f(\bar{v}) = - \lambda C + p(C) \ , \quad x \in D \ ; \quad \bar{v} > 0 \text{ on } \partial D \ .$$

If p(z) satisfies the condition

$$(6) \qquad \lim_{z \to \infty} \frac{p(z)}{z} = + \infty \ ,$$

then, for any given λ, we can find a C large enough so that $\bar{v}(x) = C$ is an upper solution. We shall assume henceforth that (6) holds.

For a lower solution, we try $\underline{v} = \epsilon\varphi_1(x)$, where ϵ is a positive constant and φ_1 is the positive fundamental eigenfunction of (2). Then

$$- \Delta\underline{v} - f(\underline{v}) = \epsilon \, \varphi_1(\lambda_1 - \lambda) + p(\epsilon \, \varphi_1) \ ; \quad \underline{v} = 0 \text{ on } \partial D \ .$$

Since $p'(0) = 0$, it is clear that for any given $\lambda > \lambda_1$, we can choose ϵ sufficiently small so that $\epsilon \, \varphi_1(\lambda_1 - \lambda) + p(\epsilon \, \varphi_1)$ is negative. Hence $\epsilon \, \varphi_1(x)$ is a lower solution. We therefore conclude, with $C > \epsilon \max \varphi_1$:

Theorem 5. If f(z) satisfies (4) and (6) and if $\lambda > \lambda_1$, then (3) has a strictly positive solution, lying between $\epsilon \, \varphi_1$ and C.

What about uniqueness? Let $\underline{u}$ and $\bar{u}$ be the minimal and maximal solutions obtained by the monotone iteration schemes mentioned after Theorem 4. To be precise, these solutions should be labeled $\underline{u}_\epsilon$ and $\bar{u}_C$ to show that they may depend on the initial approximations $\underline{v} = \epsilon \, \varphi_1$ and $\bar{v} = C$. Since $\underline{u}_\epsilon$ and $\bar{u}_C$ both satisfy (3), we obtain from Green's theorem (omitting subscripts),

$$(7) \qquad \int_D \underline{u} \, \bar{u} \, \{[p(\bar{u})/\bar{u}] - [p(\underline{u})/\underline{u}]\} \, dx = 0 \ .$$

We know that $\bar{u} \geq \underline{u}$. Suppose D' is the subdomain of D on which $\bar{u} > \underline{u}$. Then (7) holds with D' instead of D. But this yields a contradiction if p(z) satisfies

$$(8) \qquad \frac{p(z)}{z} \text{ strictly increasing for } z > 0 \ .$$

Thus $\underline{u}_\epsilon(x) \equiv \bar{u}_C(x)$. Since this is true for all sufficiently small ϵ and all sufficiently large C, it follows that there exists a unique strictly positive solution.

We recapitulate our results:

If $f(z)$ and $p(z)$ satisfy conditions (4), (6) and (8), problem (3) has one and only one strictly positive solution for $\lambda > \lambda_1$ and no solution can be positive anywhere for $\lambda \le \lambda_1$.

3. MAXIMUM PRINCIPLE AND ITS CONSEQUENCES

We shall consider problem (3) under conditions (4), (6) and (8), with $\lambda > \lambda_1$. Then there exists a unique, strictly positive solution $u(x)$ which is easily seen to satisfy $f(u(x)) \ge 0$, $x \in D$. Indeed, by (8), $p(C) = \lambda C$ has a unique solution C_λ, and $\bar{v} = C_\lambda$ is an upper solution. Thus $u(x) \le C_\lambda$ and, $f(z)$ being positive for $0 \le z \le C_\lambda$, the desired conclusion follows.

We shall now make an additional assumption on the region D. Let the boundary be of class $C^{2+\varepsilon}$, a smoothness assumption which will be relaxed later. As was shown in [8] and [13], some of the interesting optimal properties occur for noncontractible regions. We shall therefore not restrict ourselves to boundaries with positive curvature. The smoothness assumption does however imply the existence of a nonnegative constant K_o such that

$$(9) \qquad K_o \ge \max_{x \in \partial D} (-K) ,$$

where K is the average curvature of ∂D. For convex D, it is clear that we can choose $K_o = 0$.

We first obtain an upper bound on the magnitude of the gradient of $u(x)$.

Theorem 6. The following estimates hold:

$$(10) \qquad |\text{grad } u|^2 \le 2\,[F(u_M) - F(u)] + \beta(u_M - u) , \quad x \in D ;$$

$$(11) \qquad |\text{grad } u|^2 \le 2\,F(u_M) + \beta u_M , \quad x \in \partial D ;$$

where

$$F(z) = \int_o^z f(t)dt ,$$

and

$$(12) \qquad \beta = 2(n-1)K_o \left\{ (n-1)K_o u_M + [(n-1)^2 K_o^2 u_M^2 + 2F(u_M)]^{\frac{1}{2}} \right\} .$$

Remark. If the average curvature of ∂D is everywhere nonnegative, we can set $\beta = 0$ and the formulas simplify considerably.

Proof. Setting

(13) $$\Phi(x) = |\mathrm{grad}\ u(x)|^2 + 2F(u(x)) + \alpha u(x) ,$$

for an as yet unspecified $\alpha \geq 0$, we find

(14) $$\frac{\partial \Phi}{\partial x_k} = 2\ \mathrm{grad}\ u \cdot \mathrm{grad}\ \frac{\partial u}{\partial x_k} + 2f(u)\frac{\partial u}{\partial x_k} + \alpha \frac{\partial u}{\partial x_k} ,$$

and

(15) $$\Delta \Phi = \sum_{i,k} 2 \left(\frac{\partial^2 u}{\partial x_i \partial x_k}\right)^2 - 2f^2(u) - \alpha f(u) .$$

Using (14) and Schwarz's inequality on grad $u \cdot$ grad $\frac{\partial u}{\partial x_k}$, we obtain, at points where grad $u \neq 0$,

(16) $$\Delta \Phi - \frac{A \cdot \mathrm{grad}\ \Phi}{|\mathrm{grad}\ u|^2} \geq \frac{\alpha^2}{2} + \alpha f(u) \geq 0 ,$$

where A is a bounded vector. It follows that Φ assumes its maximum value either on ∂D or at a point where grad $u = 0$. If the maximum occurs at a point P on ∂D either $\Phi \equiv$ constant in D or $\frac{\partial \Phi}{\partial \nu} > 0$ at P, where ν is the outward normal. This is a direct application of Hopf's second maximum principle [3].

We now compute $\frac{\partial \Phi}{\partial \nu}$ at an arbitrary point on ∂D, and show that, for a suitable choice of α, $\frac{\partial \Phi}{\partial \nu}$ cannot be positive on ∂D. To this end we introduce normal coordinates in a neighborhood of the boundary and make use of the smoothness of ∂D to apply the differential equation at the boundary. Since u vanishes on ∂D and $f(0) = 0$, we have

$$\frac{\partial \Phi}{\partial \nu} = 2 \frac{\partial u}{\partial \nu} \frac{\partial^2 u}{\partial \nu^2} + \alpha \frac{\partial u}{\partial \nu} ,$$

and

$$\frac{\partial^2 u}{\partial \nu^2} + (n-1)K \frac{\partial u}{\partial \nu} = 0 ,$$

the last equation being just the differential equation for u after dropping terms that vanish on ∂D. Thus, on ∂D,

$$\frac{\partial \Phi}{\partial \nu} = \frac{\partial u}{\partial \nu} [\alpha - 2(n-1)K \frac{\partial u}{\partial \nu}] = q[2(n-1)(-K)q - \alpha] ,$$

where we have set

$$q = |\mathrm{grad}\ u| \text{ on } \partial D .$$

Using (9), we obtain

$$\frac{\partial \Phi}{\partial \nu} \leq q\ [2(n-1)K_o q - \alpha] .$$

Let us now choose

(17) $$\alpha = 2(n-1)K_o q_o ,$$

where q_o is the maximum (unknown) value of q on ∂D. With this choice of α, we have $\frac{\partial \Phi}{\partial \nu} \le 0$ at every point on the boundary so that either $\Phi \equiv$ constant or the maximum of Φ occurs where grad u = 0. In the latter case,

$$|\text{grad } u|^2 + 2F(u) + 2(n-1)K_o q_o u \le \max_{x \in D} [2F(u) + 2(n-1)K_o q_o u] .$$

Letting u_M stand for the maximum of u(x) and noting that F(u) is increasing for $0 < u < u_M$ we have

(18) $$|\text{grad } u|^2 + 2F(u) + 2(n-1)K_o q_o u \le 2F(u_M) + 2(n-1)K_o q_o u_M ,$$

the inequality being trivially true if $\Phi \equiv$ constant. At points on ∂D, we find

$$q^2 \le 2F(u_M) + 2(n-1)K_o q_o u_M ,$$

so that

$$q_o^2 \le 2F(u_M) + 2(n-1)K_o q_o u_M ,$$

which, when solved for q_o yields

(19) $$q_o \le (n-1)K_o u_M + [(n-1)^2 K_o^2 u_M^2 + 2F(u_M)]^{\frac{1}{2}} .$$

This proves (11). Since inequality (18) remains valid if an upper bound for q_o is used, we may substitute from (19) to obtain (10). These inequalities continue to be valid for domains with Lipschitz boundaries by resorting to approximation by smooth surfaces, as long as the curvature of the boundary is bounded from below.

Using (10), we can derive a number of integral inequalities for u(x). Let V be the volume of D and S the area of ∂D. Integration of (3) gives

(20) $$\int_D f(u)dx \le S[2F(u_M) + \beta u_M]^{\frac{1}{2}} .$$

If we multiply (3) by u and integrate over D, we find

(21) $$\int_D uf(u)dx + 2\int_D F(u)dx + \beta \int_D udx \le V[2F(u_M) + \beta u_M] ,$$

which, for convex D, reduces to

(22) $$\int_D uf(u)dx + 2\int_D F(u)dx \le 2VF(u_M) .$$

Sharper inequalities can be derived by introducing the level surfaces of u(x). Let $D_+(t)$ be the domain where $u(x) > t$; its boundary $\partial D_+(t)$ is the level surface $u = t$. Let us define

$$(23) \qquad \Omega(t) = \int_{D_+(t)} f(u(x))dx .$$

By dividing the region of integration into thin shells bounded by neighboring u-levels, we find that

$$(24) \qquad \Omega(t) = \int_t^{u_M} f(z)dz \int_{\partial D_+(z)} \frac{ds}{|\text{grad } u|} ,$$

so that

$$(25) \qquad \Omega'(t) = - f(t) \int_{\partial D_+(t)} \frac{ds}{|\text{grad } u|} .$$

Integration of (3) over $D_+(t)$ yields

$$(26) \qquad \Omega(t) = \int_{\partial D_+(t)} |\text{grad } u| ds ,$$

which, combined with (25) and (10), gives

$$- \frac{\Omega'(t)}{\Omega(t)} \geq \frac{f(t)}{2[F(u_M) - F(t)] + \beta[u_M - t]} .$$

For a convex region, β can be set equal to zero, so that

$$\frac{d}{dt} \log \Omega \leq \frac{1}{2} \frac{d}{dt} \log[F(u_M) - F(t)] ,$$

or

$$\log[\Omega(t)/\Omega(0)] \leq \frac{1}{2} \log \frac{F(u_M) - F(t)}{F(u_M)} ,$$

or

$$(27) \qquad \Omega(t) \leq \Omega(0) \left[1 - \frac{F(t)}{F(u_M)}\right]^{\frac{1}{2}} .$$

Integrating from $t = 0$ to $t = u_M$, we obtain

$$\int_0^{u_M} \Omega(t)dt \leq \Omega(0) \int_0^{u_M} \left[1 - \frac{F(t)}{F(u_M)}\right]^{\frac{1}{2}} dt ,$$

where, by (24) and a change in the order of integration, we find

$$(28) \qquad \int_D uf(u)dx \leq \int_D f(u)dx \int_0^{u_M} \left[1 - \frac{F(t)}{F(u_M)}\right]^{\frac{1}{2}} dt .$$

In the linear case, $f(u) = \lambda_1 u$, $u = \varphi_1$, the last inequality remains valid and becomes

$$(29) \qquad \int_D u^2 dx \leq \left[\int_D udx\right] \frac{\pi}{4} u_M ,$$

which actually is an equality for a one-dimensional domain. Other inequalities of the type (29) can be found in Payne and Rayner, [7]. We expect to return to the nonlinear problem in a subsequent paper with the particular goal of obtaining isoperimetric inequalities for the mean-to-peak neutron density ratio.

BIBLIOGRAPHY

1. Cohen, D. S., Arch. Rational Mech. Anal., 26, 305-315 (1967).
2. Simpson, R. B., and Cohen, D. S., J. Math. Mech., 19, 895-910 (1970).
3. Hopf, E., Proc. Amer. Math. Soc., 3, 291-293 (1952).
4. Kastenberg, W. E., and Chambré, P. L., Nucl. Sci. Eng., 31, 67-79 (1968).
5. Keller, H. B., Bull. Amer. Math. Soc., 74, 887-891 (1968).
6. Keller, H. B., J. Math. Mech., 19, 279-285 (1969-70).
7. Payne, L. E. and Rayner, M. E., ZAMP (to appear).
8. Payne, L. E. and Stakgold, I., J. Applicable Anal. (to appear).
9. Sattinger, D. H., Indiana Univ. Math. Jour., 21, 979-1000 (1972).
10. Sattinger, D. H., Topics in Stability and Bifurcation Theory, University of Minnesota Lecture Notes, 1972 - to appear in Springer Series.
11. Smets, H. B., Nukleonik, 8, 283 (1966).
12. Stakgold, I., SIAM Review, 13, 289-332 (1971).
13. Unger, H. E., J. Nucl. Energy, 25, 615-622 (1971).

SOME NON-LINEAR PROBLEMS IN STATISTICAL MECHANICS AND BIOLOGY

Colin J. Thompson
Mathematics Department
University of Melbourne
Parkville, Victoria. 3052
Australia

CONTENTS

1. INTRODUCTION

While few would dispute the claim that the laws of nature are inherently non-linear, there has not been any attempt, to my knowledge at least, to classify and treat the non-linear problems posed by physical theories. In some instances of course, most notably turbulence, and more recently in ecological studies, a great deal has been accomplished, but still much remains to be done. I will not attempt here to give a systematic treatment of these and other non-linear problems. Rather, I would like to focus attention on two subjects, statistical mechanics and biology. The presentation will by no means be complete, but I hope to give you some idea of the type of problems that occur, what has been done with such problems and what remains to be done.

To begin let us consider a set of unit mass particles with Hamiltonian

$$\mathcal{H} = \sum_i p_i^2/2 + V(q_1, q_2, \ldots). \tag{1.1}$$

p_i and q_i denote the momentum and position, respectively, of the ith particle; the first term represents the kinetic energy and the second the potential energy of the system. If the particles interact only pair-wise through a central potential ϕ, we can write

$$V(q_1, q_2, \ldots) = \sum_{i<j} \phi(|q_i - q_j|). \tag{1.2}$$

Hamilton's equations of motion

$$\frac{dp_i}{dt} = -\frac{\partial \mathcal{H}}{\partial q_i} \quad , \quad \frac{dq_i}{dt} = \frac{\partial \mathcal{H}}{\partial p_i} \tag{1.3}$$

then take the form

$$\frac{dp_i}{dt} = \sum_{(j \neq i)} F(q_i - q_j) \quad , \quad \frac{dq_i}{dt} = p_i, \tag{1.4}$$

where F is the interparticle force. Equations (1.4), which are in general non-linear, are nothing more than Newton's equations of motion.

Now in dealing with a large or macroscopic system, one is not really interested in the exact position and momenta of all the particles. Rather, one is usually only interested in certain macroscopic quantities, for example, the average energy, the temperature, the pressure exerted by the particles on the walls, the heat capacity, etc. The basic aim of statistical mechanics is to derive the macroscopic properties of matter from the basic microscopic properties governed, for example, by Equation (1.4). To fulfill this aim it is clear, first of all, that equations (1.4) should have a satisfactory time evolution.

Even granted such an evolution however, we are still left with the old problem dating back to Boltzmann: How is it possible to reconcile reversibility at the molecular level (i.e., the invariance of Equation (1.4) under inversion of time $t \to -t$ and velocities $p_i \to -p_i$) with apparent irreversibility at the macroscopic level? It was Boltzmann himself who first suggested that equilibrium could be brought about by collisions. To support his idea he derived his famous equation which is commonly felt these days to describe adequately the approach to equilibrium of a dilute gas.

2. THE APPROACH TO EQUILIBRIUM - BOLTZMANN'S EQUATION

Let us consider, as Boltzmann did, a spacially homogeneous gas of hard sphere particles with a velocity distribution function $f(v,t)$ defined by

$$f(v,t)dv = \text{the number of particles at time } t \text{ in the volume element } dv \text{ around } v. \tag{2.1}$$

($v = (v_x, v_y, v_z)$ and $dv = dv_x dv_y dv_z$ in three dimensions.) We make the following assumptions.

1. Only binary collisions occur. That is, situations in which three or more particles come together simultaneously are excluded. (Physically speaking, this is a reasonable assumption if the gas is sufficiently dilute.)

2. The distribution function for pairs of molecules is given by

$$f^{(2)}(v_1, v_2, t) = f(v_1, t)f(v_2, t). \tag{2.2}$$

That is, the number of pairs of molecules at time t, $f^{(2)}(v_1, v_2, t)dv_1dv_2$ with velocities in dv_1 around v_1 and in dv_2 around v_2, respectively, is equal to the product of $f(v_1, t)dv_1$ and $f(v_2, t)dv_2$. This is Boltzmann's famous Stosszahlansatz or assumption of molecular chaos.

Before proceeding, we note that 1 is purely a dynamical assumption and that 2 is basically a statistical assumption. Unfortunately, the latter was not clearly or adequately stressed, particularly by Boltzmann, at a time when objections to his equation were based solely on mechanical considerations. We will return to this point after first deriving here Boltzmann's equation for $f(v, t)$.

By definition, the rate of change of $f(v, t)dv$ is equal to the net gain of molecules in dv as a result of collisions; i.e.,

$$\frac{\partial f}{\partial t} = n_{in} - n_{out} , \tag{2.3}$$

where

$$n_{in(out)}dv = \text{the number of binary collisions at time } t \text{ in which one of the final (initial) molecules is in } dv. \tag{2.4}$$

Consider now a particular molecule [1] with velocity in dv_1 around v_1 and all those molecules with velocities in dv_2 around v_2 which can collide with molecule [1]. It is to be remembered that we are considering now only binary collisions of hard spheres (with diameter a).

In the relative coordinate system with molecule [1] at rest, the centre of molecule [2] must be in the "collision cylinder" as shown in Figure 1 if it is to collide with molecule [1] in the time interval δt. The collision cylinder and appropriate parameters are shown in more detail in Figure 2.

The collision cylinder has volume $bd\phi dbg\delta t$ where $b = a\cos(\theta/2)$. It follows that

$$bdbd\phi = \frac{a^2}{4}\, d\Omega \quad , \tag{2.5}$$

where $d\Omega = \sin\theta d\theta d\phi$ is the solid angle of the scattered particle.

Now from the definition of n_{out} and the Stosszahlansatz we have

$$n_{out}dv_1 = \frac{a^2}{4} \int d\Omega \int |v_2 - v_1|[f(v_2, t)dv_2][f(v_1, t)dv_1] \quad . \tag{2.6}$$

To determine n_{in} we merely look at the inverse collision $(v_1', v_2') \to (v_1, v_2)$ and use the above results to obtain

$$n_{in}dv_1 = \frac{a^2}{4} \int d\Omega \int |v_2' - v_1'|[f(v_2', t)dv_2'][f(v_1', t)dv_1']. \tag{2.7}$$

It is to be noted that the second integral in (2.6) is over v_2 and in (2.7) over v_2'.

Since we are considering perfectly elastic spheres, energy ($\frac{1}{2}mv^2$) and momentum (mv) are conserved in a collision; hence since all masses are assumed to be equal

$$v_1 + v_2 = v_1' + v_2' \quad \text{(momentum conservation)}$$

and

$$v_1^2 + v_2^2 = v_1'^2 + v_2'^2 \text{(energy conservation)} \quad . \tag{2.8}$$

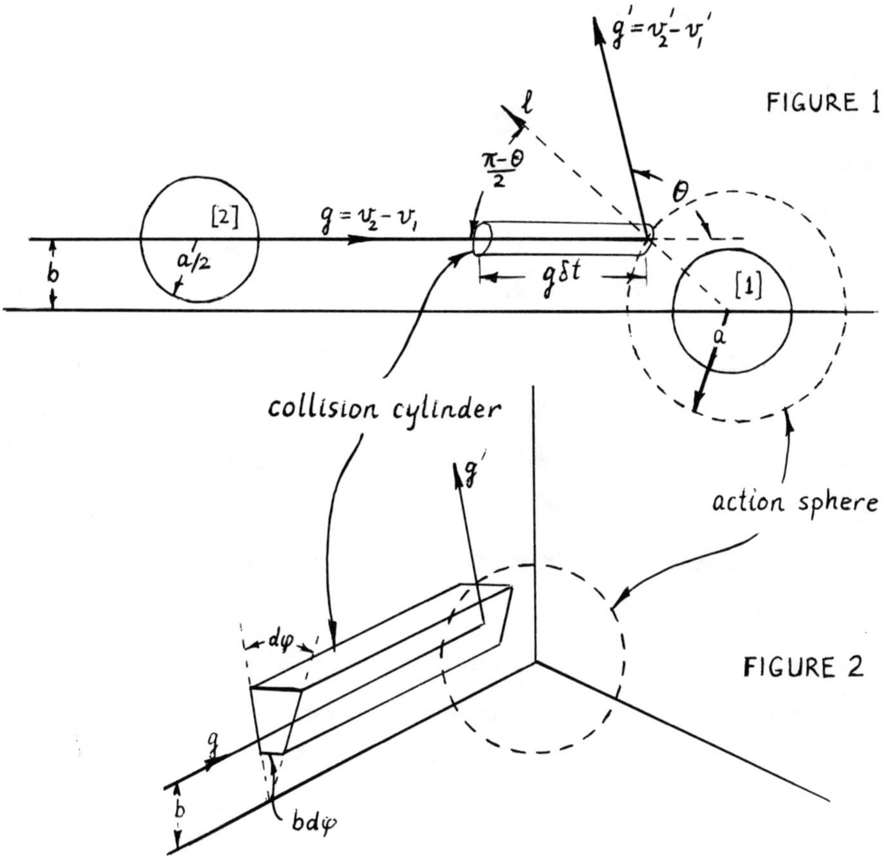

FIGURE 1

FIGURE 2

It follows almost immediately that

$$|v_2' - v_1'| = |v_1 - v_2| \quad \text{and} \quad dv_1 dv_2 = dv_1' dv_2' \ . \tag{2.9}$$

Combining the above results we obtain Boltzmann's equation

$$\frac{\partial f_1}{\partial t} = \int \frac{a^2}{4} \, d\Omega \int dv_2 |v_2 - v_1| f_1' f_2' - f_1 f_2), \tag{2.10}$$

where $f_1' = f(v_1', t)$, $f_2 = f(v_2, t)$, etc.

Equation (2.10) can be generalized in a number of ways. First, for more general interactions, we write

$$b\,db\,d\phi = I(g,\theta)d\theta \quad (g = |v_2 - v_1|), \tag{2.11}$$

which defines the "differential scattering cross-section" $I(g,\theta)$. For hard spheres $I(g,\theta) = a^2/4$; so for more general interactions $(a^2/4)d\Omega$ in (2.10) is replaced by $I(g,\theta)d\Omega$. Secondly, we can generalize (2.10) to non-uniform systems, but we will not go into the details here.

Before discussing the validity of the Boltzmann equation, let us review what is known about the equation itself.

The first problem, which was solved by Boltzmann himself, is to find the equilibrium or stationary solutions of the Boltzmann equation. If we denote the stationary solutions by $f_0(v)$ it is clear from (2.10) (in the general interaction case) that

$$\int d\Omega \int gI(g,\theta)[f_0(v_1')f_0(v_2') - f_0(v_1)f_0(v_2)]dv_2 = 0 \quad , \tag{2.12}$$

where $g = |v_2 - v_1|$ and $I(g,\theta)$ is the differential scattering cross-section defined by (2.11). Obviously, a sufficient condition on $f_0(v)$ is

$$f_0(v_1')f_0(v_2') = f_0(v_1)f_0(v_2) \quad . \tag{2.13}$$

It was a great triumph for Boltzmann that he was able to show that (2.13) is also a necessary condition for equilibrium. This is a consequence of

<u>Boltzmann's H-Theorem</u>. *If* f(v,t) *is a solution of the Boltzmann equation and* H(t) *is defined by*

$$H(t) = \int f(v,t) \log f(v,t) dv \ , \tag{2.14}$$

then

$$\frac{dH}{dt} \leqq 0 \tag{2.15}$$

with equality if and only if f(v,t) *satisfies* (2.13).

The proof of the theorem is straightforward and can be found in most books on statistical mechanics (Thompson (1972), Huang (1963), Uhlenbeck & Ford (1963)). We simply note here that the necessity of (2.13) follows from the theorem and the fact that, since

$$\frac{dH}{dt} = \int \frac{\partial f}{\partial t}(1 + \log f) dv \quad , \tag{2.16}$$

$\frac{\partial f}{\partial t} = 0$ implies that $\frac{dH}{dt} = 0$.

To show that there is only one stationary solution we note from (2.13) that

$$\log f_0(v_1') + \log f_0(v_2') = \log f_0(v_1) + \log f_0(v_2) \ , \tag{2.17}$$

which in physical language constitutes a <u>conservation law.</u> Now since we are only considering systems in which energy and momentum (and constants) are conserved, it follows that $\log f_0(v)$ must be a linear combination of v^2, the three components of v and an arbitrary constant; i.e.,

$$\log f_0(v) = \log A - B(v - v_0)^2,$$

or (2.18)

$$f_0(v) = A \exp[-B(v - v_0)^2],$$

where A, B, and the three components of v_0 are arbitrary constants. Equation (2.18) is the celebrated <u>Maxwell distribution for velocities.</u>

The arbitrary constants in (2.18) are usually expressed in terms of the density

$$\rho = \int f_0(v)dv \tag{2.19}$$

and the average kinetic energy

$$\varepsilon = \int \frac{1}{2} mv^2 f_0(v)dv[\int f_0(v)dv]^{-1} = \frac{3}{2} kT, \tag{2.20}$$

where k is Boltzmann's constant and T the absolute temperature. If it is assumed that the gas has no translational motion ("Galilean invariance"), $v_0 = 0$, and a simple calculation shows that

$$f_0(v) = \rho(\frac{m}{2\pi kT})^{\frac{3}{2}}\exp(-\frac{mv^2}{2kT}) \quad . \tag{2.21}$$

The above results are independent of the form of the "scattering force" as are the facts that the first and second moments of $f(v,t)$ are preserved in time; i.e., $\int |v|f(v,t)dv$ and $\int v^2 f(v,t)dv$ are constant.

It is to be stressed that standard derivations of the above are purely formal. In many cases the rigor has been supplied,but the results are still not as general as one would like. The initial value problem in particular has been much studied. The first result in this direction, due to Carleman (1957), is that, for hard spheres, a finite second moment for the initial distribution assures existence and uniqueness of solutions to the Boltzmann equation. Wild (1951) has obtained the same result for the (cut-off) "Maxwell gas" in which the potential is inversely proportional to r^4. The simplifying feature of the Maxwell gas is that $|v_1 - v_2|I(|v_1 - v_2|, \theta)$ is a function only of θ with a pole at $\theta = 0$. The pole at the origin is customarily removed by a cut-off so that the total cross-section $\int I(g,\theta)d\Omega$ is finite. The general interaction case remains an open question.

Another problem concerns the actual approach to equilibrium. Usually this is taken for granted, being a consequence of Boltzmann's H-theorem and the existence of a unique stationary solution. Complete rigor has been supplied for the hard sphere gas, Carleman (1957), for a wide class of cut-off potentials, Grad (1965), and for Kac's caricature of the Maxwell gas, McKean (1966). The general case is also an open question. A related problem is the rate of approach to equilibrium. This has been studied for the cases mentioned above by Grünbaum (1972), but the general case is open.

A remaining question of particular interest to physicists is the actual (asymptotic) form of the solutions for long times, in physical language, for times long compared with the time between collisions (the mean free time). One would expect the equations of classical hydrodynamics to be valid in this region (hence the term "hydrodynamical stage" in the approach to equilibrium); so what one would like to do is to derive the hydrodynamical equations from the Boltzmann equation. This was first achieved by Chapman and Enskog (See Uhlenbeck and Ford, 1963, pp. 77) with later refinements by Hilbert (1912). Hilbert's recipe was to expand f as a formal power series

$$f = f_0 + \varepsilon f_1 + \varepsilon^2 f_2 + \dots ,$$

(f_0 being the Maxwell distribution), substitute into Boltzmann's equation, solve successively for $f_1, f_2, \dots,$ and then set $\varepsilon = 1$. This process of course requires considerable faith and optimism. The rigor has not been supplied for the Boltzmann equation; only for a Boltzmann-like equation (the Telegraph equation) by McKean. So here is another problem!

The Chapman-Enskog-Hilbert solution has exactly the desired (by physicists) properties. In particular, the $n = 1 (f_1)$ solution gives the Eulerian hydrodynamical equations for the five hydrodynamical moments

$$\int v^n f_0 dv \quad , \qquad n = 0, 1, 2 \ ;$$

the $n = 2$ solution gives the Navier-Stokes equations for the hydrodynamical moments of $f_0 + f_1$, etc. The amusing thing about this solution is that the formal series for f is completely determined by the initial values of the five hydrodynamical moments (Uhlenbeck and Ford have called this Hilbert's paradox).

This completes our discussion on what is known about Boltzmann's equation. We now turn to the derivation and validity of the equation itself.

In our derivation of Boltzmann's equation we stressed that there were two distinct assumptions; one dynamical (only binary collisions are allowed) and one statistical (the Stosszahlansatz or assumption of molecular chaos). Since the main objections (historically due to Zermelo and Loschmidt) to Boltzmann's equation were based on dynamical consideration - that the H-theorem ($dH/dt \leq 0$) singles out a preferred direction in time and hence violates time reversal invariance - it is important to stress that the derivation involves an interplay between dynamics and statistics. Unfortunately this was not stressed adequately by Boltzmann, although I think it is fair to say that probabilistic considerations were in his thoughts. The problem then is when and how to introduce probabilities.

Existing "first principle" derivations of Boltzmann's equation fall roughly into two categories.

1. Begin with Hamilton's equations (or more correctly Liouville's equation) and attempt to derive the Boltzmann equation.

2. Begin with a dynamical-probabilistic model and attempt to derive the Stosszahlansatz.

Approach 1 is clearly preferable, but it is obvious that somewhere along the way a probabilistic assumption must be inserted. This has not been done in any satisfactory way as yet. Approach 2 has been carried through but it must be stressed that although it has a sounder mathematical footing than 1, it still lacks a fundamental connection with dynamics.

One probabilistic approach (2), due mainly to Uhlenbeck (1942), (Siegert (1949)), is briefly as follows. Let $p(v_1, v_2, \ldots, v_n, t)$ be the velocity distribution for a gas of n molecules with velocities $v_1, v_2, \ldots, v_n$, satisfying the "master equation"

$$\frac{\partial p}{\partial t} = \frac{2}{n} \sum_{i<j} \int [p(v_1,\ldots,v_i',\ldots,v_j',\ldots,v_n) - p(v_1,\ldots,v_i,\ldots,v_j,\ldots,v_n)] \times |v_i - v_j| I(|v_i - v_j|,\theta) \sin\theta \, d\theta d\phi \quad , \tag{2.22}$$

where the sum is over all possible pairs of molecules. The map $(v_i,v_j) \to (v_i',v_j')$ describes the result of a collision with scattering angle θ and differential scattering cross-section I. We are assuming in (2.22) that the molecules only collide in pairs. The factor $|v_i - v_j|I(|v_i - v_j|,\theta)$ describes the transition probability for the collision $(v_i,v_j) \to (v_i',v_j')$ and its inverse (so we are using microscopic reversibility). In addition, it will be assumed that the particles are indistinguishable; i.e., p is symmetric in $v_1, \ldots, v_n$.

The problem now is to derive Boltzmann's equation from (2.22). The first thing to note is that (2.22) is linear and describes the joint evolution of the n molecules. Boltzmann's (non-linear) equation, on the other hand, describes only the evolution of a single molecule. To connect the two, we define the contracted distributions

$$p^{(k)}(v_1,v_2,\ldots,v_k) = \int p(v_1,v_2,\ldots,v_k,v_{k+1},\ldots,v_n)dv_{k+1}\ldots dv_n \quad , \qquad (2.23)$$

$k = 1, 2, \ldots, n$ (note that by symmetry it does not matter which velocities are integrated). If we now integrate both sides of (2.22) over $v_2, \ldots, v_n$ and retain the v_2 integral on the right-hand side, we obtain

$$\frac{\partial p^{(1)}}{\partial t} = \int [p^{(2)}(v_1',v_2') - p^{(2)}(v_1,v_2)]|v_1 - v_2|I(|v_1 - v_2|,\theta)d\Omega dv_2, \qquad (2.24)$$

which would be Boltzmann's equation if

$$p^{(2)}(v_1,v_2,t) = p^{(1)}(v_1,t)p^{(1)}(v_2,t) \quad . \qquad (2.25)$$

Now we can always start with a "chaotic distribution"

$$p(v_1,v_2,\ldots,v_n,0) = p^{(1)}(v_1,0)p^{(1)}(v_2,0),\ldots,p^{(1)}(v_n,0), \qquad (2.26)$$

but in general it is obvious that this property will not propagate in time if there are interactions present and n is finite. Realizing this, Kac suggested that (2.25) may be valid in the limit $n \to \infty$, since in this limit, i.e., an infinite gas, a molecule after colliding with a given molecule will fly off to infinity, never to be seen again. Hence, if no correlations are present at time $t = 0$ (equation

(2.26)) they cannot develop in time. Kac was able to prove this result for his Maxwell caricature, and Alberto Grünbaum (1971) has recently proved the general result ((2.25)), but subject to certain smoothness, existence and uniqueness assumptions.

To conclude this section, let me summarize what we know and don't know about the Boltzmann equation and the approach to equilibrium.

The first problem concerns the validity of the Boltzmann equation. This is more a physical than a mathematical problem and requires some agreeable way of combining the dynamical and probability methods. A related problem which I have not discussed is the extension of the Boltzmann equation to higher densities where more than two body collisions become important. Attempts so far to obtain corrections to the Boltzmann equation at high densities have met with considerable difficulties, the most serious being that density expansions for such things as transport coefficients (viscosity, etc.) do not seem to exist. More precisely, if one attempts to obtain formal power series in the density, terms arising from more than three body events blow up. This whole area is then very much up in the air, the central question being; in what sense, if any, is Boltzmann's equation a first approximation? These problems, of course, are basically physical and should not overly concern us here.

On the mathematical side much has been accomplished. It still remains, however, to extend the existence and uniqueness theorems and, perhaps more importantly, to either justify the Chapman-Enskog-Hilbert development or find an alternative way to discuss the hydrodynamical stage in the approach to equilibrium.

3. THE GIBBS ENSEMBLES - EQUILIBRIUM STATISTICAL MECHANICS

In equilibrium statistical mechanics one seeks a prescription for calculating macroscopic thermodynamic quantities, such as pressure, etc., from the microscopic properties of the system. Such a prescription was first provided by Gibbs late last century and is now universally accepted. The only problem then is to calculate. Unfortunately, as things stand at the moment, we can calculate very few things although as a result of recent efforts we know that most things we want to calculate at least exist.

Let us begin by briefly summarizing the Gibb's postulates. Consider a system composed of N particles in a volume V with fixed energy E. The time evolution of the system is governed by Hamilton's equations

$$\frac{dp_i}{dt} = -\frac{\partial \mathcal{H}}{\partial q_i} , \quad \frac{dq_i}{dt} = \frac{\partial \mathcal{H}}{\partial p_i} , \tag{3.1}$$

where p_i and q_i are the canonical momentum and coordinates of the ith particle and $\mathcal{H}$ is the Hamiltonian of the system. We consider only Hamiltonians that do not depend on time explicitly so that

$$\mathcal{H}(p,q) = E , \tag{3.2}$$

and the time evolution of a given state, specified by a point

$$(p,q) = (p_1, \ldots, p_n; q_1, \ldots, q_n) \tag{3.3}$$

in phase or Γ-space, is represented according to (3.1) by a path in Γ-space on the energy surface specified by (3.2).

Now on the macroscopic level we are only interested in a few quantities and not the particular microscopic state of the system. We consider, therefore, an ensemble of systems composed of the original system with all possible initial conditions. At time $t = 0$ we specify a density function $\rho(p,q,t = 0)$ on this ensemble which becomes a probability density when properly normalized. After time t, $\rho(p,q,t = 0)$ becomes $\rho(p,q,t)$ defined by

$$\int_A \rho(p,q,t)dpdq = \text{the number of points in } A \text{ of } \Gamma\text{-space at time } t, \tag{3.4}$$

$\rho(p,q,t)$ being completely determined from $\rho(p,q,0)$ by Liouville's equation

$$\frac{\partial\rho}{\partial t} = \{\mathcal{H}, \rho\} \quad , \tag{3.5}$$

where the curly brackets denote Poisson bracket.

The basic postulate due to Gibbs is the following.

Postulate I. - The Gibbs Microcanonical Ensemble

The equilibrium distribution of the macroscopic state of an isolated mechanical system is the uniform distribution on the energy surface. That is,

$$D_{m.c.}(p,q) = \rho(p,q)/\int\rho(p,q)d\Gamma = \delta(\mathcal{H} - E)/S(E) \quad , \tag{3.6}$$

where $S(E)$ is the area of the energy surface

$$S(E) = \int\frac{d\sigma}{||\text{grad}\mathcal{H}||} \quad , \tag{3.7}$$

$d\sigma$ is an element of energy surface,

$$||\text{grad}\mathcal{H}||^2 = \Sigma\{(\frac{\partial\mathcal{H}}{\partial p_i})^2 + (\frac{\partial\mathcal{H}}{\partial q_i})^2\} \tag{3.8}$$

and $\delta(X)$ is the delta function.

Gibbs gave heuristic arguments to support his postulate, and although it is true that certain points in the argument can be tightened up with the aid of ergodic theorems and the like, there is still no strict derivation of (3.6). The sensible point of view might then be to take (3.6) as axiom 1 and proceed without delay to discuss its consequences. We will, however, present Gibbs' arguments and the related mathematics, not so much to justify (3.6), but more to back up a little and see what the recent theorems of Sinai tell us about non-equilibrium statistical mechanics. Before doing so we mention here, for completeness, Gibbs' second postulate or canonical ensemble which applies to more realistic non-isolated systems, e.g., systems in contact with a heat reservoir. The statement is as follows.

Postulate II.

The conditional probability density for a system S *in equilibrium with a system (heat reservoir)* R *is given by*

$$D = \exp(-\beta\mathcal{H})/\int\exp(-\beta\mathcal{H})d\Gamma \quad , \tag{3.9}$$

in the limit of infinite R, *where* $\beta = (kT)^{-1}$, k *is Boltzmann's constant,* T *the absolute temperature (of* R*),* $\mathcal{H}$ *is the Hamiltonian of the system* S *and* $d\Gamma$ *an element of its phase space. The distribution* (3.9) *is called the canonical distribution.*

Unlike the microcanonical distribution, a number of proofs, Khinchin (1949) and Grad (1952)(see also Thompson (1972) and Uhlenbeck (1963)), can be given of (3.8),granted the microcanonical postulate. All proofs, however, require a specific form for R, usually with a Hamiltonian of the form $H_1 + H_2 + \ldots + H_M$, where the H_i are "weakly coupled" to one another. We will not go into the deatils here, but will return to the canonical ensemble in the next section after completing this section with Gibbs' justification of Postulate I and the relevance of Sinai's theorem to the statistical mechanics of irreversible processes.

To begin we must first say precisely what we mean by a macroscopic variable. Unfortunately, this notion is somewhat arbitrary and depends on how detailed a description one wants of the system. Gibbs suggests that a finite number of macroscopic variables

$$y_i = f_i(p,q) \quad , \quad i = 1, 2, \ldots, m \; ,$$

used to describe the macroscopic state of a system, should satisfy the following conditions.

1. Each macroscopic state defined by a set of y_i values corresponds to a region of Γ-space, i.e., a portion of the energy surface $\mathcal{H}(p,q) = E$.

2. For large N there is one set of values $\bar{y}_1, \bar{y}_2, \ldots, \bar{y}_m$ which corresponds to a region that is overwhelmingly the largest (in area).

Requirement 1 corresponds to a "coarse graining" of the energy surface and, as stressed by the Ehrenfests, is essential for any macroscopic description. The amount of coarse graining clearly depends on the level of description demanded by the observer. Requirement 2, on the other hand, imposes restrictions on the type of description by requiring that N be large.

Granted conditions 1 and 2 on the macroscopic description, we now argue as follows. If every part of the energy surface is accessible to phase points and the time $t(A)$ a point spends in a region A is, roughly speaking, proportional to its area, we can identify the region that is overwhelmingly the largest as the (macroscopic) equilibrium state, since it follows that

(a) if the system is not in equilibrium it will almost always go into this state, and

(b) if the system is in equilibrium it will almost always stay in equilibrium.

These ideas are due to Gibbs (similar ideas although less precise and general were put forward by Boltzmann) and can be made more precise by introducing probability notions and appealing to ergodic type theorems.

First, we introduce the notion of probability by assuming that at time $t = 0$, the system is in a non-equilibrium macroscopic state specified by y_i variables and is characterized by a probability distribution $D(p,q,t = 0)$ which is non-zero only in the given region. $D(p,q,0)$ is in principle arbitrary (since we would like equilibrium to be approached independently of the initial distribution) except that it should be a reasonably smooth function of (p,q) (e.g., a constant over the initial region). This is the only probabilistic assumption. The time development of D is determined from $D(p,q,0)$ by Liouville's theorem, and one would like $D(p,q,t)$ to approach the equilibrium distribution as t increases. Gibbs' description of how this might come about is as follows. From Liouville's theorem the volume of the region where $D(p,q,t)$ is non-zero remains the same, although its shape will change drastically. As time develops the initial region will be drawn out into a thin ribbon winding over the energy surface until in a coarse grained sense the distribution becomes uniform. This is Gibbs' microcanonical postulate. The degree of uniformity and the time it actually takes to achieve uniformity clearly depend on how coarse grained a description one wants. The more coarse grained the description, the shorter the time clearly it will take to achieve the required uniform distribution.

Although this argument is extremely heuristic, it can be made more precise by appealing to the (classical) ergodic theorems. It should be stressed, however, that the basic point of the argument - the existence of a macroscopic description to begin with - is not at all solved by ergodic theory. Ergodic theorems are necessary for the validity of the Gibbs' argument, but they do not, and cannot, answer the basic question: How is a macroscopic description possible?

The ergodic theorems most commonly referred to are the classical ones, due mainly to Birkhoff. They are as follows:

1. For a bounded mechanical motion and for any phase function $y = f(p,q)$ which is integrable over the energy surface (e.g., Gibbs' macroscopic variables), the time average

$$\bar{y} = \lim_{\tau \to \infty} \tau^{-1} \int_0^\tau f(P_t)dt$$

almost always exists and is independent of the point $P_0 = (p,q)_{t=0}$ which evolves to $P_t = (p,q)_t$.

2. If the system is "metrically transitive", i.e., the energy surface cannot be decomposed (in a reasonable way) into separate parts left invariant by the flow, $\bar{y}$ is almost always equal to the ensemble average

$$\langle y \rangle = \int f(p)D_{m.c.}(p)d\sigma$$

over the energy surface (with element $d\sigma$) with respect to the uniform or microcanonical distribution $D_{m.c.}$.

Taking $f(p)$ in particular to be the characteristic function for the region A, i.e.,

$$f(p) = \begin{cases} 1 & p \in A \\ 0 & p \notin A \end{cases},$$

the ergodic theorems show that for a metrically transitive system

$$\lim_{T\to\infty} \frac{t(A)}{T} = \frac{V(A)}{V} ,$$

where $t(A)$ is the time a given point spends in A, $V(A)$ is the measure of A and V the measure of the entire energy surface.

Clearly this is a precise formulation of the basic step in the argument - that all points are accessible and the time a point spends in a region is determined by its area. A necessary prerequisite then for Gibbs' microcanonical distribution is that the system be metrically transitive. Until very recently there were no non-trivial physical examples of metrically transitive systems. Sinai, however, has recently proved that three or more hard spheres in a box form a metrically transitive system. In fact, he proved much more; i.e., for those who know the jargon, he proved that the system is a K-system (Wightman (1971)). This leads us next to consider the non-equilibrium situation.

While ergodicity is necessary for equilibrium statistical mechanics, it is probably not sufficient for either an understanding of non-equilibrium statistical mechanics or for integrals arising in the study of irreversible phenomena to exist. A common feeling these days is that something stronger is needed, e.g., Hopf's mixing flow or even stronger (Wightman (1971)).

I do not want to go into the details here except to say that Sinai proved that a finite number, larger than or equal to three, of hard spheres in a (finite) box is a mixing system, and very few physicists would doubt that most real systems have this property. As a consequence, however, we find that for such <u>finite systems</u> the normal prescription in irreversible processes gives vanishing transport coefficients (viscosity, diffusion coefficients, etc.). To save these theories we must therefore investigate <u>infinite systems</u>, beginning with Newton's equations (1.4). Such (extremely non-linear) problems do not seem to be studied in the mathematical literature. The only exception of which the author is aware is due to Oscar Lanford (1968 & 1969), who has studied Newton's equations for a number of infinite one-dimensional systems.

4. THE PROBLEM OF PHASE TRANSITIONS

The prescription in equilibrium statistical mechanics is to calculate the canonical partition function

$$Z = \int e^{-\beta \mathcal{H}} \, d\Gamma / N!, \tag{4.1}$$

which is simply the normalizing constant for the Gibbs canonical distribution (3.9) ($\beta = (kT)^{-1}$, k is Boltzmann's constant, T the absolute temperature and $\mathcal{H}$ is the Hamiltonian of the system). Once Z has been calculated, all the thermodynamic quantities of interest can be found by differentiation. For example, in terms of the free energy Ψ defined by

$$-\beta\Psi = \log Z, \tag{4.2}$$

the average energy of the system is given by

$$E = \Psi - T\left(\frac{\partial \Psi}{\partial T}\right) \qquad \text{and} \tag{4.3}$$

the specific heat (at constant volume) by

$$C_v = \frac{\partial}{\partial T} E , \tag{4.4}$$

etc.

In the real world many systems are observed to undergo what is called a phase transition. For example, if a gas is compressed at sufficiently low temperatures, it reaches a certain density and condenses at constant pressure into a liquid. Beyond a certain critical temperature T_c, however, the gas remains a gas no matter how much it is compressed. Similarly, if a magnet is heated, the spontaneous magnetization decays abruptly to zero at a critical (Curie) temperature T_c. In all cases studied experimentally, a number of interesting things happen in the neighborhood of T_c; for example, the specific heat diverges. Mathematically we would describe such situations by saying that T_c is a singular point of the free energy. If we go back to the definitions (4.1) and (4.2), however, we see immediately that for a finite system Ψ is a completely analytic function of T (Z being an integral of an analytic function over a finite domain). To have any hope then of describing mathematically what goes on in the neighborhood of a phase transition point, we must first go to the limit of an infinite system. Specifically, we are interested, for a system composed of N particles in a domain with volume V, in the thermodynamic limit

$$-\frac{\psi}{kT} = \lim_{\substack{N,V\to\infty \\ N/V=\rho \text{ fixed}}} N^{-1} \log Z \quad , \tag{4.5}$$

where ρ is the density and ψ is now the limiting free energy per particle. ψ has been shown to exist, Thompson (1972), for a large class of systems and to be a function only of ρ and T. Unfortunately, except in very special cases, very little is known about the analytic form of ψ. Perhaps the most interesting case as far as phase transitions are concerned is Onsager's celebrated solution of the two-dimensional Ising model in which it is found that the (zero field) specific heat diverges logarithmically at a finite temperature.

Since the Ising model is one of the most studied models in physics, let me briefly describe it here. Consider a lattice, for example, a two-dimensional square lattice, with "spins" $\mu_i = \pm 1$ located on the vertices. Assuming that the spins interact pair-wise with interaction constant J_{ij} between spins on the ith and jth sites and with an external magnetic field H, the Hamiltonian (or interaction energy) for the system in a given configuration $\{\mu\} = (\mu_1, \mu_2, \ldots, \mu_N)$ of spins is given by

$$\mathcal{H} = -\sum_{i<j} J_{ij}\mu_i\mu_j - H\Sigma\mu_i \quad . \tag{4.6}$$

By analogy with Equation (4.1) the partition function is given by

$$Z = \sum_{\{\mu\}} \exp(-\beta\mathcal{H}) \quad , \tag{4.7}$$

where the sum is over all (2^N) configurations of spins. The problem Onsager solved was the zero field $(H = 0)$ two-dimensional case with only nearest neighbor spins interacting ($J_{ij} = 0$ unless i and j are nearest neighbor lattice sites).

An interesting type of Ising model was constructed recently by Dyson (1969) to examine the question of the existence of a phase transition in one dimension. I mention it here because it leads to an interesting non-linear problem. The particular model, called the hierachical model, is as follows. Consider a chain of 2^N spins $\mu_i = \pm 1$ with Hamiltonian

$$\mathcal{H} = - \sum_{p=1}^{N} C_p \sum_{r=1}^{2^{N-p}} (s_{p,r})^2 - H \sum_{i=1}^{2^N} \mu_i \quad , \tag{4.8}$$

where $s_{p,r}$ is the sum of the rth block of 2^p spins,

$$s_{N,1} = \sum_{i=1}^{2^N} \mu_i, \quad s_{N-1,1} = \sum_{1}^{2^{N-1}} \mu_i, \quad s_{N-1,2} = \sum_{2^{N-1}+1}^{2^N} \mu_i \text{ , etc.} \tag{4.9}$$

The model is shown pictorially in the figure below for 8 spins.

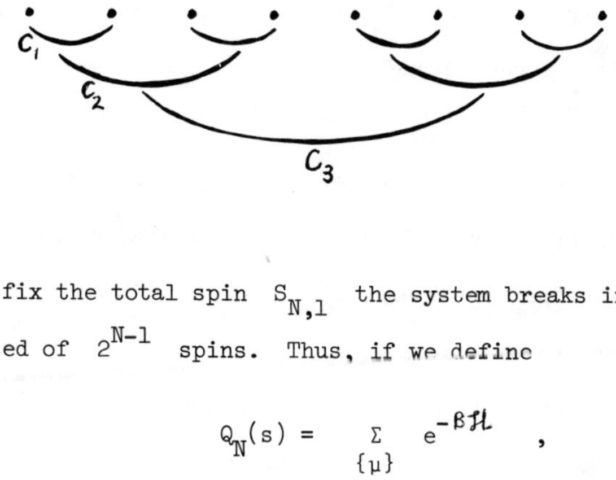

Obviously, if we fix the total spin $S_{N,1}$ the system breaks into two identical subsystems composed of 2^{N-1} spins. Thus, if we define

$$Q_N(s) = \sum_{\substack{\{\mu\} \\ s_{N,1}=s}} e^{-\beta \mathcal{H}} \quad , \tag{4.10}$$

it follows from (4.8) that

$$Q_N(s) = e^{\beta C_N s^2} \sum_{s'} Q_{N-1}(s') Q_{N-1}(s - s') \quad . \tag{4.11}$$

Using the identity

$$e^{\frac{a^2}{2}} = (2\pi)^{-\frac{1}{2}} \int_{-\infty}^{\infty} e^{-\frac{x^2}{2} + ax} dx \quad , \tag{4.12}$$

with $a = s(2\beta C_N)^{\frac{1}{2}} = sd_N$ in (4.11), we obtain

$$Q_N(s) = (2\pi)^{-\frac{1}{2}} \int_{-\infty}^{\infty} e^{-\frac{x^2}{2}} \sum_{s'} Q_{N-1}(s')e^{xd_N s'} Q_{N-1}(s - s')e^{xd_N(s-s')} \quad . \tag{4.13}$$

Summing $Q_N(s)$ over s gives the partition function $Z_N(h,\beta)$ for the system with $h = \beta H$. A simple calculation then gives

$$Z_N(h,\beta) = (2\pi)^{-\frac{1}{2}} \int_{-\infty}^{\infty} e^{-\frac{x^2}{2}} [Z_{N-1}(h + xd_N,\beta)]^2 dx \quad . \tag{4.14}$$

Of particular interest is the case $C_p = 2^{-\alpha p}$ where Dyson has shown that a limiting free energy exists, that there is a phase transition when $h = 0$ and $1 < \alpha < 2$ and no phase transition when $\alpha > 2$. It would be very nice indeed to see this directly from (4.14) and to deduce the analytic form of ψ in the neighborhood of its singular point without actually solving the non-linear recursion relation (4.14).

An equation similar to (4.14) has recently been derived by Wilson (1971) for a different class of spin systems. This equation, unlike (4.14), is only approximate, but since it has caused such a stir it would be well worth a detailed investigation.

It may be that the ultimate framework for the study of phase transitions will be something non-linear. Certainly a number of approaches of this type have appeared in recent years. Apart from Wilson's work there is an approach due to Grmela (1971) and others which starts with a kinetic type equation and emerges with a phase transition point as a point of bifurcation. The difficulty or draw-back with this approach, of course, is that the starting point is very difficult to justify. The correct starting point, without a doubt, should be the Gibbs canonical distribution. Another approach does precisely that, incorporating along the way some of Thom's structural stability ideas (in his theory of morphogenesis). The deficiency in this approach is that the picture is only a qualitative one and therefore has no hope of telling us what we really want to know - the analytic form of the free energy in the neighborhood of its singular point.

At this point we leave statistical mechanics and conclude with some non-linear problems in biology.

5. SOME NON-LINEAR PROBLEMS IN BIOLOGY

In biology we are confronted with complex systems with complex interactions: substances combining together to give other substances, species competing for survival and so forth. In view of the complexity of real systems, it is virtually out of the question to construct theories from first principles, even assuming we know what the first principles are. The only thing to do then is to construct model systems and investigate their behavior. We are then at the mercy of the biologists who more often than not claim (and rightly so) that the models do not take account of all the facts. Nevertheless the hope is that by studying oversimplified model systems some insight and understanding may be gained into the behavior of real systems.

In competing and cooperating systems we are usually led to consider non-linear rate equations of the general form

$$\frac{dc_j}{dt} = k_j c_j + \sum_{\ell,m} k^{(j)}_{\ell,m} c_\ell c_m - c_j \sum_\ell k^{(j)}_\ell c_\ell \quad , \tag{5.1}$$

where c_j, $j = 1, 2, \ldots, N$ represent the concentrations (or numbers) of objects, which may be chemical substances undergoing a chain of chemical reactions or species competing for survival, etc. The first term in Equation (5.1) represents the rate at which the jth object grows without interacting with the remaining objects, the second term represents the growth of j through interactions with all ℓ and m which might produce j and the third term, the loss of j through interactions with the remaining objects.

A clasic example of (5.1) was formulated by Volterra (Goel, Maitra & Montroll (1971)), after discussions with his friend D'Ancona on statistics of fish catches in the Adriatic. It was observed that the population of two species of fish varied with the same period, but somewhat out of phase. The larger species of fish, which we will label 1, ate the smaller one, 2, grew and multiplied until there were insufficient numbers of small fish available for eating. The larger ones then diminished in numbers, the smaller ones prospered and the whole cycle began afresh. Volterra described this situation by two equations

$$\frac{dN_1}{dt} = -\alpha_1 N_1 + \lambda_1 N_1 N_2 \ ,$$
$$\frac{dN_2}{dt} = \alpha_2 N_2 - \lambda_2 N_1 N_2 \ , \qquad (5.2)$$

where N_1 and N_2 represent the population of large (predator) and small (prey) fish,respectively. The term $\lambda_1 N_1 N_2$ represents the gain rate of the large fish due to "collisions" with the small ones and $-\lambda_2 N_1 N_2$ the loss rate of the small ones through the same collisions. In this model, species 1 would die out exponentially in the absence of species 2, and 2 would grow exponentially in the absence of species 1. Prior to Volterra, Lotka had investigated these equations in the theory of autocatalytic chemical reactions as well as in the theory of competing species.

If we denote the stationary solutions of (5.2) by q_1 and q_2, we immediately obtain

$$q_1 = \frac{\alpha_2}{\lambda_2} \text{ and } q_2 = \frac{\alpha_1}{\lambda_1} . \qquad (5.3)$$

Defining v_j by

$$N_j = q_j \exp(v_j) \quad j = 1, 2 \qquad (5.4)$$

and substituting in (5.2) gives

$$\frac{dv_1}{dt} = \alpha_1 (e^{v_2} - 1) \quad , \qquad (5.5a)$$

$$\frac{dv_2}{dt} = -\alpha_2 (e^{v_1} - 1) \quad . \qquad (5.5b)$$

If we then multiply (5.5a) by $\alpha_2(e^{v_1} - 1)$ and (5.5b) by $\alpha_1(e^{v_2} - 1)$ and add, we obtain

$$\alpha_2(e^{v_1} - 1)\frac{dv_1}{dt} + \alpha_1(e^{v_2} - 1)\frac{dv_2}{dt} = \frac{d}{dt}[\alpha_2(e^{v_1} - v_1) + \alpha_1(e^{v_2} - v_2)] = 0 .$$

It follows that the quantity G defined by

$$G = \alpha_1(e^{v_2} - v_2) + \alpha_2(e^{v_1} - v_1) \qquad (5.6)$$

is a constant (of the motion). In terms of the quantities $f_i = N_i/q_i = e^{v_i}$, (5.6) becomes

$$u_1 u_2 = \text{const.} \quad , \tag{5.7}$$

where

$$u_i = (f_i e^{-f_i})^{\frac{1}{\alpha_i}} \quad . \tag{5.8}$$

Given condition (5.7), it is now an easy matter to show the periodic character of the solution. This is most easily done by drawing a few pictures.

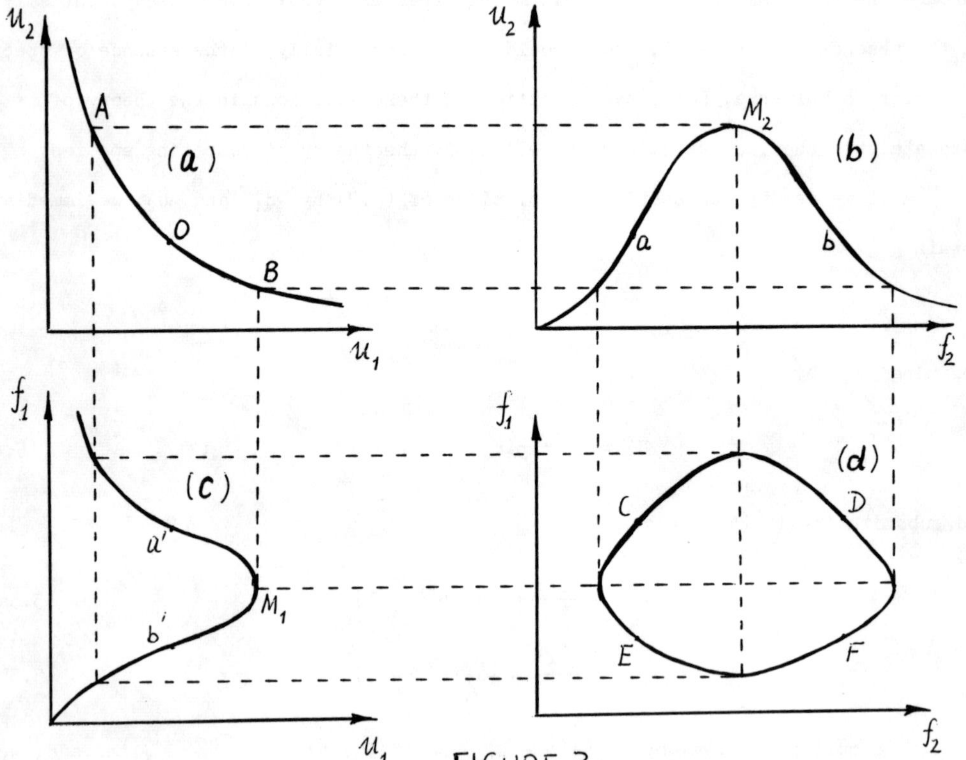

FIGURE 3

In Figure 3a we have plotted the hyperbola (5.7). Figures b and c show the behavior of u_2 and u_1 as functions of f_2 and f_1, respectively. It is important to note that u_2 and u_1 attain maxima, identified by M_2 and M_1, in Figures b and c. It follows that the relevant region of a is bounded by the points A and B. Note also that the point 0 between A and B corresponds to two values of f_2(a and b) and two values of f_1 (a' and b'). Hence in Figure d, which relates f_1 and f_2, the point 0 corresponds to four points, C, D, E, and F.

As one goes from A to B in Figure a, one traces out the closed curve in Figure d.

From (5.2) and the definition of q_1 and q_2 (5.3), it follows that the initial conditions $N_i = 0$ and $N_i(0) = q_i$ yield populations which remain at their initial values for all time. The first set is clearly of no interest. The second set however, represents a true equilibrium state of the system. Any other set of initial populations yields periodic population variations, typically shown in Figure 4.

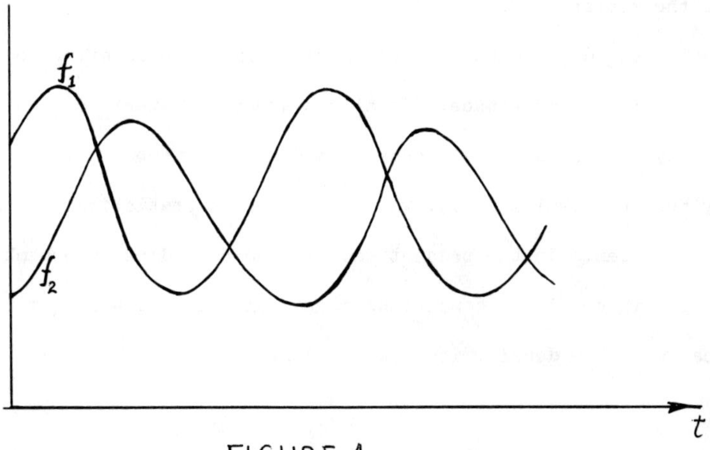

FIGURE 4.

For n competing species, Volterra considered the generalization of (5.2),

$$\frac{dN_i}{dt} = k_i N_i + \beta_i^{-1} \sum_{j=1}^{n} a_{ij} N_i N_j \quad . \tag{5.9}$$

Since one species gains and another one loses during a collision, it is clear that a_{ij} and a_{ji} have opposite signs. The positive quantities β_i^{-1} were called equivalence numbers by Volterra. During binary collisions of i and j, the ratio of i's lost (or gained) per unit time to j's gained (or lost) is $\beta_i^{-1}/\beta_j^{-1}$. With this definition

$$a_{ij} = -a_{ji} \quad i \neq j \quad . \tag{5.10}$$

We will further assume that $a_{ii} = 0$ so that the matrix with entries a_{ij} is antisymmetric. If $q_1 \ldots q_n$ are non-zero positive stationary solutions of (5.9) and $v_j = \log(N_j/q_j)$ (Eq. (5.4)), an elementary calculation shows that

$$\frac{d}{dt} \sum_j \beta_j q_j (e^{v_j} - v_j) = \sum_{i,j} a_{ji} q_j q_i (e^{v_j} - 1)(e^{v_i} - 1) \quad . \tag{5.11}$$

Since $a_{ij} = -a_{ji}$, the right-hand side of (5.11) vanishes, and it follows that

$$G = \sum_j \beta_j q_j (e^{v_j} - v_j) \tag{5.12}$$

is a constant of the motion (c.f. (5.6)).

The fact that G is a sum of individual terms, each relating to a different species, is of considerable importance. It allows a natural specification of the components of the system in the usual sense of statistical mechanics. This fact was exploited by Kerner (1961 & 1971), who constructed a statistical mechanics of the Volterra-Lotka system. In the present context one considers an ensemble of systems with all possible initial conditions compatible with $G = G(0) = \text{const.}$ The microcanonical ensemble density is then given by

$$D_{m.c.} = \delta(G - G(0))/S(G(0)) \ , \tag{5.13}$$

where $S(G(0))$ is the area of the surface $G = G(0)$. One can justify this postulate in the same way as before and proceed to calculate averages, fluctuations, etc. This is all very nice, but what is lacking is a microscopic description which we saw in statistical mechanics was necessary before contemplating a macroscopic description. We would be on safe grounds if G were in some sense a Hamiltonian, but this is definitely not the case. In fact it is clear that $G = \text{const.}$ is an accidental feature of the Volterra-Lotka system. If we introduce a damping term into the Volterra equations so that a population will not explode in the absence of others, we would presumably have a more realistic description or model (such models have in fact been considered - see Goel, Maitra & Montroll (1971)). The introduction of a damping term, however, in general destroys the constancy of G and along with it the whole statistical mechanical structure. The basic difficulty of course is that a system with a constant of motion is not structurally stable; i.e., it is extremely sensitive tc small perturbations in the structure of the equations.

We conclude this section with another set of non-linear rate equations proposed recently by Manfred Eigen (1971) to describe the "Selforganization of Matter and the Evolution of Biological Macromolecules." Eigen's idea was to start with a set of objects that reproduce themselves and mutant copies of themselves and show that under certain conditions only one object will survive or select out of the system.

If we let x_i, $i = 1, 2, \ldots, N$ represent the concentrations of the N objects in question, Eigen's phenomenological equations for the evolution of the system are

$$\frac{dx_i}{dt} = (F_i - R_i)x_i + \sum_{\ell \neq i} \phi_{i\ell} x_\ell \quad . \tag{5.14}$$

F_i is the formation rate for self-reproduction, R_i a removal rate, and the third term comes from imprecise copying of the remaining objects. F_i and R_i may (and will) depend on $x_1, x_2, \ldots, x_N$. To simplify (5.14) Eigen writes

$$F_i = A_i Q_i \quad , \tag{5.15}$$

where A_i is a rate constant describing all formations from i, and Q_i $(0 \leq Q_i \leq 1)$ is a quality factor which counts the fraction of processes leading to precise copying of i. We further write

$$R_i = D_i + \phi_{oi} \quad , \tag{5.16}$$

where D_i represents a decomposition rate and ϕ_{oi} a dilution rate. It will be assumed that

$$\phi_{oi} = \Phi_o / \sum_1^N x_i \tag{5.17}$$

when Φ_o is the "total dilution flow", which is chosen to compensate for the overall excess production, i.e.,

$$\Phi_0 = \sum_{k=1}^{N} (A_k - D_k) x_k \quad . \tag{5.18}$$

Combining the above definitions, (5.14) becomes

$$\frac{dx_i}{dt} = (W_i - E)x_i + \sum_{\ell \neq i} \phi_{i\ell} x_\ell \ , \tag{5.19}$$

where

$$W_i = A_i Q_i - D_i \qquad \text{("selective value")}$$

$$E = \sum_{k=1}^{N} E_k x_k \Big/ \sum_{k=1}^{N} x_k \qquad \text{("mean productivity")} \tag{5.20}$$

$$E_k = A_k - D_k \qquad \text{("excess productivity")} \ .$$

To obtain "selective strains" Eigen imposes the following constraints on the system, termed constant overall organization,

(a) $$\sum_{k=1}^{N} x_k = n \quad \text{(constant)} \tag{5.21}$$

(b) The concentrations of the "energy rich" units being fed into the system are held fixed.

We will take (b) to mean that W_i and E_i are constants (independent of the x_i's and time). Note that since (5.19) is homogeneous in the x_i's, we can, without loss of generality, take $n = 1$ in Eq. (5.21).

The problem now is to find out under what conditions a "master" will emerge or select out at the expense of the rest. That is, under what conditions will, for some m,

$$x_m \to 1 \ , \quad x_i \to 0 \qquad i \neq m \quad \text{as} \quad t \to \infty \ .$$

Eigen asserts that this will indeed be the case if $W_m = \max_{1 \leq i \leq N} W_i$; i.e., the object with the largest selective value will win out, regardless of the mutation rates $\phi_{i\ell}$, as long as they are "sufficiently small". Unfortunately, Eigen's assertion is only correct under very special conditions, in fact, under conditions that are much too severe to be of interest. We will see in a moment, however, that the assertion is almost correct - the object with the largest selective value predominates, but not at the expense of the rest.

To see what is going on, let us suppose that $\bar{x}_1, \bar{x}_2, \ldots, \bar{x}_N$ is a stationary solution of (5.19) with $\Sigma\bar{x}_i = 1$, i.e.,

$$(W_i - \sum_{k=1}^{N} E_k\bar{x}_k)\bar{x}_i + \sum_{\ell\neq i} \phi_{i\ell}\bar{x}_\ell = 0 \qquad i = 1, 2, \ldots, N \ . \tag{5.22}$$

If $\bar{x}_\ell = 0 \quad \ell \neq m$, setting $i = m$ in (5.22) gives

$$(W_m - E_m\bar{x}_m)\bar{x}_m = 0 \ . \tag{5.23}$$

Since $\Sigma\bar{x}_i = \bar{x}_m = 1$, it follows that

$$W_m - E_m = A_m(Q_m - 1) = 0, \text{ i.e., } Q_m = 1 \quad , \tag{5.24}$$

or m is a perfect copier. Setting $i \neq m$ in (5.22) we obtain $\phi_{im} = 0, \quad i \neq m$, which is obvious since m is perfect. The interesting case is clearly when no $Q_i = 1$.

Returning to (5.19) we see that if we sum on i and use the fact that $\Sigma x_i = 1$,

$$\sum_{i=1}^{N} (W_i - E_i)x_i + \sum_{\ell\neq 1}\sum \phi_{i\ell}x_\ell = 0 \ . \tag{5.25}$$

It follows that if

$$\sum_{\substack{\ell=1 \\ (\ell\neq i)}}^{N} \phi_{\ell i} = E_i - W_i = A_i(1 - Q_i) \quad , \tag{5.26}$$

(5.25) will be satisfied, and, moreover, if $x_i(t)$ is any solution of (5.19) satisfying $\Sigma x_i = 1$ at $t = 0$, $\Sigma x_i = 1$ for all t.

We now derive an exact solution of (5.19). If we write

$$x_i(t) = e^{-\int_0^t E(\tau)d\tau} z_i(t) \tag{5.27}$$

and substitute into (5.19), we obtain

$$\frac{dz_i}{dt} = \sum_{\ell=1}^{N} A_{i\ell}z_\ell \quad , \tag{5.28}$$

where

$$A_{i\ell} = \begin{cases} W_i & i = \ell \\ \phi_{i\ell} & i \neq \ell \end{cases} \quad . \tag{5.29}$$

(5.28) can be solved immediately and we obtain

$$z_i = \sum_{\ell=1}^{N} a_{i\ell} e^{\lambda_\ell t} \quad , \tag{5.30}$$

where $\lambda_1, \lambda_2, \ldots, \lambda_N$ are the eigenvalues of the matrix $A = (A_{ij})$, assumed for convenience to be distinct, and $a_{i\ell}$ are constants to be determined from the initial conditions $(z_i(0) = x_i(0))$. Having found $z_i(t)$, we return to (5.19) and write

$$E = \sum_{k=1}^{N} E_k x_k = x_i \sum_{k=1}^{N} E_k \left(\frac{x_k}{x_i}\right) = x_i \sum_{k=1}^{N} E_k z_{ki}(t) \quad , \tag{5.31}$$

where from (5.20)

$$Z_{ki}(t) = \frac{x_k}{x_i} = \frac{z_k}{z_i} \tag{5.32}$$

is a <u>known</u> function of t.

Similarly, we can write

$$\sum_{\ell \neq i} \phi_{i\ell} x_\ell = x_i \sum_{\ell \neq i} \phi_{i\ell} Z_{\ell i} \quad . \tag{5.33}$$

The functions

$$E_i(t) = \sum_{k=1}^{N} E_k Z_{ki}(t), \quad \Phi_i(t) = \sum_{\ell = i} \phi_{i\ell} Z_{\ell i}(t) \tag{5.34}$$

are known functions of t, and (5.19) can be written in the form

$$\frac{dx_i}{dt} - g_i(t)x_i + E_i(t)x_i^2 = 0, \tag{5.35}$$

where

$$g_i(t) = W_i + \Phi_i(t) \tag{5.36}$$

and $E_i(t)$ is defined by (5.34).

(5.35) is a Bernoulli equation with solution

$$x_i(t) = \{e_i(t) \int^t E_i(t)(e_i(t))^{-1} dt\}^{-1} \quad , \tag{5.37}$$

where

$$e_i(t) = \exp(-\int^t g_i(\tau)d\tau) \quad , \tag{5.38}$$

and the lower limits of integration are to be fixed by the initial conditions.

To investigate (in a rough way) the asymptotic behavior of the x_i's, let us assume that $W_1 > W_2 > \ldots W_N$ and that $\phi_{i\ell} > 0$. In the extreme case when all $\phi_{i\ell} = 0$ (which is of no interest), it is clear from (5.28) and (5.29) that

$$z_i = e^{W_i t} \quad . \tag{5.39}$$

In the general case, from (5.30)

$$z_i \sim a_{i1} e^{\lambda_1 t} \quad ,$$

where λ_1 is the maximum eigenvalue of the matrix (5.29). Clearly, when the ϕ's are small, $\lambda_1 \sim W_1$, $a_{i1} << a_{11}$, and in a rough sense a_{i1}/a_{11} will be a measure of how much i mutates compared with 1. It is easy to see from (5.32) that

$$\frac{x_i}{x_1} \sim \frac{a_{i1}}{a_{11}} (<<1 \quad i \neq 1) \quad \text{as} \quad t \to \infty \tag{5.40}$$

and that unless $a_{i1} = 0$, species i will not die out. This all very rough, but it is clear what is going on.

1. The maximum eigenvalue of A (Eq. (5.29)) determines which species will predominate (for small ϕ's this will correspond to the one with the largest selection value W_i - as asserted by Eigen).

2. Except under very special circumstances (which remain to be investigated thoroughly), no species will completely die out, the concentration relative to the "master" being determined by mutation rates which are related through (5.26) to quality factors.

REFERENCES

Carleman, T., "Problèmes Mathématiques dans la Théorie Cinétique des Gaz" (Almquist and Wiksells, Uppsala, 1957).

Dyson, F.J., Commun. Math. Phys. 12, 91, 212 (1969).

Eigen, M., Die Naturwissenschaften 58, 33 (1971).

Goel, N.S., Maitra S.C., and Montroll, E.W., Rev. Mod. Phys. 43, 231 (1971).

Grad, H., Symp. Appl. Math. 17, 154 (1965).

Grad, H., Commun. Pure Appl. Math. 5, 455 (1952). See also Thompson, C.J. (1972) and Uhlenbeck, C.J. (1972).

Grünbaum, F.A., Trans. Amer. Math. Soc. 165, 425 (1972).

Grünbaum, F.A., Arch. Rat. Mech. & Anal. 42, 323 (1971).

Grmela, M., J. Stat. Phys. 3, 347 (1971).

Hilbert, D., "Gründzuge einer allgemeinen Theorie der linearen Integralgleichugen" (Teubner, 1912) p.27.

Huang, K., "Statistical Mechanics" (Wiley, 1963).

Kerner, E.H., Bull. Math. Biophys. 12, 121 (1957); 21, 217 (1959); 23, 141 (1961); "Gibbs Ensembles: Biological Ensembles" (Gordon Breach, 1971).

Khinchin, A.I., "Mathematical Foundations of Statistical Mechanics" (Dover, 1949).

Lanford, O., Commun. Math. Phys. 9, 176 (1968); 11, 257 (1969).

McKean, H.P., Arch. Rat. Mech. & Anal. 21, 343 (1966).

Siegert, A.J.F., Phys. Rev. 76, 1708 (1949).

Thompson, C.J. "Mathematical Statistical Mechanics" (Macmillan, New York, 1972).

Uhlenbeck, G.E., Phys. Rev. 62, 11, 467 (1942).

Uhlenbeck, G.E., and Ford, G., "Lectures in Statistical Mechanics" (AMS, Providence, R.I., 1963).

Wightman , A.S., in "Statistical Mechanics at the Turn of the Decade", E.D.G. Cohen, Ed. (Dekker, 1971).

Wild, E., Proc. Camb. Phil. Soc. 47, 602 (1951).

Wilson, K., Phys. Rev. 4B, 3184 (1971). See Equations (3.40)-(3.42).

STABILITY PROPERTIES AND PERIODIC BEHAVIOR OF CONTROLLED BIOCHEMICAL SYSTEMS

Charles Walter
Departments of Biomathematics and Biochemistry
University of Texas M. D. Anderson Hospital
Houston, Texas 77025

INTRODUCTION

Recent discoveries about the detailed biochemical operation of certain metabolic control mechanisms provide a sound basis for studying the dynamics of complex biochemical systems. Perhaps the most fundamental discovery is that highly specific control mechanisms operate through numerous feedback devices to regulate the concentrations of metabolites or the activities of enzymes. The establishment of specific closed loops in the biochemical organization of cells provides the detail necessary for an understanding of metabolic control mechanisms and for studies of their dynamical properties.

Metabolic sequences consist of a series of consecutive, reversible, enzyme-catalyzed chemical reactions. In such a set of reactions the formation of substance S_{i+1} from S_i is catalyzed by enzyme E_i via one or more enzyme-substrate compounds. If the S exchange with the "exterior" but the enzyme species are confined to the "interior", the system is said to be "open" with respect to the S but "closed" with respect to the catalysts. A non-equilibrium stationary state for the interior system can be maintained by a large external reservoir of S_o (the zeroth substrate). Under these conditions the concentration of S_o in the interior is maintained constant.

In general, the rate of an enzyme-catalyzed reaction depends on many things. At constant temperature and pressure the rate may still depend on the concentration of practically any component in the environment. Fortunately, most enzymes are relatively insensitive to most of their usual environmental chemicals. However, in the case of every enzyme that has been studied so far, the activity of the enzyme depends on the concentration of the substrate of the catalyzed reaction. Usually it is found that there is a limiting rate obtained at very high substrate concentration. If this asymptotic rate is also the highest possible rate, it is often called the "maximum velocity" or "Vm". One possible mathematical representation of these general findings is to express the rate of enzyme-catalyzed reactions as a ratio of polynomials in the substrate concentrations (Botts, 1958). Since enzyme-catalyzed rates are usually asymptotic at high substrate concentrations, the degree of the polynomial in the numerator is greater than the degree of the denominator polynomial; since enzymes cannot convert substrates unless the substrate is present, there is usually no constant term associated with the numerator polynomial:

$$\text{rate} = F(S) = \frac{\sum_{i=1}^{p} e_i S^i}{1 + \sum_{i=1}^{q} c_i S^i} \quad . \tag{1}$$

For many enzyme-catalyzed reactions these polynomials can be abbreviated, and the rate can be approximated by the familiar hyperbolic function of the substrate concentration,

$$F(S) = \frac{e_1 S}{1 + c_1 S} \quad , \tag{1a}$$

where e_1 is the ratio of Vm to the so-called "Michaelis constant" ("Km") and $c_1 = 1/Km$ (Briggs and Haldane, 1925). Further simplification of the equation describing the relationship between the reaction rate and the substrate concentration can be accomplished by linearizing $F(S)$:

$$F(S) = K\,S \quad . \tag{1b}$$

In equation (1b), K is $\frac{\partial F}{\partial S}$ evaluated at the concentration of S where the linearization is to be used. For sufficiently small S the linearization will have a relatively large neighborhood, and $K = e_1$. For hyperbolic rates $K \leq e_1$, but for rates wherein F(S) experiences an inflection or an extremum, K could be larger than e_1 .

If we suppose (1) that the rate of loss of each S_i except S_n to the exterior is slow compared to the rate of conversion to S_{i+1} and (2) that the rate of each catalyzed step is unaffected by any component other than its substrate, then the dynamical description of a metabolic sequence of enzyme-catalyzed reactions is

$$\begin{aligned} \dot{S}_1 &= F_o(S_o) - F_1(S_1) \\ &\vdots \\ \dot{S}_i &= F_{i-1}(S_{i-1}) - F_i(S_i) , \quad \underline{i} = 2, \ldots, \underline{n} . \end{aligned} \tag{2}$$

Since S_o is constant,

$$\begin{aligned} \dot{S}_1 &= K_o - F_1(S_1) \\ &\vdots \\ \dot{S}_i &= F_{i-1}(S_{i-1}) - F_i(S_i) , \quad \underline{i} = 2, \ldots, \underline{n} . \end{aligned} \tag{2a}$$

Using the definition for the linearized reate (F(S) in equation 1b), we obtain

$$\begin{aligned} \dot{S}_1 &= K_o - K_1 S_1 \\ &\vdots \\ \dot{S}_i &= K_{i-1} S_{i-1} - K_i S_i , \quad \underline{i} = 2, \ldots, \underline{n} . \end{aligned} \tag{2b}$$

As long as the rate constants in equations 2b are non-negative real numbers, these equations 2b cannot exhibit instabilities, nor can the variables experience sustained oscillations (Walter, 1972).

A number of metabolic sequences include a catalyzed step that is subject to feed back by a downstream product. For our mechanisms we choose the step catalyzed by E_o as the step at which the feedback occurs and S_n as the product that mediates the feedback. Then, instead of equations 2b we obtain

$$\begin{aligned} \dot{S}_1 &= f(S_n) - K_1 S_1 \\ &\vdots \\ \dot{S}_i &= K_{i-1} S_{i-1} - K_i S_i , \qquad \underline{i} = 2,\ldots,\underline{n} , \end{aligned} \tag{3}$$

where $f(S_n)$ is the nonlinear function that describes the effect of S_n on the step catalyzed by E_o. Equations 3 provide a sound basis for studies of the stability and dynamical properties of a wide class of metabolic control systems (Walter, 1970, 1972).

<u>Asymptotic Stability in the Small of Linearized Biochemical Control Systems</u>

If equations 3 describe real chemical systems, there must be at least one real non-negative singular point (S*). Linearization in the neighborhood of S* leads to the abridged equations,

$$\begin{aligned} \dot{\xi}_1 &= a\,\xi_n - K_1 \xi_1 \\ &\vdots \\ \dot{\xi}_i &= K_{i-1} \xi_{i-1} - K_i \xi_i , \qquad \underline{i} = 2,\ldots,\underline{n} , \end{aligned} \tag{4}$$

where $\xi = S - S^* = 0$ and a is the linearized feedback term. If $a < 0$ the feedback is referred to as negative, but if $a > 0$ we call the feedback positive. Obviously, this distinction between linearized negative and positive feedback should be used only in the neighborhood where the linearization has been carried out.

The characteristic equation for the abridged equations is

$$\prod_{i=1}^{n} (K_i + \mu) - a \prod_{i=1}^{n-1} K_i = 0 , \tag{5}$$

If we set the characteristic exponents, $\mu=0$, we obtain the "equation of zero roots":

$$K_n = a. \tag{6}$$

If we set $\mu = i\omega$ for various $\underline{n}$, we obtain the series of "neutrality equations":

For $\underline{n} = 2$: $\quad K_1 + K_2 = 0,$

For $\underline{n} = 3$: $\quad (K_1 + K_2)(K_1 + K_3)(K_2 + K_3) + K_1K_2a = 0,$ (7)

etc. (Walter, 1969). With the aid of these equations we can partition the plane of the parameters, a and K_n, into areas for which the real part of at least one characteristic exponent changes sign. In Figure 1 appear the "line of zero roots" (solid lines) and the "neutrality line" (dashed lines) for $\underline{n}$ = 2, 3, 4 and 5. In the shaded sections of Figure 1, the real part of at least one characteristic exponent is positive; the shaded sections therefore denote areas wherein the linearized system is unstable. For values of a and K_n in the unshaded areas, the singularity ξ = 0 is asymptotically stable in the small provided the real parts of all the other characteristic exponents are also negative.

Since K_n is a non-negative rate constant, we can limit our attention to the first and fourth quadrants of Figure 1. For $\underline{n}$ = 2 there is no unshaded area in the fourth quadrant. Thus, negative feedback systems are always asymptotically stable in the neighborhood of ξ = 0 when there are only two kinetically important steps. There is, however, an area of instability for positive feedback systems; the size of this area is independent of the number of variables in the control system.

For $\underline{n}$ = 3 there is an area of instability in the fourth quadrant. This means that negative feedback systems could have unstable singularities when there are three kinetically important steps. As $\underline{n}$ gets larger, the area of instability in the fourth quadrant also gets larger. This suggests that the potential for oscillatory behavior in negative feedback systems might become greater as the number of kinetically important steps gets larger.

The Global Asymptotic Stability of Biochemical Control Systems with Nonlinear Feedback

1. Derivation of the Transformed Equations

The Lur'e transformation (Lur'e, 1951) of equations 3 can be obtained by the methods outlined by Letov (1961). The canonical variables will be determined with the aid of the equations

$$\eta_i = \sum_{j=2}^{n} \beta_j^{(i)} \xi_j + \xi_1, \qquad \underline{i} = 2,3,\ldots,\underline{n}. \qquad (8)$$

In order to obtain the equations involving the new variables in the canonical form

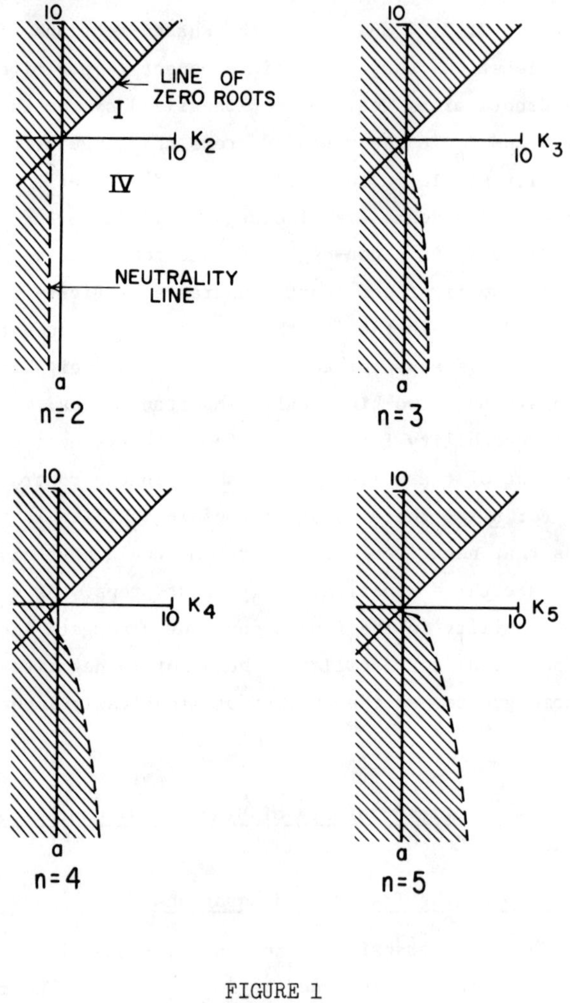

FIGURE 1

<u>Figure 1</u>. Partition of the plane of the linear feedback term, a , and K_n into areas of instability (shaded portion) and asymptotic stability (unshaded portion). The solid line is the line of zero roots (equation 6), and the dashed line is the neutrality line (for example, equations 7 - 8) .

$$\dot{\eta}_i = -G(\xi_n) - K_i\eta_i, \qquad \underline{i} = 1,2,\ldots,\underline{n}, \qquad (9)$$

the transformation constants must satisfy the relations

$$-K_i\beta_k^{(i)} = \sum_{j=2}^{n} B_{jk}\,\beta_j^{(i)}, \qquad \underline{k},\underline{i} = 2,3,\ldots,\underline{n}, \qquad (10)$$

$$K - K_i = K_1\beta_1^{(i)}, \qquad \underline{i} = 2,3,\ldots,\underline{n}, \qquad (11)$$

where $B_{jk} = -K_k$ if $j=k$, $B_{jk} = K_k$ if $j=k+1$ and all other $B_{jk} = 0$ $G(\xi_n)$ is any function satisfying the conditions

$$G(\xi_n) = 0 \text{ for } |\xi_n| \le \xi_n^{\#}$$
$$\xi_n\, G(\xi_n) > 0 \text{ for } |\xi_n| < \xi_n^{\#}. \qquad (12)$$

In chemical systems $-G(\xi_n)$ are functions that include the type describing negative chemical feedback; whereas $G(\xi_n)$ are functions usually associated with positive feedback.

For the chemical systems under consideration here equations 10 and 11 can always be solved provided the K_i are different; consequently the transformation described by equations 8 exists and is not singular. Because the transformation is not singular, it is possible to solve equations 8 for ξ_i. In particular, to complete the reduction of the initial equations to the canonical form, it is necessary to express ξ_n in terms of the new variables. After this step the canonical equations assume the final form

$$\dot{\eta}_i = G(\xi_n) - K_i\eta_i, \qquad \underline{i} = 1,2,\ldots,\underline{n},$$
$$\xi_n = \sum_{i=1}^{n} \frac{\prod_{j=1}^{n-1} K_j}{\prod_{j=1}^{n} (K_i - K_j),\ i \ne j}\, \eta_i \qquad (13)$$

where $\xi_1 = \eta_1$.

Equations 3 have the non-negative real solution S = S*. The corresponding solutions for the transformed equations are $\xi = \xi^* = 0$ and $\eta = \eta^* = 0$. These solutions have a one-to-one correspondence which can always be established through the transformation equations. The asymptotic stability of the equations involving the transformed variables means that the solution of the equations involving the original variables is a stationary solution that cannot experience sustained oscillations.

2. Construction of the Lyapunov Function

Methods for the construction of Lyapunov functions for the Lur'e (1951) transformed canonical equations have been described elsewhere (Letov, 1961). In what follows this general method is applied to the particular chemical control systems of interest here. In order to investigate the stability of the solution of equations 3, consider the Lyapunov function

$$V = \sum_{i=1}^{n} \sum_{j=1}^{n} \frac{a_j a_i}{K_j + K_i} \eta_j \eta_i + \int_0^{\xi_n} G(\xi_n)\, d\xi_n . \tag{14}$$

From the obvious equality

$$\frac{1}{K_j + K_i} = \int_0^{\infty} e^{-(K_j + K_i)\tau}\, d\tau , \tag{15}$$

it is clear that the first term on the r.h.s. of equation 14 is

$$\sum_{i=1}^{n} \sum_{j=1}^{n} \left(a_j a_i \int_0^{\infty} e^{-(K_j + K_i)\tau}\, d\tau \right) \eta_j \eta_i =$$

$$\int_0^{\infty} \left[\sum_{j=1}^{n} a_j \eta_j e^{-K_j \tau} \right]^2 d\tau. \tag{16}$$

For the chemical control systems under discussion here the integrand on the r.h.s. of equation 16 is the square of a real number provided the a_i are real numbers. Therefore, for arbitrary real a_i the first term on the r.h.s. of equation 14 is sign invariant and non-negative for all possible values of the η_i. Furthermore, if the K_i are all different, then for arbitrary real a_i, this term is positive definite for all possible values of the η_i.

3. Stability of the General Nonlinear System

In order to examine the stability of equations 3, we shall use the positive definite Lyapunov function in equation 14. According to Lyapunov's second theorem, the solution of equations 3 is asymptotically stable if the total derivative of V with respect to time is negative definite:

$$\frac{dV}{dt} = \sum_{j=1}^{n} \sum_{i=1}^{n} \frac{a_j a_i}{K_i + K_j} \left[\eta_i (G - K_j \eta_j) + \eta_j (G - K_i \eta_i) \right]$$

$$-G \sum_{i=1}^{n} \frac{K_i \prod_{j=1}^{n-1} K_j}{\prod_{j=1}^{n} (K_i - K_j),\ i \neq j} \eta_i \,. \tag{20}$$

Following the derivation suggested by Letov (1961), one obtains

$$\frac{dV}{dt} = - \left[\sum_{i=1}^{n} a_i \eta_i \right]^2 + G \left[- \frac{K_i \prod_{j=1}^{n-1} K_j}{\prod_{j=1}^{n} (K_i - K_j),\ i \neq j} + 2a_i \sum_{j=1}^{n} \frac{a_j}{K_i + K_j} \right] \eta_i \,. \tag{21}$$

Therefore dV/dt is negative definite if

$$- \frac{K_i \prod_{j=1}^{n-1} K_j}{\prod_{j=1}^{n} (K_i - K_j),\ i \neq j} + 2a_i \sum_{j=1}^{n} \frac{a_j}{K_i + K_j} = 0 \,, \qquad \underline{i} = 1,2,\ldots,\underline{n} \,. \tag{22}$$

Thus if equations 22 are satisfied by real a_i, it is possible to construct a sign definite Lyapunov function, V, having a sign definite Eulerian derivative, dV/dt, opposite in sign to V. Under these conditions it is possible to guarantee absolute asymptotic stability of the solution of equations 3. If the solution of a set of differential equations possesses global aymptotic stability, sustained oscillations about that solution are not possible. Thus, satisfaction of equations 22 by real a_i guarantees that sustained concentration oscillations will not occur in the chemical control systems described by equations 3.

4. Stability of the Nonlinear System: n = 2

For $\underline{n} = 2$, the canonical equations are

$$\dot{\eta}_1 = G(\xi_2) - K_1\eta_1$$
$$\dot{\eta}_2 = G(\xi_2) - K_2\eta_2 \qquad (13a)$$
$$\xi_2 = \frac{K_1}{K_1 - K_2}(\eta_2 - \eta_1).$$

In this case there are two sufficiency criteria:

$$-\frac{K_1^2}{K_1 - K_2} + 2a_1\left(\frac{a_1}{2K_1} + \frac{a_2}{K_1 + K_2}\right) = 0$$
$$\frac{K_1K_2}{K_1 - K_2} + 2a_2\left(\frac{a_2}{2K_2} + \frac{a_1}{K_1 + K_2}\right) = 0. \qquad (22a)$$

Addition of equations 22a gives the condition

$$K_1 > 0; \qquad (23)$$

whereas multiplying the first equation by K_1, the second by K_2, and adding gives the condition

$$K_1 + K_2 > 0. \qquad (24)$$

Both conditions, inequalities 23 and 24, are always met for the chemical control systems involving negative feedback.

However, for functions of the positive feedback type, $G(\xi_2)$, the first condition is inverted:

$$K_1 < 0. \qquad (25)$$

Inequality 25 will never be satisfied in chemical systems since rate constants are always non-negative numbers.

These results illustrate that for any function of the negative feedback type, it is always possible to choose real

$$a_1 = \frac{\pm K_1\sqrt{K_1(K_1 + K_2)}}{K_2 - K_1}$$

and

$$a_2 = \frac{\pm K_2\sqrt{K_1(K_1+K_2)}}{K_1-K_2}$$

such that equations 22a are satisfied. Thus, for chemical control systems of the type described by equations 3, $\underline{n} = 2$, global asymptotic stability of the stationary state is guaranteed and sustained concentration oscillations cannot occur. However, for the same chemical systems involving positive feedback, no guarantee about the stability of the stationary state can be made.

5. Stability of the Nonlinear System: n = 3

In this case the canonical equations are

$$\begin{aligned} \dot{\eta}_1 &= G(\xi_3) - K_1\eta_1 \\ \dot{\eta}_2 &= G(\xi_3) - K_2\eta_2 \\ \dot{\eta}_3 &= G(\xi_3) - K_3\eta_3 \end{aligned} \qquad (13b)$$

$$\xi_3 = \frac{K_1K_2}{(K_1-K_2)(K_1-K_3)(K_2-K_3)}\left[(K_3-K_2)\eta_1 + (K_1-K_3)\eta_2 + (K_2-K_1)\eta_3\right].$$

For $\underline{n} = 3$ there are three sufficiency criteria:

$$-\frac{K_1^2K_2}{(K_1-K_2)(K_1-K_3)} + 2a_1\left(\frac{a_1}{2K_1} + \frac{a_2}{K_1+K_2} + \frac{a_3}{K_1+K_3}\right) = 0$$

$$-\frac{K_1K_2^2}{(K_2-K_1)(K_2-K_3)} + 2a_2\left(\frac{a_2}{2K_2} + \frac{a_1}{K_1+K_2} + \frac{a_3}{K_2+K_3}\right) = 0 \qquad (22b)$$

$$-\frac{K_1K_2K_3}{(K_3-K_1)(K_3-K_2)} + 2a_3\left(\frac{a_3}{2K_3} + \frac{a_1}{K_1+K_3} + \frac{a_2}{K_2+K_3}\right) = 0.$$

In this case we multiply the first equation by K_1, the second by K_2, and the third by K_3. Addition of the resulting equations yields the condition

$$K_1K_2 > 0 \qquad (26)$$

which can be fulfilled in chemical systems.

If, on the other hand, we divide the first equation by K_1, the second by K_2, and the third by K_3 and add the resulting equations, we obtain

$$-a_1 = \frac{K_1}{K_2} a_2 + \frac{K_1}{K_3} a_3 \quad . \tag{27}$$

Addition of equations 22b provides another relationship

$$-a_1 = \frac{\frac{a_2}{K_2} + \frac{4a_3}{K_2 + K_3}}{\frac{a_1}{K_1} + \frac{4a_2}{K_1 + K_2}} a_2 + \frac{\frac{a_3}{K_3} + \frac{4a_1}{K_1 + K_3}}{\frac{a_1}{K_1} + \frac{4a_2}{K_1 + K_2}} a_3 . \tag{28}$$

Substitution from equation 27 into equation 28 yields

$$\frac{a_2}{a_3} = - \frac{K_2(K_1 - K_3)}{K_3(K_1 - K_2)} \left[A \pm \sqrt{A^2 - \frac{K_1 + K_2}{K_1 + K_3}} \right] \tag{29}$$

where

$$A = \frac{K_2(K_1 + K_3) + K_3(K_1 + K_2)}{(K_1 + K_3)(K_2 + K_3)} \quad .$$

We are free to choose real a_i provided

$$A^2 > \frac{K_1 + K_2}{K_1 + K_3} . \tag{30}$$

But inequality 30 cannot be satisfied for any set of non-negative K_1, K_2 and K_3. Thus, the sufficiency criteria for global asymptotic stability can never be satisfied by real a_i for three-component chemical systems involving negative feedback. This means that limit cycle behavior cannot be excluded on this basis for any set of K_1, K_2 and K_3 for all nonlinear negative feedback functions.

For functions of the positive feedback type, the condition in inequality 30 remains the same, but the condition in inequality 26 is inverted. This means that the sufficiency criteria are not met for chemical systems involving three components and positive feedback.

6. Numerical Simulations of Systems with Nonlinear Negative Feedback

Morales and McKay (1967) suggested using the function

$$f(S_n) = \frac{K_o}{1 + \alpha\left[S_n\right]^{\rho}} \qquad (31)$$

to describe metabolic end-product inhibition of the Yates-Pardee (1956) type. We have used this function in equations 3 to obtain numerical solutions for a wide range of $\underline{n}$, ρ, K_o, α, and K_n. The results are summarized in Figure 2. Sustained oscillatory behavior was not observed for systems involving only two variables, but limit cycles were found for certain systems involving three or more variables. Furthermore, if a limit cycle arose for a given set of $\underline{n}$, ρ, α, and K_o, then a limit cycle also occurred for larger values of any one of these parameters provided none of the remaining three were decreased.

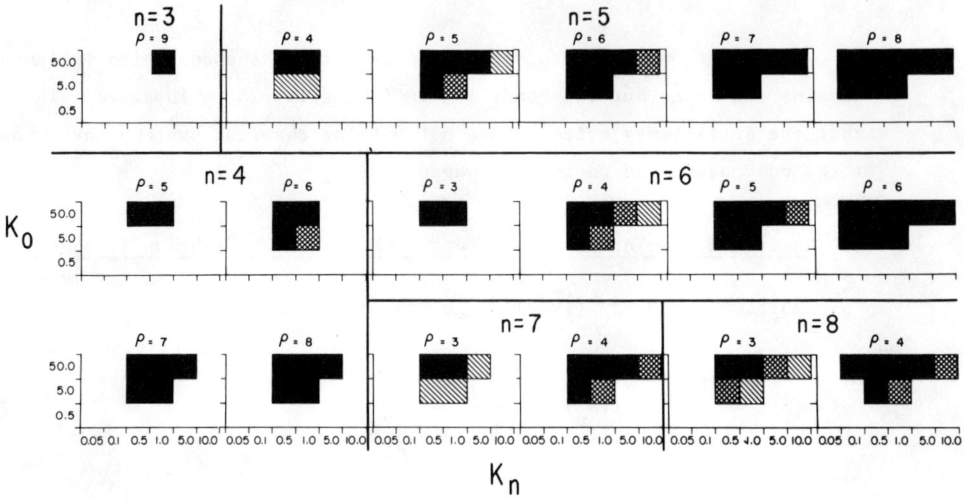

FIGURE 2

Figure 2. A summary of the results obtained from the numerical solutions (digital simulations) of equations 3. The squares at the intersection of specific values of K_o = 0.5, 5.0 or 50.0 (vertical axis) and specific values of K_n = 0.05, 0.1, 0.5, 1.0, 5.0 or 10.0 (horizontal axis) are filled in if a limit cycle arises in the control systems described by equations 3. If the square at the intersection is filled in with slashes leaning to the left, the limit cycle was found only when $\alpha = 1.0$; if the intersection is filled with slashes leaning to the left and right (diamonds), limit cycles were found when $\alpha = 0.1$ and $\alpha = 1.0$; if the intersection is filled with solid darkening, limit cycles were found when $\alpha = 0.01$, $\alpha = 0.1$ and $\alpha = 1.0$. If the intersection is not filled, the systems involving the indicated values of K_o and K_n were simulated for $\alpha = 0.01$, $\alpha = 0.1$, $\alpha = 1.0$ and the indicated values of n and ρ, but a limit cycle did not arise. In each of these cases equations 3 possess global asymptotic stability. The values of n and ρ used in each case are identified as inserts on the figure; for each value of n all 54 possible combinations of the indicated values of K_o, K_n and α were simulated for all possible (integer) values of ρ less than those indicated on the figure. The results obtained from these lower values of ρ were that no limit cycles arose, and equations 3 possessed global asymptotic stability.

This research was supported in part by Grant GB 20612 from the National Science Foundation.

References

Botts, J.,(1958). Trans. Faraday Soc. 54, 593.

Briggs, G. E.,and Haldane, J. B. S. (1925). Biochem. J. 19, 383.

Letov, A. M.,(1961). "Stability in Nonlinear Control Systems". Princeton Univ. Press, Princeton, N. J.

Lur'e, A. L.,(1951). "Certain Nonlinear Problems in the Theory of Automatic Control". Gostekhizdat, Moscow.

Morales, M.,and McKay, D. (1967). Biophys. J. 7, 621.

Walter, C.,(1969). Biophys. J. 9, 863.

Walter, C., (1970). J. Theor. Biol. 27, 259.

Walter, C., (1972). "Kinetic and Thermodynamic Aspects of Biological and Biochemical Control Mechanisms" in "Biochemical Regulatory Mechanisms in Eukaryotic Cells" (E. Kun and S. Grisolia, editors), John Wiley and Sons, Inc., New York, Ch. 11.

Yates, R. A.,and Pardee, A. B. (1956). J. Biol. Chem. 221, 757.

Lecture Notes in Mathematics

Comprehensive leaflet on request

Vol. 146: A. B. Altman and S. Kleiman, Introduction to Grothendieck Duality Theory. II, 192 pages. 1970. DM 18,–

Vol. 147: D. E. Dobbs, Cech Cohomological Dimensions for Commutative Rings. VI, 176 pages. 1970. DM 16,–

Vol. 148: R. Azencott, Espaces de Poisson des Groupes Localement Compacts. IX, 141 pages. 1970. DM 16,–

Vol. 149: R. G. Swan and E. G. Evans, K-Theory of Finite Groups and Orders. IV, 237 pages. 1970. DM 20,–

Vol. 150: Heyer, Dualität lokalkompakter Gruppen. XIII, 372 Seiten. 1970. DM 20,–

Vol. 151: M. Demazure et A. Grothendieck, Schémas en Groupes I. (SGA 3). XV, 562 pages. 1970. DM 24,–

Vol. 152: M. Demazure et A. Grothendieck, Schémas en Groupes II. (SGA 3). IX, 654 pages. 1970. DM 24,–

Vol. 153: M. Demazure et A. Grothendieck, Schémas en Groupes III. (SGA 3). VIII, 529 pages. 1970. DM 24,–

Vol. 154: A. Lascoux et M. Berger, Variétés Kähleriennes Compactes. VII, 83 pages. 1970. DM 16,–

Vol. 155: Several Complex Variables I, Maryland 1970. Edited by J. Horváth. IV, 214 pages. 1970. DM 18,–

Vol. 156: R. Hartshorne, Ample Subvarieties of Algebraic Varieties. XIV, 256 pages. 1970. DM 20,–

Vol. 157: T. tom Dieck, K. H. Kamps und D. Puppe, Homotopietheorie. VI, 265 Seiten. 1970. DM 20,–

Vol. 158: T. G. Ostrom, Finite Translation Planes. IV. 112 pages. 1970. DM 16,–

Vol. 159: R. Ansorge und R. Hass. Konvergenz von Differenzenverfahren für lineare und nichtlineare Anfangswertaufgaben. VIII, 145 Seiten. 1970. DM 16,–

Vol. 160: L. Sucheston, Constributions to Ergodic Theory and Probability. VII, 277 pages. 1970. DM 20,–

Vol. 161: J. Stasheff, H-Spaces from a Homotopy Point of View. VI, 95 pages. 1970. DM 16,–

Vol. 162: Harish-Chandra and van Dijk, Harmonic Analysis on Reductive p-adic Groups. IV, 125 pages. 1970. DM 16,–

Vol. 163: P. Deligne, Equations Différentielles à Points Singuliers Reguliers. III, 133 pages. 1970. DM 16,–

Vol. 164: J. P. Ferrier, Seminaire sur les Algebres Complètes. II, 69 pages. 1970. DM 16,–

Vol. 165: J. M. Cohen, Stable Homotopy. V, 194 pages. 1970. DM 16,–

Vol. 166: A. J. Silberger, PGL_2 over the p-adics: its Representations, Spherical Functions, and Fourier Analysis. VII, 202 pages. 1970. DM 18,–

Vol. 167: Lavrentiev, Romanov and Vasiliev, Multidimensional Inverse Problems for Differential Equations. V, 59 pages. 1970. DM 16,–

Vol. 168: F. P. Peterson, The Steenrod Algebra and its Applications: A conference to Celebrate N. E. Steenrod's Sixtieth Birthday. VII, 317 pages. 1970. DM 22,–

Vol. 169: M. Raynaud, Anneaux Locaux Henséliens. V, 129 pages. 1970. DM 16,–

Vol. 170: Lectures in Modern Analysis and Applications III. Edited by C. T. Taam. VI, 213 pages. 1970. DM 18,–

Vol. 171: Set-Valued Mappings, Selections and Topological Properties of 2^X. Edited by W. M. Fleischman. X, 110 pages. 1970. DM 16,–

Vol. 172: Y.-T. Siu and G. Trautmann, Gap-Sheaves and Extension of Coherent Analytic Subsheaves. V, 172 pages. 1971. DM 16,–

Vol. 173: J. N. Mordeson and B. Vinograde, Structure of Arbitrary Purely Inseparable Extension Fields. IV, 138 pages. 1970. DM 16,–

Vol. 174: B. Iversen, Linear Determinants with Applications to the Picard Scheme of a Family of Algebraic Curves. VI, 69 pages. 1970. DM 16,–

Vol. 175: M. Brelot, On Topologies and Boundaries in Potential Theory. VI, 176 pages. 1971. DM 18,–

Vol. 176: H. Popp, Fundamentalgruppen algebraischer Mannigfaltigkeiten. IV, 154 Seiten. 1970. DM 16,–

Vol. 177: J. Lambek, Torsion Theories, Additive Semantics and Rings of Quotients. VI, 94 pages. 1971. DM 16,–

Vol. 178: Th. Bröcker und T. tom Dieck, Kobordismentheorie. XVI, 191 Seiten. 1970. DM 18,–

Vol. 179: Seminaire Bourbaki – vol. 1968/69. Exposés 347-363. IV. 295 pages. 1971. DM 22,–

Vol. 180: Séminaire Bourbaki – vol. 1969/70. Exposés 364-381. IV, 310 pages. 1971. DM 22,–

Vol. 181: F. DeMeyer and E. Ingraham, Separable Algebras over Commutative Rings. V, 157 pages. 1971. DM 16.–

Vol. 182: L. D. Baumert. Cyclic Difference Sets. VI, 166 pages. 1971. DM 16,–

Vol. 183: Analytic Theory of Differential Equations. Edited by P. F. Hsieh and A. W. J. Stoddart. VI, 225 pages. 1971. DM 20,–

Vol. 184: Symposium on Several Complex Variables, Park City, Utah, 1970. Edited by R. M. Brooks. V, 234 pages. 1971. DM 20,–

Vol. 185: Several Complex Variables II, Maryland 1970. Edited by J. Horváth. III, 287 pages. 1971. DM 24,–

Vol. 186: Recent Trends in Graph Theory. Edited by M. Capobianco/ J. B. Frechen/M. Krolik. VI, 219 pages. 1971. DM 18.–

Vol. 187: H. S. Shapiro, Topics in Approximation Theory. VIII, 275 pages. 1971. DM 22,–

Vol. 188: Symposium on Semantics of Algorithmic Languages. Edited by E. Engeler. VI, 372 pages. 1971. DM 26,–

Vol. 189: A. Weil, Dirichlet Series and Automorphic Forms. V, 164 pages. 1971. DM 16,–

Vol. 190: Martingales. A Report on a Meeting at Oberwolfach, May 17-23, 1970. Edited by H. Dinges. V, 75 pages. 1971. DM 16,–

Vol. 191: Séminaire de Probabilités V. Edited by P. A. Meyer. IV, 372 pages. 1971. DM 26,–

Vol. 192: Proceedings of Liverpool Singularities – Symposium I. Edited by C. T. C. Wall. V, 319 pages. 1971. DM 24,–

Vol. 193: Symposium on the Theory of Numerical Analysis. Edited by J. Ll. Morris. VI, 152 pages. 1971. DM 16,–

Vol. 194: M. Berger, P. Gauduchon et E. Mazet. Le Spectre d'une Variété Riemannienne. VII, 251 pages. 1971. DM 22,–

Vol. 195: Reports of the Midwest Category Seminar V. Edited by J.W. Gray and S. Mac Lane.III, 255 pages. 1971. DM 22,–

Vol. 196: H-spaces – Neuchâtel (Suisse)- Août 1970. Edited by F. Sigrist, V, 156 pages. 1971. DM 16,–

Vol. 197: Manifolds – Amsterdam 1970. Edited by N. H. Kuiper. V, 231 pages. 1971. DM 20,–

Vol. 198: M. Hervé, Analytic and Plurisubharmonic Functions in Finite and Infinite Dimensional Spaces. VI, 90 pages. 1971. DM 16.–

Vol. 199: Ch. J. Mozzochi, On the Pointwise Convergence of Fourier Series. VII, 87 pages. 1971. DM 16,–

Vol. 200: U. Neri, Singular Integrals. VII, 272 pages. 1971. DM 22,–

Vol. 201: J. H. van Lint, Coding Theory. VII, 136 pages. 1971. DM 16,–

Vol. 202: J. Benedetto, Harmonic Analysis on Totally Disconnected Sets. VIII, 261 pages. 1971. DM 22,–

Vol. 203: D. Knutson, Algebraic Spaces. VI, 261 pages. 1971. DM 22,–

Vol. 204: A. Zygmund, Intégrales Singulières. IV, 53 pages. 1971. DM 16,–

Vol. 205: Séminaire Pierre Lelong (Analyse) Année 1970. VI, 243 pages. 1971. DM 20,–

Vol. 206: Symposium on Differential Equations and Dynamical Systems. Edited by D. Chillingworth. XI, 173 pages. 1971. DM 16,–

Vol. 207: L. Bernstein, The Jacobi-Perron Algorithm – Its Theory and Application. IV, 161 pages. 1971. DM 16,–

Vol. 208: A. Grothendieck and J. P. Murre, The Tame Fundamental Group of a Formal Neighbourhood of a Divisor with Normal Crossings on a Scheme. VIII, 133 pages. 1971. DM 16,–

Vol. 209: Proceedings of Liverpool Singularities Symposium II. Edited by C. T. C. Wall. V, 280 pages. 1971. DM 22,–

Vol. 210: M. Eichler, Projective Varieties and Modular Forms. III, 118 pages. 1971. DM 16,–

Vol. 211: Théorie des Matroïdes. Edité par C. P. Bruter. III, 108 pages. 1971. DM 16,–

Please turn over

Vol. 212: B. Scarpellini, Proof Theory and Intuitionistic Systems. VII, 291 pages. 1971. DM 24,-

Vol. 213: H. Hogbe-Nlend, Théorie des Bornologies et Applications. V, 168 pages. 1971. DM 18,-

Vol. 214: M. Smorodinsky, Ergodic Theory, Entropy. V, 64 pages. 1971. DM 16,-

Vol. 215: P. Antonelli, D. Burghelea and P. J. Kahn, The Concordance-Homotopy Groups of Geometric Automorphism Groups. X, 140 pages. 1971. DM 16,-

Vol. 216: H. Maaß, Siegel's Modular Forms and Dirichlet Series. VII, 328 pages. 1971. DM 20,-

Vol. 217: T. J. Jech, Lectures in Set Theory with Particular Emphasis on the Method of Forcing. V, 137 pages. 1971. DM 16,-

Vol. 218: C. P. Schnorr, Zufälligkeit und Wahrscheinlichkeit. IV, 212 Seiten 1971. DM 20,-

Vol. 219: N. L. Alling and N. Greenleaf, Foundations of the Theory of Klein Surfaces. IX, 117 pages. 1971. DM 16,-

Vol. 220: W. A. Coppel, Disconjugacy. V, 148 pages. 1971. DM 16,-

Vol. 221: P. Gabriel und F. Ulmer, Lokal präsentierbare Kategorien. V, 200 Seiten. 1971. DM 18,-

Vol. 222: C. Meghea, Compactification des Espaces Harmoniques. III, 108 pages. 1971. DM 16,-

Vol. 223: U. Felgner, Models of ZF-Set Theory. VI, 173 pages. 1971. DM 16,-

Vol. 224: Revètements Etales et Groupe Fondamental. (SGA 1). Dirigé par A. Grothendieck XXII, 447 pages. 1971. DM 30,-

Vol. 225: Théorie des Intersections et Théorème de Riemann-Roch. (SGA 6). Dirigé par P. Berthelot, A. Grothendieck et L. Illusie. XII, 700 pages. 1971. DM 40,-

Vol. 226: Seminar on Potential Theory, II. Edited by H. Bauer. IV, 170 pages. 1971. DM 18,-

Vol. 227: H. L. Montgomery, Topics in Multiplicative Number Theory. IX, 178 pages. 1971. DM 18,-

Vol. 228: Conference on Applications of Numerical Analysis. Edited by J. Ll. Morris. X, 358 pages. 1971. DM 26,-

Vol. 229: J. Väisälä, Lectures on n-Dimensional Quasiconformal Mappings. XIV, 144 pages. 1971. DM 16,-

Vol. 230: L. Waelbroeck, Topological Vector Spaces and Algebras. VII, 158 pages. 1971. DM 16,-

Vol. 231: H. Reiter, L¹-Algebras and Segal Algebras. XI, 113 pages. 1971. DM 16,-

Vol. 232: T. H. Ganelius, Tauberian Remainder Theorems. VI, 75 pages. 1971. DM 16,-

Vol. 233: C. P. Tsokos and W. J. Padgett. Random Integral Equations with Applications to Stochastic Systems. VII, 174 pages. 1971. DM 18,-

Vol. 234: A. Andreotti and W. Stoll. Analytic and Algebraic Dependence of Meromorphic Functions. III, 390 pages. 1971. DM 26,-

Vol. 235: Global Differentiable Dynamics. Edited by O. Hájek, A. J. Lohwater, and R. McCann. X, 140 pages. 1971. DM 16,-

Vol. 236: M. Barr, P. A. Grillet, and D. H. van Osdol. Exact Categories and Categories of Sheaves. VII, 239 pages. 1971, DM 20,-

Vol. 237: B. Stenström. Rings and Modules of Quotients. VII, 136 pages. 1971. DM 16,-

Vol. 238: Der kanonische Modul eines Cohen-Macaulay-Rings. Herausgegeben von Jürgen Herzog und Ernst Kunz. VI, 103 Seiten. 1971. DM 16,-

Vol. 239: L. Illusie, Complexe Cotangent et Déformations I. XV, 355 pages. 1971. DM 26,-

Vol. 240: A. Kerber, Representations of Permutation Groups I. VII, 192 pages. 1971. DM 18,-

Vol. 241: S. Kaneyuki, Homogeneous Bounded Domains and Siegel Domains. V, 89 pages. 1971. DM 16,-

Vol. 242: R. R. Coifman et G. Weiss, Analyse Harmonique Non-Commutative sur Certains Espaces. V, 160 pages. 1971. DM 16,-

Vol. 243: Japan-United States Seminar on Ordinary Differential and Functional Equations. Edited by M. Urabe. VIII, 332 pages. 1971. DM 26,-

Vol. 244: Séminaire Bourbaki - vol. 1970/71. Exposés 382-399. IV, 356 pages. 1971. DM 26,-

Vol. 245: D. E. Cohen, Groups of Cohomological Dimension One. V, 99 pages. 1972. DM 16,-

Vol. 246: Lectures on Rings and Modules. Tulane University and Operator Theory Year, 1970-1971. Volume I. X, 661 pages. DM 40,-

Vol. 247: Lectures on Operator Algebras. Tulane University Rin Operator Theory Year, 1970-1971. Volume II. XI, 786 pages. DM 40,-

Vol. 248: Lectures on the Applications of Sheaves to Ring Th Tulane University Ring and Operator Theory Year, 1970-1971 ume III. VIII, 315 pages. 1971. DM 26,-

Vol. 249: Symposium on Algebraic Topology. Edited by P. J. H VII, 111 pages. 1971. DM 16,-

Vol. 250: B. Jónsson, Topics in Universal Algebra. VI, 220 p 1972. DM 20,-

Vol. 251: The Theory of Arithmetic Functions. Edited by A. A. and D. L. Goldsmith VI, 287 pages. 1972. DM 24,-

Vol. 252: D. A. Stone, Stratified Polyhedra. IX, 193 pages. DM 18,-

Vol. 253: V. Komkov, Optimal Control Theory for the Dampi Vibrations of Simple Elastic Systems. V, 240 pages. 1972. DM

Vol. 254: C. U. Jensen, Les Foncteurs Dérivés de lim et leurs plications en Théorie des Modules. V, 103 pages. 1972. DM

Vol. 255: Conference in Mathematical Logic - London '70. Edite W. Hodges. VIII, 351 pages. 1972. DM 26,-

Vol. 256: C. A. Berenstein and M. A. Dostal, Analytically Uni Spaces and their Applications to Convolution Equations. VII, pages. 1972. DM 16,-

Vol. 257: R. B. Holmes, A Course on Optimization and Best proximation. VIII, 233 pages. 1972. DM 20,-

Vol. 258: Séminaire de Probabilités VI. Edited by P. A. Meyer 253 pages. 1972. DM 22,-

Vol. 259: N. Moulis, Structures de Fredholm sur les Variétés bertiennes. V, 123 pages. 1972. DM 16,-

Vol. 260: R. Godement and H. Jacquet, Zeta Functions of Si Algebras. IX, 188 pages. 1972. DM 18,-

Vol. 261: A. Guichardet, Symmetric Hilbert Spaces and Related pics. V, 197 pages. 1972. DM 18,-

Vol. 262: H. G. Zimmer, Computational Problems, Methods, Results in Algebraic Number Theory. V, 103 pages. 1972. DM

Vol. 263: T. Parthasarathy, Selection Theorems and their Applicati VII, 101 pages. 1972. DM 16,-

Vol. 264: W. Messing, The Crystals Associated to Barsotti- Groups: with Applications to Abelian Schemes. III, 190 pages. 1 DM 18,-

Vol. 265: N. Saavedra Rivano, Catégories Tannakiennes. II, pages. 1972. DM 26,-

Vol. 266: Conference on Harmonic Analysis. Edited by D. Gu and R. L. Lipsman. VI, 323 pages. 1972. DM 24,-

Vol. 267: Numerische Lösung nichtlinearer partieller Differential- Integro-Differentialgleichungen. Herausgegeben von R. Ansorge W. Törnig, VI, 339 Seiten. 1972. DM 26,-

Vol. 268: C. G. Simader, On Dirichlet's Boundary Value Problem. IV pages. 1972. DM 20,-

Vol. 269: Théorie des Topos et Cohomologie Etale des Schér (SGA 4). Dirigé par M. Artin, A. Grothendieck et J. L. Verdier. 525 pages. 1972. DM 50,-

Vol. 270: Thèorie des Topos et Cohomologie Etle des Schém Tome 2. (SGA 4). Dirige par M. Artin, A. Grothendieck et J. L. Ver V, 418 pages. 1972. DM 50,-

Vol. 271: J. P. May, The Geometry of Iterated Loop Spaces. IX, pages. 1972. DM 18,-

Vol. 272: K. R. Parthasarathy and K. Schmidt, Positive Definite nels, Continuous Tensor Products, and Central Limit Theorem Probability Theory. VI, 107 pages. 1972. DM 16,-

Vol. 273: U. Seip, Kompakt erzeugte Vektorräume und Analysis 119 Seiten. 1972. DM 16,-

Vol. 274: Toposes, Algebraic Geometry and Logic. Edited by. F Lawvere. VI, 189 pages. 1972. DM 18,-

Vol. 275: Séminaire Pierre Lelong (Analyse) Année 1970-1971 181 pages. 1972. DM 18,-

Vol. 276: A. Borel, Représentations de Groupes Localement C pacts. V, 98 pages. 1972. DM 16,-

Vol. 277: Séminaire Banach. Edité par C. Houzel. VII, 229 pa 1972. DM 20,-